Smart Materials for Ranging Systems

NATO Science Series

A Series presenting the results of scientific meetings supported under the NATO Science Programme.

The Series is published by IOS Press, Amsterdam, and Springer in conjunction with the NATO Public Diplomacy Division

Sub-Series

I. **Life and Behavioural Sciences**	IOS Press
II. **Mathematics, Physics and Chemistry**	Springer
III. **Computer and Systems Science**	IOS Press
IV. **Earth and Environmental Sciences**	Springer

The NATO Science Series continues the series of books published formerly as the NATO ASI Series.

The NATO Science Programme offers support for collaboration in civil science between scientists of countries of the Euro-Atlantic Partnership Council.The types of scientific meeting generally supported are "Advanced Study Institutes" and "Advanced Research Workshops", and the NATO Science Series collects together the results of these meetings. The meetings are co-organized by scientists from NATO countries and scientists from NATO's Partner countries – countries of the CIS and Central and Eastern Europe.

Advanced Study Institutes are high-level tutorial courses offering in-depth study of latest advances in a field.
Advanced Research Workshops are expert meetings aimed at critical assessment of a field, and identification of directions for future action.

As a consequence of the restructuring of the NATO Science Programme in 1999, the NATO Science Series was re-organized to the four sub-series noted above. Please consult the following web sites for information on previous volumes published in the Series.

http://www.nato.int/science
http://www.springer. com
http://www.iospress.nl

Series II: Mathematics, Physics and Chemistry – Vol. 226

Smart Materials for Ranging Systems

edited by

Jaap Franse

Van der Waals – Zeeman Instituut,
Universiteit van Amsterdam,
The Netherlands

Victor Eremenko

Institute for Low Temperature Physics,
National Academy of Sciences of Ukraine,
Kharkov, Ukraine

and

Valentyna Sirenko

Institute for Low Temperature Physics,
National Academy of Sciences of Ukraine,
Kharkov, Ukraine

Springer

Published in cooperation with NATO Public Diplomacy Division

Proceedings of the NATO Advanced Research Workshop on
Smart Materials for Ranging Systems
Krasnoyarsk, Russia
28 August – 1 September 2004

A C.I.P. Catalogue record for this book is available from the Library of Congress.

ISBN-10 1-4020-4645-6 (PB)
ISBN-13 978-1-4020-4645-2 (PB)
ISBN-10 1-4020-4644-8 (HB)
ISBN-13 978-1-4020-4644-5 (HB)
ISBN-10 1-4020-4646-4 (e-book)
ISBN-13 978-1-4020-4646-9 (e-book)

Published by Springer,
P.O. Box 17, 3300 AA Dordrecht, The Netherlands.

www.springer.com

Printed on acid-free paper

Printed in the Netherlands.

Table of Contents

Preface vii

List of Contributors xi

Acknowledgement xv

Contributions:

1. J. Bartolomé, F. Bartolomé, L.M. García, F. Luis, F. Petroff, V. Cros, H. Jaffrès, and A. Vaurès
MAGNETIC DYNAMICS OF CO NANOSPHERES: ORIGIN OF THE ENHANCED ANISOTROPY 1

2. Janis Kliava and René Berger
NANOPARTICLES IN OXIDE GLASSES: MAGNETIC RESONANCE STUDIES 27

3. G. Petrakovskii, M. Popov, V. Zinenko, B. Roessli, J. Schefer, M. Boehm and U. Staub
EFFECT OF MAGNETIC FIELD ON THE MAGNETIC STATE OF COPPER METABORATE 49

4. V.V. Eremenko and V.A. Sirenko
INTRINSICALLY INHOMOGENEOUS MAGNETIC STATES IN ANTIFERROMAGNETS IN THE VICINITY OF PHASE TRANSITIONS INDUCED BY MAGNETIC FIELD 67

5. M. Chiba, H. Nakamura, Y. Fujii, H. Kikuchi, Y. Yamamoto, H. Hori, G. Petrakovskii, M. Popov and L. Bezmaternikh
NMR STUDY OF THE SUCCESSIVE PHASE TRANSITIONS IN CuB_2O_4 101

6. A. Sherman
MAGNETIC PROPERTIES OF CUPRATE PEROVSKITES 115

7. J.J.M. Franse
QUANTUM CRITICAL PHENOMENA AND THE ANOMALOUS LOW-TEMPERATURE PROPERTIES OF THE U(Pt,Pd) SYSTEM 129

8. Yu.A. Izyumov and D.S. Alexeev
THE PERIODIC ANDERSON MODEL IN THE GENERATING FUNCTIONAL APPROACH 153

9. I.O. Troyanchuk, V.A. Khomchenko and H. Szymczak
ORBITAL ORDERING AND PHASE SEPARATION PHENOMENA IN LIGHTLY DOPED MANGANITES 165

10. V. R. Shah, Christian Schanzer, Peter Böni and Hans-Benjamin Braun
INTERFACE AND MAGNETIC CHARACTERIZATION OF FM/AF/FM MULTILAYERS 179

11. S. Bondarenko and N. Nakagawa
SQUID-BASED MAGNETIC MICROSCOPE 195

12. V.V. Kruglyak and R.J. Hicken
FEMTOSECOND ELLIPSOMETRY: A NEW TOOL FOR THE MEASUREMENT OF HOT ELECTRON MOMENTUM RELAXATION TIMES IN METALS 203

13. A.N. Vasiliev, R.Z. Levitin and V.V. Khovaylo
FERROMAGNETIC SHAPE MEMORY ALLOYS $Ni_{2+x}Mn_{1-x}Ga$ 223

14. H. Szymczak and R. Szymczak
MAGNETOSTRICTION OF MANGANITES AND COBALTITES 245

15. V.V. Eremenko, S.B. Feodosyev, I.A. Gospodarev, V.A. Sirenko, M.Yu. Shvedun, M. Tovar and W. McCallum
STRUCTURAL STUDY OF MAGNETOSTRICTIVE SUPERCONDUCTING COMPOUNDS 259

Subject index 271

PREFACE

The problem of determining the location of an object, which is usually called ranging, attracts at present much attention in many different areas of applications, among them in ecological and safety devices. Electromagnetic waves along with sound waves are widely used for this purpose. Familiar examples of ranging systems are radar, sonar, GPS positioning, speed meters, etc.. Most are echo-type of devices, generating a wave and interpreting its echo from the object of interest. GPS is a cooperative system, in which the receiver observes timing signals from sources at known locations, and locates itself in reference to them. Passive ranging makes use of waves generated by the object to be located that are picked up by an observer. As indicated, there are three kinds of ranging systems, successively described as echo, cooperative and passive systems. Echo ranging is by far the most common method in practice. The observer at a certain point emits a wave of some physical nature at a certain time. When the outgoing wave front strikes the object, a scattered wave front is launched, which is detected at the observer point a certain time interval later. In special cases, the observer is able to determine the direction of the received wave. This, of course, is the principle of radar and echoscopy. The observed object in echo ranging is passive and takes no active part in the procedure. In cooperative ranging, from the other hand, the object takes an active part. In one form of cooperative ranging, the object emits waves that contain time information. Two fixed stations with clocks synchronized with that of the object note the time delay of the received wave and from the two distances thereby determined the object can be located. If needed, the stations can radio back the evaluated position. In the second form, the object receives waves from two fixed stations and subsequently determines its position from the distances to the two stations. Of course, in three dimensions, three distances are necessary. The GPS system uses the second option, with the peculiar circumstance that the 'fixed' stations are actually moving in orbits. But their positions can be accurately evaluated at any time. Cooperative ranging places great demands on clocks, especially when carried out with electromagnetic waves. Timing sound waves is less critical. Passive ranging, finally, makes use of waves generated by the object to be located that are picked up by an observer. The wave used in this application may be generated in the normal action of the object, and not necessarily generated for location purposes. An example is the location of a boat from its propeller noise, or a whale from its singing. The fundamental data in these applications are the relative times, the time delays, at which the signals are picked up at the different stations. Smart material devices in which sensing, actuating, and control functions are integrated are urgently needed for the realization of practical devices. Magnetic field measurements and the imaging of its spatial distribution are valuable tools in the related range of applications, including navigation systems, resource exploration drilling, spacecraft attitude control systems and assessments of the impact of 'space weather' caused by cosmic particles. In addition to these macroscopic applications, the microscopic approach should be mentioned. The position of micro particles, even on an atomic scale can be observed by scanning tunnelling microscopy. Examples are the atomic force microscope (AFM) and the magnetic force microscope (MFM).

Application of the underlying ideas needs the design of appropriate "smart" materials as a first step. This ARW outlines magnetic research that hopefully will help to transform scientific ideas into practical system designs. The workshop's focus was on:

- an up-to-date knowledge of the latest results of solid state physics which is indispensable for a deep insight in material properties;
- advanced techniques for material preparation, for nano-size samples in particular, which is the key for many modern applications;
- the integration of different features in one composite which marks the new generation of smart devices;
- sophisticated detection and measuring techniques which open new horizons for advanced applications, in particular on the propagation of magnetic waves;
- the dynamics of magnetic structures in materials with unusual magnetic and superconducting structures.

The different aspects of new materials with specific magnetic, electric and elastic properties were discussed during the workshop in order to obtain a general and consistent view on the potentialities for application in design and manufacturing of smart materials. As a result of the workshop we mention the progress that was reported in the fabrication and understanding of in-situ formation and characterization of solid state structures with specified properties, as well as in the observation and study of the mobility of magnetic structures and of the kinetics of magnetic ordering transitions. Looking from a different perspective, one of the outcomes of the ARW is the emphasis on the important role that collective phenomena like spin waves in systems with a magnetically ordered ground state, or critical currents in superconductors could play at the design of magnetic-field sensitive sensor materials. The lectures presented covered the following relevant topics:

1. Nano-technology problems were addressed in talks by J. Bartolomé (Magnetic dynamics of Co nanospheres) and J. Kliava (Nanoparticles in oxide glasses).
2. Strongly correlated electron systems including high-T_c superconductors were discussed in talks by S. Ovchinnikov (Theory for transition metal oxides), G. Petrakovskii (Magnetism of 2D spin systems),

S. Maleyev (Theory of cubic magnets with Dzyaloshinskii-Moriya interaction), V. Eremenko (Intrinsic inhomogeneity aspect of sensor units), M. Chiba (NMR in copper metaborate), A. Sherman (Magnetism of copper perovskites), V. Mitskan (Singlet formation of 2D magnets), M. Korshunov (Superconductivity of cuprates), J. Franse (Quantum criticality in the $U(Pt,Pd)_3$ system), Yu. Izyumov (The periodic Anderson model) and I. Lubutin (Magnetism of iron oxides under pressure).

3. Magnetism of molecular magnets in view of new perspectives for electronic applications was presented in a talk by G. Romanenko (Phase transitions in hetero-spin systems).

4. Materials with colossal magnetoresistance for application in different types of sensors in manganites in particular were addressed in talks by I. Troyanchuk and V. Shah (Magnetism of multilayered magnetic films).

5. SQUID sensors as high-sensitivity devices for a variety of applications were presented in a talk by S. Bondarenko (SQUID-based magnetic microscope).

6. Optical properties of electronic and magnetic systems were discussed in talks by R. Hicken (Electron relaxation in metals by femtosecond ellipsometry) and R. Pisarev (Nonlinear optical spectroscopy).

7. Ferromagnetic materials exhibiting shape memory effects and superelasticity under action of a magnetic field turn out to have a great potential for applications. This problem was discussed in a talk by A. Vasiliev (Shape memory alloys of the Ni-Mn-Ga type).

8. Magnetostrictive materials with potentialities for sensing unit applications were considered in presentations of H. Szymczak (magnetostriction of manganites) V.Eremenko and V.Sirenko (Structural study of magnetostrictive superconducting compounds)

The meeting was well organised and took place in the House of Scientists of the Academy of Science in Krasnoyarsk, situated in a spacious area of an outlying district of the city of Krasnoyarsk where both the Academy of Science and the University of Krasnoyarsk are located. In total, thirty participants took part in this NATO-ARW. It was a great experience for most participants of the workshop to discover a new centre of academic research that was until recent years closed for visitors from the West and that turned out to be well equipped and involved in sophisticated research in the field of interest of the workshop.

The editors

LIST OF CONTRIBUTORS (IN ORDER OF PUBLICATION)

J. Bartolomé, F. Bartolomé, L.M. García, F. Luis
Instituto de Ciencia de Materiales de Aragón, CSIC-Universidad de Zaragoza, E-50009 Zaragoza, Spain

F. Petroff, V. Cros, H. Jaffrès, A. Vaurès
Unité Mixte de Physique. CNRS/ Thales. 91404 Orsay, France and Université Paris-Sud. 91405 Orsay, France.

Janis Kliava and René Berger
CPMOH, UMR Université Bordeaux I-CNRS 5798, 33405 Talence Cedex, France

G. Petrakovskii, M. Popov, V. Zinenko
Institute of Physics SB RAS, 660036 Krasnoyarsk, Russia

B. Roessli and J. Schefer
Laboratory for Neutron Scattering, ETH Zurich & Paul Scherrer Institute, CH-5232Villigen, Switzerland

M. Boehm
Laboratory for Neutron Scattering, ETH Zurich & Paul Scherrer Institute, CH-5232Villigen Switzerland and Institut Laue-Langevin, 38042 Grenoble, France

U. Staub
Swiss Light Source, Paul Scherrer Institute, CH-5232 Villigen, Switzerland

V. Eremenko and V. Sirenko
Institute for Low Temperature Physics & Engineering National Academy of Sciences of Ukraine

M. Chiba, H. Nakamura, Y. Fujii and H. Kikuchi
Department of Applied Physics, University of Fukui, Fukui 910-8507, Japan

Y. Yamamoto and H. Hori
Physics Department, School of Materials Science, JAIST, Ishikawa 923-1292, Japan

L. Bezmaternikh
Institute of Physics, SBRAS, 660036, Krasnoyarsk, Russia

A. Sherman
Institute of Physics, University of Tartu, Riia 142, 51014Tartu, Estonia

J.J.M. Franse
Van der Waals – Zeeman Instituut Valckenierstraat 65, 1018 XE Amsterdam, The Netherlands

Yu.A. Izyumov and D.S. Alexeev
Institute for Metal Physics of the RAS, Ural Division, 620219 Ekaterinburg,Russia

I.O. Troyanchuk and V.A. Khomchenko
Institute of Solid State and Semiconductor Physics, NAS, P. Brovka str. 17, 220072Minsk, Belarus

H. Szymczak
Institute of Physics, Polish Academy of Sciences, Lotnikow 32/46, 02-668 Warsaw, Poland

V. R. Shah, Christian Schanzer, Peter Böni
Physik Department E21, Technische Universität München, James-Franck Strasse, D-85747 Garching, Germany

Hans-Benjamin Braun
Institut Laue-Langevin, 6 rue Jules Horowitz, BP 156-38042 Grenoble Cedex 9, France

S. Bondarenko
Institute for Low Temperature Physics & Engineering National Academy of Sciences of Ukraine

N. Nakagawa
Center of Non Destructive Evaluation, Iowa State University, Ames, USA

V.V. Kruglyak and R.J. Hicken
School of Physics, University of Exeter, Stocker road, Exeter, EX4 4QL, UK

A.N. Vasiliev, R.Z. Levitind and V.V. Khovailo
PhysicsFaculty,MoscowStateUniversity,Moscow119899,Russia

V.V. Khovailo
Institute of Radio Engineering and Electronics of RAS, Moscow125009, Russia

R. Szymczak
Institute of Physics, Polish Academy of Sciences, Lotnikow 32/46, 02-668 Warsaw, Poland

S.B. Feodosyev, I.A. Gospodarev and M.Yu. Shvedun
Institute for Low Temperature Physics & Engineering National Academy of Sciences of Ukraine

M. Tovar
BENSC, Hahn-Meitner-Institute, Berlin, Germany

W. McCallum
Ames Lab., Iowa State University, Ames, USA

ACKNOWLEDGEMENT

This workshop constitutes a continuation of a series of ARW meetings on advanced magnetic and superconducting materials. The organisers acknowledge the generous financial support from NATO side and express their hope for further cooperation within the NATO Scientific Program.

The organisers

MAGNETIC DYNAMICS OF Co NANOSPHERES: ORIGIN OF THE ENHANCED ANISOTROPY

J. Bartolomé, F. Bartolomé, L.M. García, F. Luis
Instituto de Ciencia de Materiales de Aragón
CSIC-Universidad de Zaragoza
E-50009 Zaragoza, Spain
barto@unizar.es

F. Petroff, V. Cros, H. Jaffrès, A. Vaurès
Unité Mixte de Physique
CNRS/ Thales. 91404 Orsay, France and Université Paris-Sud. 91405 Orsay, France.

Abstract: The present work deals with the observation of enhancement of the magnetic anisotropy of Co nanoparticles and its origin. The samples were granular multilayer samples prepared by sequential deposition, by sputtering, of amorphous Al_2O_3 and Co layers on a Si substrate. Co nanoparticles are self-organized in a quasi-regular spatial order of approximately hexagonal close-packed symmetry. The particles studied range in average diameter $\langle D \rangle$ between 0.7 nm and nearly 5 nm, with a narrow size distribution. This well controlled morphology has enabled us to circumvent ambiguities in sample configuration and, by means of a simple model for fluctuating moments, explain the dynamics of the Co particle moments in terms of an activation energy with contributions from anisotropy K_{eff}, dipole-dipole interactions E_{dip}, and a bias magnetic field H. The anisotropy is enhanced by one to two orders of magnitude with respect to the bulk fcc Co due to strong pinning of the surface Co magnetic moments anisotropy, and increase as $1/D$ as the particle diameter decreases. The origin of this enhancement is related to an increase of the orbital magnetic moment at the surface atoms. Capping the Co nanospheres with a Cu film increases further the particle anisotropy and the orbital magnetic moment of the surface atoms.

Keywords: nanoparticles, magnetic anisotropy, orbital moment, size effects, superparamagnetism, magnetic dynamics.

J. Franse et al. (eds.), Smart Materials for Ranging Systems, 1–25.

1. Introduction

It is evident that our globalized world is based on informatics and communications, and both technologies are supported by the existence of fast computers. However, it is often overseen that an essential part of the computer is its memory. A computer memory should be able to store huge amounts of data; moreover, they must be accessible in an easy and fast way and occupy a reduced volume. An information bit recorded in a memory must be able to remain unspoiled for about 10 years, and be read or written reversibly in a minimum of time. At present, the only materials that fulfill all these requirements are magnetic materials.

Magnetic materials have a long tradition in the recording technology. Already in the first electrical recorder (1898) a steel wire with embedded ferrite particles was used. Since then, and save for technological details, the physical principles remain the same. The medium on which we want to record the information, be it a disk, a tape or a card, consists of a supporting material on which the magnetic material is fixed. The recording medium moves with respect to the read/write head. The head generates a magnetic field sufficiently intense to magnetize a region of the medium in a given direction. After a time delay to allow the magnetized region to be out of the magnetizing influence of the head, a new region can be magnetized and a new bit recorded. The maximum quantity of information that can be recorded depends, thus, on the minimum surface of the magnetic region that contains a bit, and the maximum speed of recording, that is, the magnetic reversal speed of the magnetic material [1].

The principles of magnetic recording are, thus, based, on the technological capacity of perturbing a localized region of the magnetic medium in a controlled way, and sensing at a later occasion the remanence of that perturbed region localized in position. The two steps are typical processes in ranging systems, and thus, are of interest for the present workshop.

The strategies to increase the amount of recorded data are manifold. While remaining within the concept described above of recording on a surface, the trend has been to reduce the size of the magnetic region supporting a bit. In Dynamic Random Access Memories (DRAM) the memory density has increased by a factor of four every three years (Moore's Law). To date the most compact ultra high density commercial recording media is capable of storing 5 Gbits in a 1 inch diameter disk, and a prototype has been demonstrated to read/write on a magnetic media of 101 Gbit/in^2 [2]. It could seem feasible to reduce the size of the memory elements progressively to particles of smaller diameters till we reach the size of the atom as memory unit. However, there are physical limits to this trend. If the magnetically oriented particle is so small as to be single domain it loses its polarization spontaneously if the temperature is above a threshold temperature, the blocking temperature T_B, and with it also the information is lost. Unluckily T_B becomes lower the smaller the particle, and for each magnetic material there is a threshold size below which it cannot be used as a recording material. The parameter that determines the ability of a magnetic bit to store information is the magnetic anisotropy, which represents the energy per unit volume that must be supplied to the memory unit to invert its magnetic moment direction. The conclusion is that, in order to use magnetic particles as recording media, we should either cool the material or increase the magnetic anisotropy, or both things altogether.

An alternative strategy, albeit still in its starting stages, is to store information in three dimensions (3D). So far this has been already achieved by optical means, by burning regions of a polymeric matrix by convergent beams. However, the information so stored cannot be modified. Holographic methods are also being tested for the same purpose. It could be envisaged that magnetic recording in 3D could also be achieved if spin polarized currents can be mastered spatially, since they could induce reversal of the magnetic moment of the individual particle in the recording media.

In this paper we review our work on a very special type of magnetic particles, Co nanospheres, that can be dispersed in a matrix regularly in three dimensions and retain its magnetic polarization at liquid helium temperatures. Though evidently not suitable for commercial magnetic recording, they have provided very interesting results on the dynamics of nanoparticles and on effects induced by their nanometric sizes. Our work is exploratory but it provides some corner stones in the development of the field. The first section deals with properties of single particle, the second with the interactions with the embedding matrix, and the third with the interparticle interactions.

2. The Co particles

2.1 Morphology

The samples were prepared by sequential deposition by sputtering of amorphous Al_2O_3 and Co layers on a Si substrate. The Co layer is formed by clusters that are well dispersed in the Al_2O_3 layer. By varying the amount of Co deposited on each layer, determined by the deposition time or the nominal thickness t_{Co} that the layer would have if it were continuous, one gets Co aggregates with average diameter $\langle D \rangle$, which can be changed in the range from $0.7 < \langle D \rangle < 5$ nm. On top of each Co layer a layer of Al_2O_3 of variable thickness is deposited. The particles are spherical on average, free from oxidation and those with $D < 4$-5 nm, are well crystallized at the core in the fcc phase [3]. Within a layer the Co clusters order quite regularly, with six nearest neighbors on average, and the edge to edge distance between particles, about 2 nm, is independent of the amount of deposited Co. A conspicuous property is that the clusters of a layer nucleate preferentially in the valleys left by the previous layer. We denote by N the number of Al_2O_3/Co bilayers in a sample. We may consider the sample as a self-organized set of periodically arranged Co spheres in a topology that varies progressively from two-dimensional (N = 1) to three-dimensional for increasing N. Loosely speaking, the self-organized cluster array in the Al_2O_3 matrix resembles that of a hexagonal close-packed structure (Fig. 1).

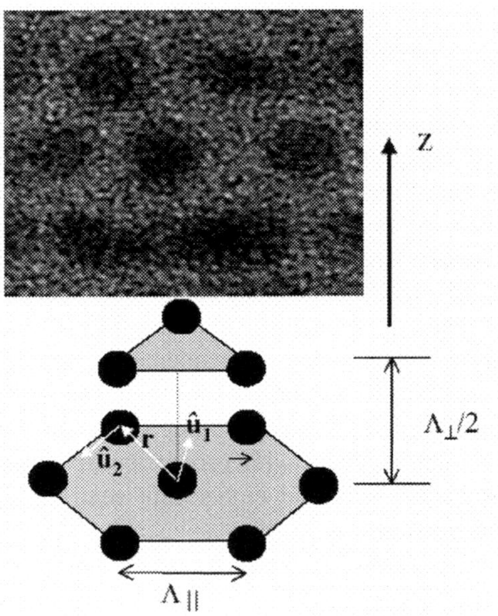

Figure 1. Top panel, TEM cross section of a multilayer, showing the quasiordered arrangements of the Co cluster particles (dark circles). Lower panel, 3-dimensional schematic view of two layers.

An issue that needs some attention is the fact that the amount of Co deposited on the surface is larger than the Co mass that forms the clusters, as observed by transmission electron microscopy (TEM). However, this discrepancy diminishes for the larger t_{Co}; that is, the amount of Co that forms non-aggregated atoms or very small clusters decreases with increasing size of the observed spherical clusters. The unavoidable dispersed non-aggregated Co in the matrix shows a different physical behavior than the clusters. By studying the magnetic response of the samples in a wide range of T and H it is possible to discern the contributions of the two subsets in most experiments, as we shall show below.

Evidently, all spherical particles in a sample do not have the same diameter but are distributed around an average dimension. The TEM study permitted to determine the average diameter, and that it depends linearly with t_{Co}. Moreover, the distribution was found to be Gaussian with a distribution parameter $\sigma \approx 0.26 \langle D \rangle$.

All the samples measured showed characteristic superparamagnetic behavior with a blocking temperature T_B. An independent method of determining the parameters of the particle size distribution *g(D)* is by means of the analysis of magnetic measurements under equilibrium conditions, i.e. at temperatures above the superparamagnetic blocking temperature T_B. For this purpose we performed magnetization measurements as a function of field *M(H)* at different temperatures [4,5].

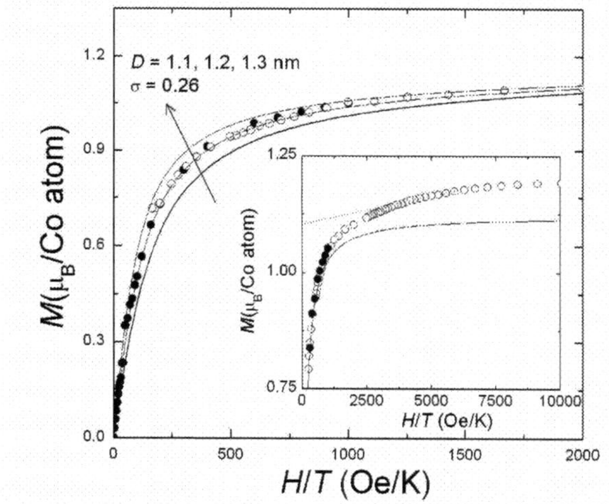

Figure 2. Magnetization curves of the t_{Co} = 0.2 nm sample fit by averaging a Langevin curve over a Gaussian distribution of sizes with σ =0.26. Inset, dotted line, fit to the paramagnetic contribution of non-aggregated Co above 2500 Oe/K, thick line, contribution due to the Co clusters.

The *M(H)* curves measured well above T_B merge into a common curve when plotted as a function of H/T, indicating that the effect of the anisotropy is weak at these temperatures (Fig. 2) [5]. The saturation magnetization shows a Curie tail at the lowest temperatures. This contribution is likely caused by the non-aggregated Co atoms. The *M(H)* curves may be used to estimate the fraction x_{para} of the total Co atoms that do not form clusters. The *M(H/T)* curve can be fitted by the sum of a pure Langevin curve, indicating that the size distribution of the particles is quite narrow, plus a paramagnetic contribution due to the non-aggregated Co. The

Co magnetization per Co atom in the cluster is assumed to be $M_{sb} = 1.7\ \mu_B$. At high enough field (see Fig. 2, inset) the cluster contribution is saturated while the paramagnetic contribution increases linearly with *H/T*. By extrapolation to $H = 0$ of a fitted straight line in this region we eliminate the

Table 1. Parameters of the Gaussian distribution of particle sizes obtained by TEM (Briatico et al., 1999) (a) and from the fit of the magnetization and susceptibility data (b). The width σ is given in units of $\langle D \rangle$. x_{para} is the fraction of non-aggregated Co.

	a			b		
t_{Co}(nm)	$\langle D \rangle$ (nm)	σ	x_{para}	$\langle D \rangle$ (nm)	σ	x_{para}
0.1				0.8(1)	0.35(5)	0.7(2)
0.2	0.83(20)	0.3		1.3(1)	0.3(1)	0.7(2)
0.3	1.4(3)	0.3		1.4(1)	0.32(5)	0.22(4)
0.4	1.4(3)	0.22	0.77	2.2(1)	0.2(1)	0.5(1)
0.7	2.9(6)	0.23	0.45	3.1(3)	0.2(1)	0.25(5)
1	4.2(8)	0.27	0.27	5.2(3)	0.25(5)	0.13(3)

paramagnetic contribution and obtain the saturation magnetization of the cluster contribution $M_s = (1 - x_{para})1.7\ \mu_B$. Solving the equation we obtain x_{para}. The results for x_{para} are shown in Table 1, where the values obtained from the TEM measurements are included for comparison. The x_{para} values found in TEM are systematically larger than those from the magnetic

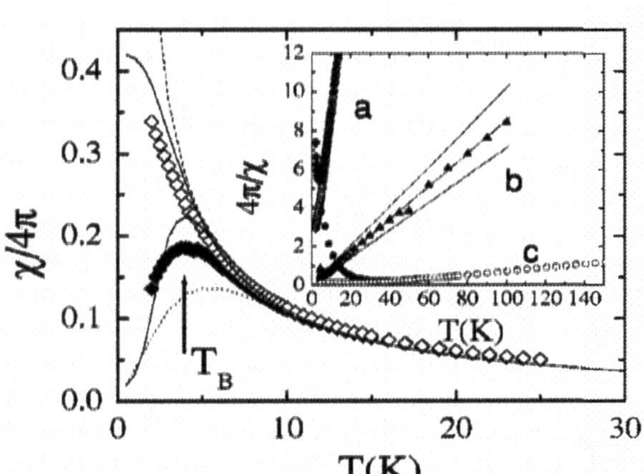

Figure 3. dc susceptibility of a multilayer with $t_{Co} = 0.1$ nm and N = 100 and H=0.01 T: open symbols, FC; solid symbols: ZFC. Lines are explained in the text. Inset, inverse susceptibility for three multilayers. a) $t_{Co} = 0.1$ nm and N = 100, b) $t_{Co} = 0.3$ nm and N = 40, c) $t_{Co} = 0.7$ nm and N = 30. The lines correspond to $1/\chi_{eq}$ for $\langle D \rangle$=1.4 nm and three σ's. Best fit with σ=0.3.

measurements, which implies that more clusters give contributions to the magnetization than are visible by TEM. The final fit of the Langevin curve plus paramagnetic contribution is very satisfactory and yields quite precisely the average diameter $\langle D \rangle$ of the particles in the sample. This fit is especially sensitive to the average value of the particle diameter, as can be seen in Fig. 2. It is also noteworthy that $\langle D \rangle$ is quite independent of the value of N.

A fit taking into account a Gaussian size distribution function $g(D)$ with width σ was also performed. Our results are in reasonable agreement with the TEM ones [3]. The correction due to the paramagnetic contribution is applied hereafter in all the experiments considered.

In contrast to the $M(H)$ curves above T_B, that are rather insensitive to the size distribution function, the high-temperature equilibrium susceptibility is quite sensitive to the contribution due to the largest particles. Using this fact the fit of $1/\chi$ vs T (see Fig. 3 inset) yields a rather accurate estimate of the distribution width σ. This is favorable since the paramagnetic contribution can be neglected in this range. The values of $\langle D \rangle$ and σ obtained form the fits are included in Table 1. The width of the distribution is found to be rather constant and in good agreement with the values found by TEM. In fact, σ increases slightly with decreasing t_{Co}, as was also observed in the TEM experiments. Also of importance is the result that the average diameter and distribution width are practically identical for mono-or multilayers with the same t_{Co}.

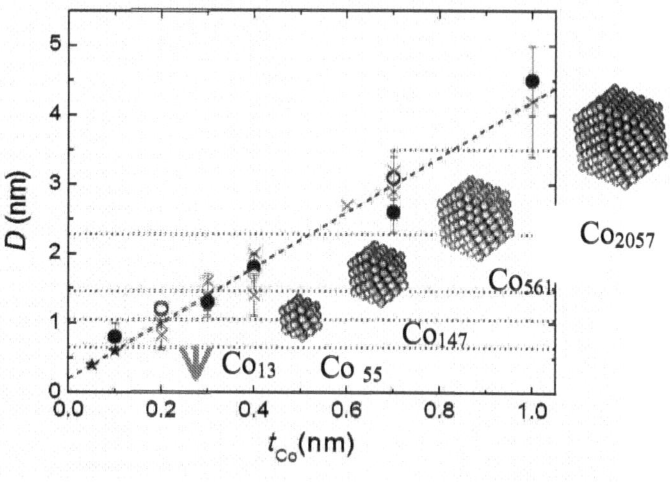

Figure 4. Average particle diameter as a function of nominal thickness, as determined from TEM and magnetic measurements. Included are closed shell clusters for comparison purpose. (×) TEM results, (other symbols) Magnetic measurements.

2.2 Size-dependent anisotropy

To determine the magnetic anisotropy of the clusters we measured dc susceptibility χ and ac susceptibility, with components χ' in phase, and χ'' out-of-phase, as a function of temperature and frequency. (Torres et al., 2002).

The temperature at the maximum in the zero field dc susceptibility measurement determines the blocking temperature T_B (Fig. 3). It was found to increase with the average diameter of the clusters of a given sample. The simplest model to explain superparamagnetism assumes that each particle has a uniaxial anisotropy with a direction independent of that of the other particles. The energy that is needed to reverse the magnetization U, determines the relaxation time of this process,

$$\tau = \tau_0 \exp(U / k_B T) \tag{1}$$

Here, $\tau_0 \approx 10^{-10} - 10^{-13}$ s is the inverse attempt frequency that depends on the damping of the magnetic moments by the phonons. The superparamagnetic blocking occurs when τ equals the measuring time of each experimental point, t_e, therefore $T_B \cong \alpha U / k_B \ln(t_e / \tau_0)$, where α is a constant that depends on the width of the particle size distribution.

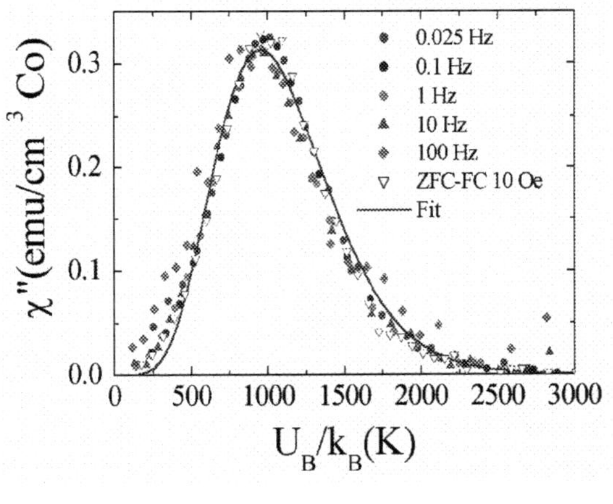

Figure 5. χ'' of a multilayer sample N=30 and t_{Co} = 0.7 nm, vs scaling parameter U_B. Solid line, fit to eq. 2 with a Gaussian distribution $g(D)$.

Since we have observed that T_B increases with t_{Co} the approximation of $U = K_{eff} V$ has been considered. Here K_{eff} is an effective anisotropy that encompasses all possible contributions, such as intrinsic and surface anisotropies [5].

To determine the effective anisotropy constant we need to compare the activation energy of particles with a given size (i.e. $D = \langle D \rangle$) with their volume $V(\langle D \rangle)$. We introduce the distribution of activation energies $f(U)$ that can be determined from the $\chi''(T)$. The simplifying assumption that those particles with $U > U_b$ are blocked and the rest are in equilibrium is applied. This leads to

$$\chi'' \cong \frac{\pi}{2} k_B T \chi_{eq}(T, U_B) f(U_B) \qquad (2)$$

where $U_B = k_B T \ln(1/\omega\tau_0)$ is the activation energy of those particles having exactly $\tau = t_e = 1/\omega$ at a given temperature and $\chi_{eq} = M_{sb}^2 V / 3k_B T$, is the equilibrium susceptibility [6]. The frequency dependent ac susceptibility curve should then coalesce into a single curve proportional to $f(U)$ if plotted versus the scaling variable U_B. This is nicely achieved in Fig. 5, from which we determine the activation energy $U(\langle D \rangle)$. K_{eff} is then obtained as the ratio $K_{eff} = U(\langle D \rangle) / (\pi \langle D \rangle^3 / 6)$.

The K_{eff} obtained for samples with varying average diameters are plotted in Fig. 6. In the same plot the anisotropy constant for fcc Co $K_{eff} \approx$

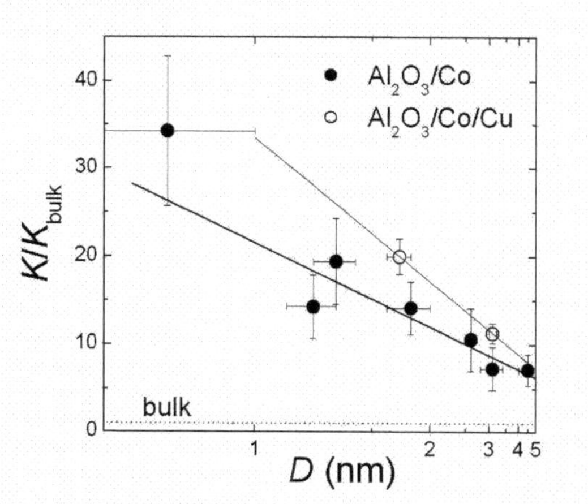

Figure 6. Size dependence(shown in reciprocal scale of diameters) of the relative effective anisotropy for the different average diameters for Al_2O_3/Co and Cu capped $Al_2O_3/Co/Cu$. multilayers. Lines are guides to the eye.

6.5 x 10^4 J/m^3 is included for comparison purposes. We arrive to this value taking into account that for cubic anisotropy $K_{eff} = K/4$ [7] where K is the second order intrinsic anisotropy constant, and that $K = 2.6 \times 10^5$ J/m^3 [8]. Therefore, as Fig. 6 shows, K_{eff} of nanoclusters with $\langle D \rangle < 5$ nm is within one to two orders of magnitude larger than the value of the bulk. Moreover, K_{eff} increases with decreasing $\langle D \rangle$ according to the expression

$$K_{eff} = K_{\infty} + \frac{6K_s}{\langle D \rangle} \tag{3}$$

with $K_{\infty} = 5(\pm 2) \times 10^4$ J/m^3 and $K_S = 3.3(\pm 0.5) \times 10^{-1}$ mJ/m^2. The first term corresponds to the bulk intrinsic contribution, and the second is inversely proportional to the average diameter. If multiplied by the volume to obtain the activation energy, this last term gives a contribution proportional to the surface atoms. We conclude from our experiments that the surface contributes strongly to the net anisotropy of the clusters. Similar results had been obtained earlier for Co particles of comparable diameters [9,10].

With this result in hand, we revisited the dc χ and ac $\chi'(T)$ of the smaller particles and found that the measurements could be beautifully fit when taking into account the preeminence of the surface anisotropy (Fig. 3). The implication is that the distribution function can be written as $f(U) \propto D^2 g(D)$ and as a result its width becomes smaller than in the case of volume anisotropy. The conclusion is that for small nanoparticles the detailed behavior of the ac susceptibility in the proximity of its maximum is strongly dependent on the surface atoms of the clusters. Moreover, one has to avoid using the naive method of ascribing T_B to the maximum of the susceptibility to draw conclusions on the activation energy.

2.3 Origin of the anisotropy

Once that the enhancement of the anisotropy due to the nanometric dimensions of the particle has been proven in our samples beyond doubt, the pertinent question is whether we can find the physical origin of this enhancement. A number of possible mechanisms can give rise to this strong variation in anisotropy: size effects due to the increasing ratio of surface to bulk atoms, electronic confinement within the cluster leading to 3d band narrowing, surface oxidation, stress induced anisotropy, or even shape anisotropy due to a departure of sphericity. The two latter ones can be readily disregarded. The interatomic Co distance is estimated from the analysis of EXAFS spectra of these clusters. They are shorter than the bulk fcc Co and comparable to that found for free Co clusters [11,12]. Therefore it seems that the amorphous Al_2O_3 does not exert tensions on the Co particle as would be expected from mismatch effects were the matrix crystalline. Besides, the maximum asphericity (a needle) yields a shape anisotropy that is one half of the experimental value found for the smallest particles. Therefore, also this source of anisotropy can be ruled out since the observed asphericity is extremely small [3].

In the EXAFS and Electron Energy Loss Spectroscopy (EELS) spectra [11] show no trace of Co oxide, however XAS spectra of the smallest particles present characteristic peaks of CoO bonds in the L_3 contribution. The fraction of Co bonded to oxygen atoms is proportional to the fraction of paramagnetic atoms, or very small clusters. However, for $\langle D \rangle > 2$ nm this contribution disappears and the XAS spectrum is metallic-like. We performed X-ray Magnetic Circular Dichroism (XMCD) at the $L_{2,3}$ Co edges on multilayer samples for the sake of increased signal to noise ratio [13,14]. The resulting spectra, scaled to a common L_2 peak, show unambiguously an increase of the difference at the L_3 edge with decreasing size (Fig. 7).

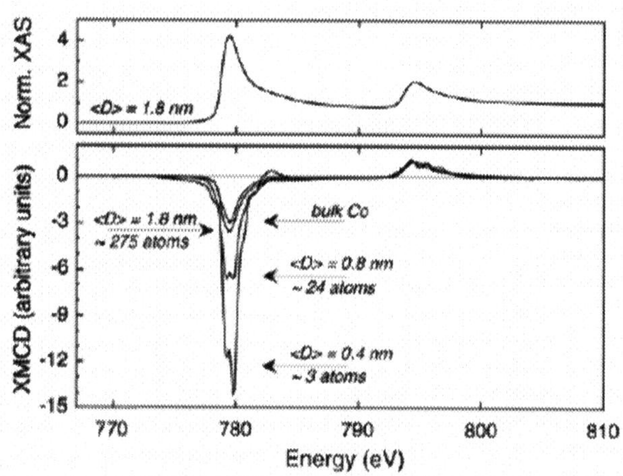

Figure 7. Upper panel: XAS spectra of a N=25 multilayer with D=1.8 nm. Lower panel:XMCD spectra of several Al_2O_3/Co granular multilayers with different $\langle D \rangle$, and a 3 nm thick Co layer as bulk reference.

An analysis of the integrated XMCD intensities gives the ratio of the angle averaged orbital to angle averaged spin magnetic moment $\langle m_L \rangle / \langle m_S \rangle$, that is depicted in Fig. 8 as a function of $\langle D \rangle$. This ratio is independent of the effective number of holes n, which may affect the value

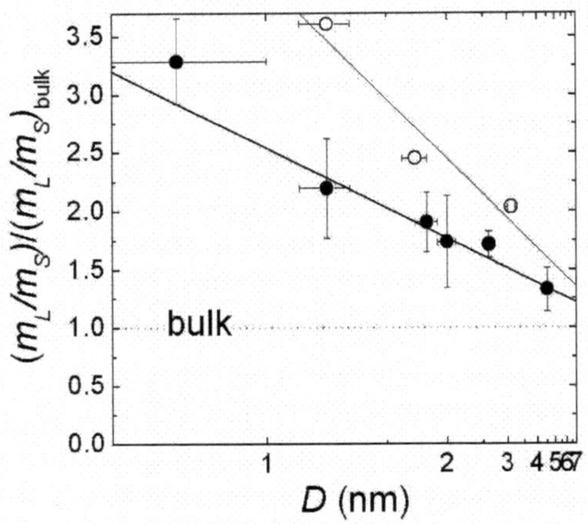

Figure 8. Size dependence (shown in reciprocal scale of diameters) of the ratio $\langle m_L \rangle / \langle m_S \rangle$ scaled to the bulk. Lines are guide to the eyes.

of $\langle m_S \rangle$. Furthermore, we expect this change to be small since it has been shown experimentally that *n* differs by only 1% between bulk and Co adatoms on Pt [15]. Therefore, the increase in $\langle m_L \rangle / \langle m_S \rangle$ probably reflects an increase in the average orbital moment at the particle. As happens with K_{eff}, $\langle m_L \rangle$ is also inversely proportional to $\langle D \rangle$, though the $\langle m_L \rangle$ variation is smaller by a factor of ten (Fig. 8). In spite of this difference, a clear relation between both quantities is shown to exist. An increase in the perpendicular anisotropy of a Co thin film has been related to the non-quenched orbital moment of the Co moments at the surface [16,17]. We conclude that the same effect, although more intense, takes place at the surface of our Co nanoparticles. A rough estimation gives $\langle m_L \rangle \approx 0.39\ \mu_B$ for the surface atoms, that is, approximately three times that of the bulk. The increase in $\langle m_L \rangle$ can give rise to the enhanced anisotropy that is known to originate from the difference in spin-orbit energy between the orientations perpendicular and parallel to the surface.

Besides, in the same L_3 spectra it can be observed that the lines narrow and split into two as the diameter decreases. A possible origin of split lines is the presence of CoO in the paramagnetic atoms. Indeed, the split lines can be understood in terms of the 3d electrons in the smallest clusters (less than 10 atoms) tending to atomic like (discrete) spectra, similar to the features observed in the absorption spectra of magnetic impurities embedded in alkali metal films [14,19].

We conclude in this section that we are dealing with quite spherical, metallic Co nanoparticles with a very narrow size distribution, that give rise to superparamagnetic behavior. The effective anisotropy of the particles increases by one to two orders of magnitude with respect to the bulk with decreasing size. The origin of this anisotropy enhancement can be traced to the anisotropy of the orbital moment and the increase of the absolute orbital moment of the Co surface atoms, at least for $\langle D \rangle > 2$ nm.

3. Interactions with the embedding media

The Co nanoparticles are embedded in an insulating matrix of amorphous Al_2O_3. As mentioned above, this matrix does not affect the metallic character of the particle since the Co atoms at the surface do not seem to form Co oxides even for the smallest particles. Stress anisotropy can also be disregarded as source of anisotropy. However, the surface Co atoms play the most important role in generating the large anisotropy described above. Besides, it is known from Co/X multilayer studies [20] that a metal in direct contact with a Co thin film induces perpendicular (X = Pt, Au and Ir), or parallel (X = Cu, Ag and Mo) anisotropy, on the one hand, and that the thin film dimensionality may even modify the sign of the anisotropy. The reduced thickness dimensionality gives rise to confinement of the 3d electrons, narrowing of the 3d bands and modification of the DOS at the surface atoms indicating an increase of the orbital moment anisotropy.

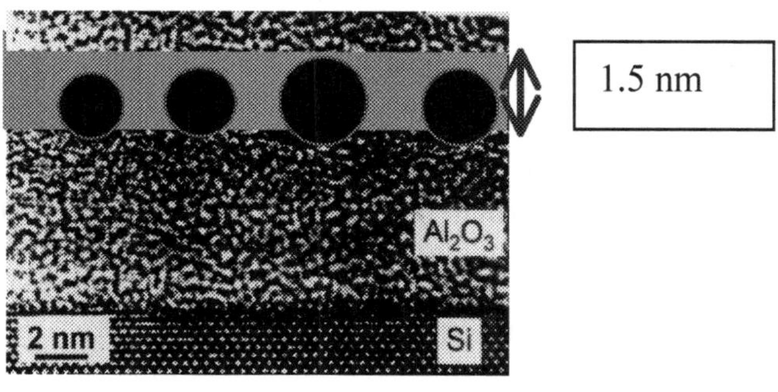

Figure 9. Transversal scheme of a Cu capped Al_2O_3/Co/Cu trilayer.

So, it seemed plausible that the nature of the embedding matrix could induce similar modifications of the anisotropy as observed in thin films. A way to do so, while maintaining the spherical Co cluster morphology, was to cap each Co by a Cu layer, thus creating $Co/Cu/Al_2O_3$ trilayers. The so called capping Cu layer has a thickness of 1.5 nm and covers partially the spheres (Fig. 9). In this case it was not possible to determine by TEM the size of the nanoparticles. Therefore we relied on the magnetic measurements, as described in the former section, to determine the average diameter and distribution width of the capped spheres. It was gratifying to verify that the particle morphology was practically identical for the same Co depositing nominal thickness t_{Co}.

On the other hand, for multilayers with N>10 (3-D limit) and the same $\langle D \rangle$ the χ'' shifts clearly to higher temperatures (Fig. 10), a clear indication that the capping modifies the magnetic properties of the Co particles.

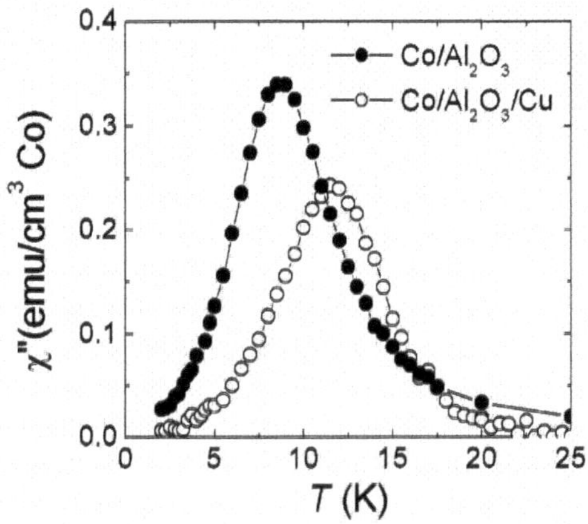

Figure 10. χ'' as a function of temperature for two samples, uncapped Al_2O_3/Co multilayer (N = 100) and Cu capped (N =30), with t_{Co} =0.4 nm, $\langle D \rangle$ = 1.8(1) nm, at an excitation frequency of ν=10 Hz.

Following the scaling method described above, K_{eff} was determined for three multilayers with particle diameters ranging between 1 and 3 nm. The poor signal to noise ratio did not allow to get a clear result for the smallest particle set. However, what we found in the other two samples that could be measured was that the anisotropy becomes about 20% larger than that of the uncapped ones (Fig. 6). Moreover, the trend of increasing anisotropy for decreasing $\langle D \rangle$ is also fulfilled as in the uncapped case, therefore, the effect is clearly related to the surface atoms being affected by the Cu layer. This exciting result implies that the electronic properties of the surface Co atoms are modified by the contact with the connecting Cu layer, thus allowing the possibility of controlling, to some extent, the effective anisotropy.

To deepen the understanding of this effect we have performed an X-ray Magnetic Circular Dichroism (XMCD) study of the L_3 and L_2 Co thresholds of the capped samples, identical to that for the uncapped ones. We show in Fig. 11 the results for the capped and uncapped D = 2.8 nm samples, where an increase in the L_3 XMCD signal can be clearly discerned, if normalized to a common L_2 signal. The relative area of the L_3 peak, with respect to the L_2 one, is definitely larger for the capped clusters. This result proves through the sum rules analysis that the $\langle m_L \rangle / \langle m_S \rangle$ ratio increases with the Cu capping.

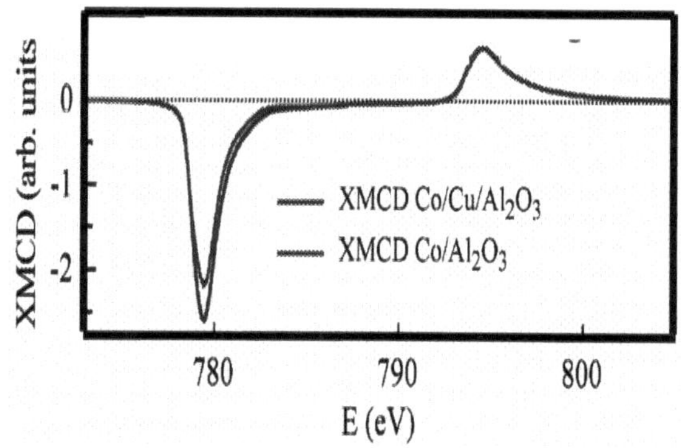

Figure 11. X-ray magnetic circular dichroism of uncapped Al_2O_3/Co and Cu capped granular multilayers, with $\langle D \rangle$ = 2.8 nm, normalized to the L_2 peak.

In Fig. 8 we point out that the $\langle m_L \rangle / \langle m_S \rangle$ ratio is much larger in the cluster array than in bulk Co, the increase being proportional to $1/\langle D \rangle$. This increase in $\langle m_L \rangle / \langle m_S \rangle$ is not as strong as the dependence found for K_{eff}, again, as for the uncapped Co particles, K_{eff} is modified tenfold more than the $\langle m_L \rangle / \langle m_S \rangle$ ratio with decreasing particle diameter. The enhancement of K_{eff} and $\langle m_L \rangle / \langle m_S \rangle$ when Co is put in contact with Cu has been observed also by XMCD in Co thin films deposited on Cu [21]. It was argued that this effect originated from the hybridization of the Co 3d electron states with the Cu conduction band, that modifies the energy and the occupation numbers of the bands.

The Co nanoparticles are preformed on the Al_2O_3 substrate, and only later they are capped by Cu. This procedure seems to let the spheres unscathed by the Cu layer, while the Co atoms do not alloy with the Co atoms at the surface. The modification of the anisotropy by capping with a metallic layer may open a simple way of tailoring the magnetic properties of the particles.

4. Interparticle interactions and magnetic relaxation

The interparticle distance in our samples amounts to a few nanometers. This distance is sufficiently small and the magnetic moments of the particles are large enough for dipolar interaction between particles to be effective in modifying the activation energy *U*. So far we have dealt with their morphology, and interaction with the supporting media. We now turn our attention to the variation of the magnetic relaxation of the particles due to interparticle dipolar interactions.

In the last few years this subject has attracted considerable interest as the question can be of relevance in the behavior of high-density recording media. Despite this extensive work, the question of whether interactions tend to increase or decrease *U* is a matter of debate. Some authors propose an increase of *U* with the number of neighboring particles [22], whereas others suggest the opposite [23,24]. In addition, collective behavior of sets of particles have been reported [25] as due to these interactions for high enough particle density. Experimental results are hampered by ambiguities in the samples, for example, clustering of the particles in ferrofluids.

Our regularly dispersed Co particle samples enable a better control of the sample morphology and, therefore circumventing some of the difficulties [12]. First, the self organization of the clusters below 5 nm diameter guarantee the absence of clustering. In addition, as was described above, the particles in a given sample have a very narrow size distribution, although it cannot be considered as negligible. In a monolayer each Co particle has, on average, six neighbors at similar distance; e.g. in the t_{Co} = 0.7 nm ($\langle D \rangle \approx 3$ nm) samples the center to center distance is $\Lambda_{\|} \approx 5.4$ nm. In a multilayer the average number of neighbors is three at the upper adjacent layer, and three more at the lower adjacent one (Fig. 1).

The interlayer distance can be modified by increasing the thickness of Al_2O_3. In the following paragraphs we shall deal with a series of samples with *N* ranging between 1 and 20, with the same distance between particles in adjacent layers, $\Lambda_{\perp} \approx 4.5$ nm. For comparison purpose we prepared a sample with a thicker Al_2O_3 layer $\Lambda_{\perp} \approx 10.3$ nm. With this series of samples we could study how the magnetic relaxation rates are modified with respect to that of a two-dimensional array of Co particles in a monolayer as we progressively add new layers to reach the three-dimensional limit. We first checked, using *M(H)* and $\chi(T)$ curves measured above T_B, that the samples with different *N* had the same $\langle D \rangle$ and σ. The maximum in χ' and χ'' increased with increasing number of layers, which already solves the

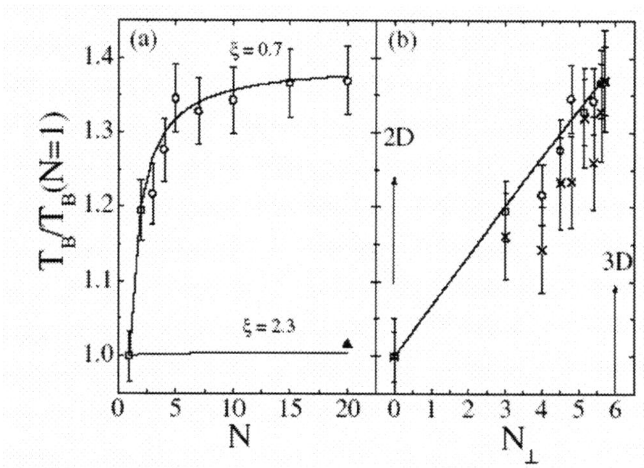

Figure 12. Left panel, Variation of T_B with the number of layers *N*. Open circles t_{Al2O3} =3 nm, solid triangle, t_{Al2O3} = 10 nm. Right panel, T_B's and activation energies (×) as a function of the average number of nearest neighbors in adjacent layers.

question of whether the activation energy increases with increasing number of neighbors. We compare in Fig. 12 the T_B 's determined from χ' measurements at 10 Hz for the samples with different *N* layers, and the one with $N = 20$ and thick alumina spacer layers. The stronger relative increase takes place with the addition of one layer, implying an increase of 20%, thereafter the increase is more moderate and tends to a stable value for $N >$ 10, that is 30% higher than for the monolayer. The average number of neighbors at adjacent layers in a hcp cluster lattice is $N_{\perp}$=6 (N-1)/N. When plotted against $N_{\perp}$ the increase in T_B is approximately linear. In contrast, the T_B of the sample with $N = 20$ and thicker layers of alumina only increases by 2%, i.e. the intralayer interactions remain dominant in this sample, and the multilayer sample shows the same behavior as the two-dimensional monolayer.

The activation energy of each of the samples was determined from the Arrhenius law which is fulfilled by the T_B's (Fig. 13)

$$\ln(1/\omega) = \ln(\tau_0) + \frac{U_{eff}}{k_B T} \quad (4)$$

U_{eff}, the effective activation energy, has contributions due to the magnetic anisotropy *(U_0)* and the dipole-dipole interactions *(U_{int})*. When the U_{eff}'s are plotted against $N_{\perp}$ we note again a roughly linear dependence (Fig. 12, right panel).

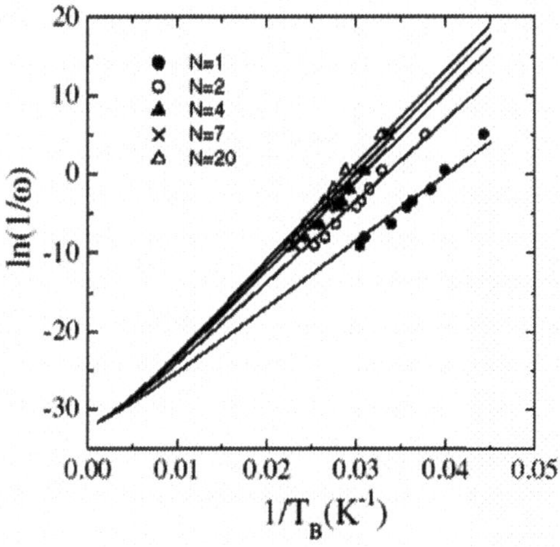

Figure 13. Arrhenius plot of the measuring time ($1/\omega$) as a function of T_B, as obtained from χ' and ZF dc $\chi(T)$ measurements, for multilayers with different N.

It is evident that the dipolar interaction increases the activation energy and slows the relaxation time of moment reversal. To explain the dependence quantitatively a simple model was proposed [26], inspired by the references contained in [22], where the relaxation time considered is for independent clusters and the dipolar interaction modifies U_{eff}. In the model, a test cluster of uniaxial anisotropy is surrounded by clusters with their anisotropy axis randomly oriented. The susceptibility is dominated by the largest particles in a sample since it is proportional to $\int V^2 g(D)dD / \int Vg(D)dD$. Then, for a sample with a Gaussian distribution around $\langle D \rangle = 2.6$ nm the particles that contribute most to χ' have $D \approx 3.5$ nm (maximum in V^2D). Approximately 90% of the clusters in the sample are smaller than these, and, therefore, relax faster. We may then assume safely that if the magnetic moment of the test cluster reverses direction the neighboring moments have time to relax to their respective equilibrium orientation. As a result, the energy of the ensuing state is always lower than that of the reversed state. The relaxation time is then always longer than for non-interacting clusters. The expressions for the intralayer interaction energy $U_{\parallel}$ and interlayer interaction $U_{\perp}$ are:

$$U_{\parallel} = N_{\parallel} \left\langle x \tanh\left(\frac{x}{k_B T}\right)\right\rangle \tag{5}$$

$$U_{\perp} = N_{\perp}\left\langle x \tanh\left[\frac{x}{k_B T\left(\xi^2 + 1/3\right)^{3/2}}\right]\right\rangle + \varepsilon\left(1/r^6\right)$$

where $x = [3(ur)(ru') - uu']\mu^2 / \Lambda_{\parallel}^2$. μ is the magnetic moment of a cluster, $N_{\parallel} \approx 6$ is the average number of neighbors in the same plane, and $\xi = \Lambda_{\perp}/2\Lambda_{\parallel}$. The $\langle\ \rangle$ indicate average over random orientations of the directional vectors of the moments, u and u', and r is the position vector of one cluster with respect to the test one. The assumption that the moments obey Ising statistics is also taken to simplify the model without losing the important ingredients. The contributions to the activation energy are proportional to the number of neighbors, as observed in the experiment, and decay with $1/r^6$. Therefore only interactions with the nearest neighboring clusters need to be accounted for. This dependence also explains why the multilayer sample with $\xi = 2.3$ behaves as the monolayer, since in this case the adjacent layer contribution has practically died off. With these equations the T_B and U_{eff} were fit, finding $U_0 = 560$ K, $U_{\parallel} \approx 250$ K and $U_{\perp} \approx 300$ K. We conclude that U_{int} may be as high as 50% of the total U_{eff}.

The application of a bias external magnetic field to a Co nanoparticle array has strong effects in the relaxational behavior since the hindering

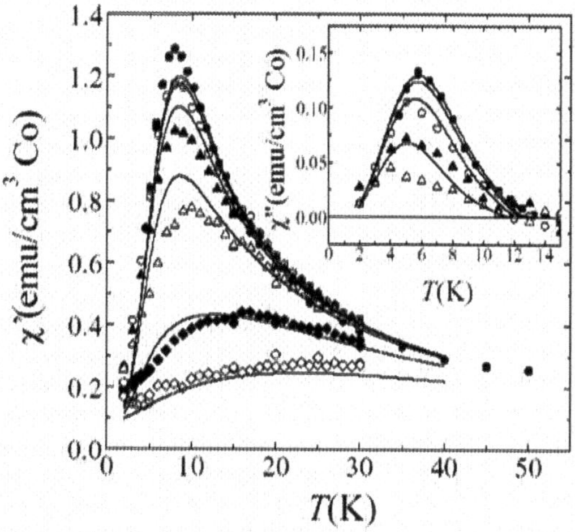

Figure 14. χ' and χ'' (inset) of sample with $\langle D \rangle = 1.4$ nm, and fixed frequency $\nu = 1$ Hz, for (●, ○, ▲, △, ◆ and ◇) H=0, 50, 100, 200, 500 and 1000 Oe, respectively. Full curves are theoretical predictions (see text).

barriers to magnetic reversal are modified by it. Two samples were studied to this end, with $\langle D \rangle$ =1.4 and 3.1 nm, because the size difference is sufficient to represent the cases of weak and strong dipole-dipole interaction between the embedded Co nanoparticles.

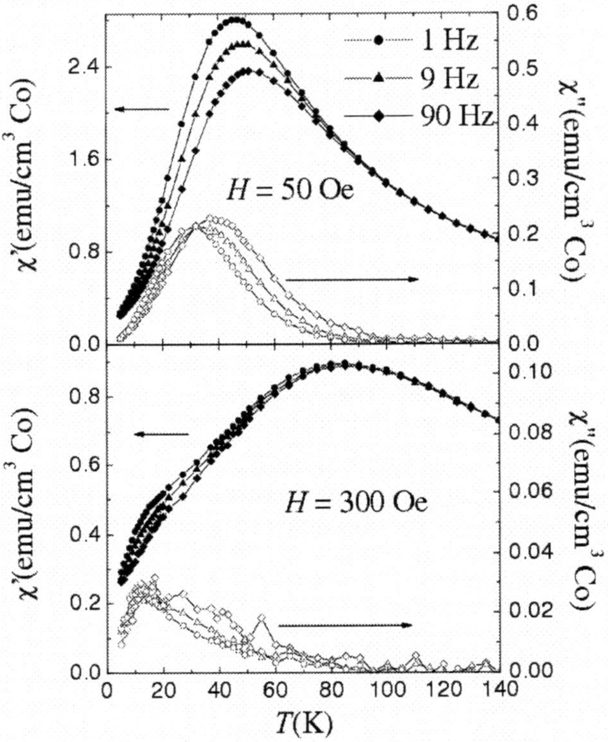

Figure 15. χ' (solid symbols) and χ'' (open symbols) of the sample with $\langle D \rangle$= 3.1 nm, for two different bias fields and three different frequencies.

Ac susceptibility measurements performed on the sample with the smallest clusters in a bias field are shown in Fig. 14. χ' has a maximum that shifts to higher temperature as the bias field increases, while that of χ'' decreases. This apparently contradicts the classical behavior since the bias field reduces the activation energy. This bizarre feature is clarified by the measurements on the larger particles at sufficiently high field (300 Oe), since in that case the maximum, shifted to higher temperature, is not frequency dependent (Fig. 15). Under these conditions, the χ' response is dominated by the equilibrium susceptibility, and the temperature at the maximum is the temperature of the equilibrium susceptibility $T_{eq} \approx 2\mu H/k_B$, related to the saturation of the magnetic moments by H, not to the superparamagnetic blocking temperature [27]. This result is important as a note of caution in the incorrect, but extended use of relating directly zero field χ' maximum to the dynamical freezing at T_B (Fig. 15) [28].

In contrast, the χ'' component of the susceptibility reflects directly the relaxation behavior under the applied field. For both samples, the maxima of χ'', measured at a given frequency, shift to lower temperature for increasing bias field. It also shifts to lower temperature for decreasing excitation frequency, and fixed field. Incidentally, this behavior indicates that the relaxation follows the classical predictions, and that no quantum tunneling relaxation is discernible since the trend would be of a shift to higher temperatures with increasing field. The dependence on H that we have adopted is

$$U(H,\psi) = U(H=0)\left[1 - {}^{H}\!/_{H_c(\psi)}\right]^{\eta(\psi)} \tag{6}$$

Here ψ is the angle between the magnetic field and the anisotropy axis. $H_c(\psi) = H_c(0)\left[\sin(\psi)^{2/3} + \cos(\psi)^{2/3}\right]^{-3/2}$, $H_c(0) = 2U(H=0)/\mu$ is the anisotropy field, and the exponent $\eta(\psi) = 0.86 + 1.14 H_c(\psi)/H_c(0)$. The relaxation time $\tau = \tau_0 \exp\, (U(H)/k_BT)$; i.e. the simple model for independent clusters, with the zero field activation energy modified by the dipole-dipole interactions, is applied again to explain the data. The prefactor τ_0 may reflect some field dependence, however, the limits of our experimental accuracy do not allow us to derive conclusions about it. The ac susceptibility $\chi'$and χ'' were calculated as a function of temperature, for fixed field and fixed frequency, using the anisotropy constants found in the first section, that is $U(H=0) \approx K_s D^2$ with very satisfactory agreement with the experiment at zero field and at the maximum field applied. At intermediate field values the trends observed in the evolution of the susceptibility maximum and the height of the susceptibility are qualitatively properly predicted by the model, although quantitatively the agreement is worse.

There are quite some strong approximations in the proposed model of independent clusters that may explain the departure of the model; in particular it implies that the proposed activation energy overestimates the value of τ at finite H.

The simple model described in the previous section is also capable of explaining these results. As said, near T_B the susceptibility response is dominated by the largest particles. When one of these particles flips the

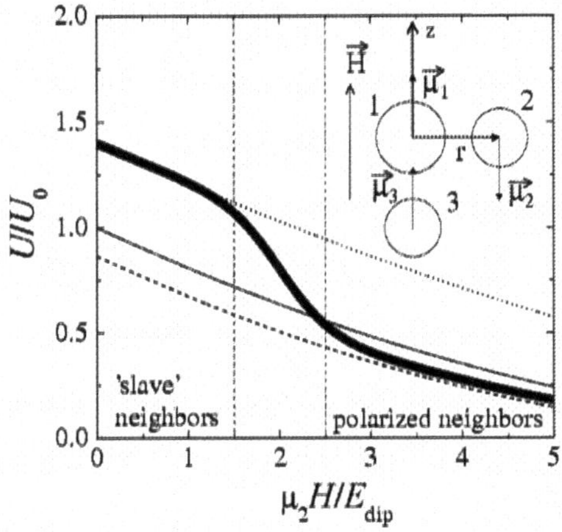

Figure 16. Field dependence of the activation energy U with respect to the anisotropy contribution U_0.. Thin full lines stand for the isolated cluster; thick curves stand for the interacting case.

surrounding magnetic moments reorient to minimize the dipolar energy and because of these "slave" particle flipping the activation energy $U(H = 0)$ becomes higher than in the non-interacting case. When the bias field is strong enough the slave moments are polarized more easily than the large moment and the influence of interparticle interaction between particles on T_B is reduced. Therefore, there is a crossover in the behavior caused by the applied field. In Fig. 16 the variation of the activation energy with applied field calculated for three clusters, the larger with D_1 = 3.55 nm and the two slave ones with $D_{2,3}$ = 3.1nm is shown. For comparison purpose, the field dependence of non-interacting particles with U_0 = 450 K is also included. At low fields the dipole-dipole interaction enhances U_{eff} by nearly 50%, and the rate of change is similar to the non-interacting case till the field reaches values $H / E_{dip} > 1.5$ ($E_{dip} \equiv \mu_1\mu_2/r^3$ the dipolar energy). When this ratio becomes larger than 2.5, the activation energy becomes close to the non-interacting case. The conclusion is that the "freezing" of the fluctuations of superparamagnetic spins by the bias field is the cause of the decrease of the effect of dipole-dipole interactions [29].

5. Conclusions

We have studied the magnetic anisotropy and relaxation phenomena of samples containing Co nanoclusters. The system of self-organized Co nanospheres in regularly ordered layers separated by an Al_2O_3 matrix has enabled us to show that the nanometric size of the particles dominate the magnetic anisotropy due to the predominance of the surface Co atoms. Due to loss of symmetry at the surface there is a lower quenching of the orbital moment of the surface Co atoms than in the bulk and, therefore, an increase in the orbital moment anisotropy of the surface Co atoms. For the samples with $\langle D \rangle < 2$ nm there is a predominance of the contribution due to the paramagnetic atoms which bond with oxygen, giving rise to splitting of the L_3 line. The properties of the larger Co clusters may be modified by capping with Cu which, in turn, increase the surface anisotropy by hybridization between the Cu and Co electrons 3d bands.

The possibility of modifying the anisotropy, either by capping with a metallic layer, change of the matrix, or even of the particle material seems an interesting field in trying to reach higher blocking temperatures, ideally over room temperature for applications. The method of production of the self-organized particles allows to obtain surfaces, which is important for applications. However, an improvement in the regularity of the self-organization of the particles, either by physical or chemical means would also improve the chances of this material to become competitive as recording media in the nanometric scale. The proven possibility of synthesizing periodical three-dimensional particle arrays is also exciting in view of the prospect of creating three dimensional magnetic memories.

Dipole-dipole interactions, together with magnetic anisotropy are the contributions to the activation energy that govern magnetization reversal of these systems. Interestingly, a bias field has the effect of dominating the dipolar interactions rendering the relaxation of the system identical to that of non-interacting particles.

Acknowledgments

We are grateful to D. Babonneau, J. Carrey and J.L. Maurice for the GIXAXS and TEM experiments and many fruitful discussions. The early magnetic measurements of J.M. Torres-Bruna and J. Stanckiewicz are also acknowledged. This work has been financed by the MAT03/166 MCYT project. We acknowledge the European Synchrotron Radiation Facility for provision of synchrotron radiation facilities and we would like to thank Peter Bencok and Nick Brookes for assistance in using beamline ID08."

References

[1] F. Jorgenson, *The complete Handbook of Magnetic Recording*. Tab Books, 3d ed (1988).
[2] Seagate, News+Info. Nov. (2004) .
[3] J.L. Maurice, J. Briatico, J. Carrey, F. Petroff, L.F. Schelp, A. Vaurès, Philos. Mag. **A79**, 2921 (1999).
[4] J.M. Torres, F. Luis, L.M. García, J. Bartolomé, J. Stanckiewicz, F. Petroff, F. Fettar and A. Vaurès, J. Magn. & Magn. Mat. **242-245**, 575 (2002).
[5] F. Luis, J.M. Torres, L.M. García, J. Bartolomé, J. Stanckiewicz, F. Petroff, F. Fettar, J.L. Maurice and A. Vaurès, Phys. Rev. B, **65**, 094409 (2002).
[6] M.I. Shliomis and V.I. Stepanov, Adv. Chem. Phys. **87**, 1 (1994).
[7] J.I. Gittleman, B. Abeles, S. Bosovski, Phys. Rev. B, **9**, 3891 (1974).
[8] W.A. Sucksmith and J.E. Thompson, Proc. R. Soc. London, Ser. **A 225**, 362 (1954).
[9] J. P. Chen, C.M. Sorensen, K.J. Klarbunde and G.C. Hadjipanajis, Phys. Rev. B, **51**, 11527 (1995).
[10] M. Respaud, J.M. Broto, H. Rakoto, A.R. Fert, L. Thomas, B. Barbara, M. Verelst, E. Snoeck, P. Lecante, A. Mosset, J. Osuna, T. Ould Ely, C. Amiens, B. Chaudret, Phys. Rev. B, **57**, 2925 (1998).
[11] J. Briatico, J.L. Maurice, J. Carrey, D. Imhoff, F. Petroff and A. Vaurès, Eur. Phys. J. D **9**, 517 (1999).
[12] D. Babonneau, F. Petroff, J.L. Maurice, F. Fettar, A. Vaurès, Appl. Phys. Lett. **76**, 2892 (2000).
[13] J. Bartolomé, F. Luis, F. Petroff, F. Bartolomé, L.M. García, V. Cros and H. Jaffres, Phys. Metals & Metallography. To be published. (2005).
[14] F. Bartolomé, F. Luis, F. Petroff, L.M. García, J. Bartolomé, V. Cross and H. Jaffrès, J. Magn. & Magn. Mater. **272**, e1275 (2004).
[15] P.S. Gambardella, Rusponi, M. Veronese, S. S. Dhesi, C. Grazioli, A. Dallmeyer, I. Cabria, R. Zeller, P. H. Dederichs, K. Kern, C. Carbone, and H. Brune, Science **300**, 1130 (2003).
[16] P. Bruno, Phys. Rev. B, **39**, R865 (1989).
[17] J.G. Van der Laan, Phys. Cond. Matter, **10**, 3239 (1998).
[18] T.J. Regan, H. Ohldag, C. Stamm, F. Notling, J. Lüning, J. Stöhr, R.L. White, Phys. Rev. B, **64**, 214422 (2001).
[19] P. S. Gambardella, S. Dhesi, S. Gardonio, C. Grazioli, P. Ohresser and C. Carbone, Phys. Rev. Lett. **88**, 047202 (2002).
[20] F.J.A. den Broeder, W. Hoving and P.J.H. Bloemen, J. Magn. & Magn. Mat. **93**, 562 (1991).
[21] M. Tisher, O. Hjortstam, D. Arvanitis, J. Hunter Dunn, F. May, K. Baberschke, J. Trygg, J. M. Wills, B. Johansson, and O. Eriksson, Phys. Rev. Lett. **75**, 1602 (1995).
[22] 22. J.L. Dormann, D. Fiorani and E. Tronc, J. Magn. & Magn. Mater. **202**, 251 (1999).
[23] S. Mørup and E. Tronc, Phys. Rev. Lett. **72**, 3278 (1994).
[24] J.L. Garcia-Palacios, Europhysics Letters **55**, 41826 (2001).
[25] C. Djurberg, P. Svedlingh, P. Nordblad, M.F. Hansen, F. Bodker and S. Mørup, Phys. Rev. Lett. **79**, 5154 (1997).
[26] F. Luis, F. Petroff, J.M. Torres, J.M. García J. Bartolomé, J. Carrey and A. Vaurès, Phys. Rev. Lett. **88**, 217205 (2002); Luis F., F. Petroff, J.M. Torres, J.M. García, J. Bartolomé, J. Carrey and A. Vaurès. Phys. Rev. Lett. **90**, 59706 (2003).

[27] J.L. García-Palacios, Adv. Chem. Phys. **87**, 1 (2000).
[28] F. Luis, J. Bartolomé, F. Petroff, L.M. García, A. Vaurès and J. Carrey, J. Appl. Phys. **93**, 7032 (2003).
[29] F. Luis, F. Petroff and J. Bartolomé, J. Phys.: Condens. Matter **16**, 5109 (2004).

NANOPARTICLES IN OXIDE GLASSES: MAGNETIC RESONANCE STUDIES

Janis Kliava and René Berger
CPMOH, UMR Université Bordeaux I-CNRS 5798, 33405 Talence Cedex, France
jkliava@cpmoh.u-bordeaux1.fr

Abstract: We overview the application of the electron magnetic resonance spectroscopy assisted with computer simulations to a new phenomenon observed in annealed iron-containing glasses, viz. the superparamagnetic resonance arising from magnetically ordered nanoparticles. The characteristics of the nanoparticle size and shape distribution in glass are obtained by numerical simulation of the resonance spectra. This phenomenon provides a basis of a novel method of morphological analysis of the magnetic nanoparticles.

Keywords: magnetic resonance, oxide glasses, iron, superparamagnetism, nanoparticles, computer simulations.

1. Introduction

An unusual and exciting phenomenon is observed in certain iron containing oxide glasses under a thermal treatment above the glass transition temperature T_g. The well-known feature at $g_{eff} \approx 4.3$ in the X-band EPR spectrum, characteristic of isolated ion Fe^{3+} ions, gradually disappears and a new signal at $g_{eff} \approx 2.0$ begins to grow. In borate glasses at lower treatment temperatures this signal consists of a sharp line which is transformed to a two-line pattern and then, as the anneal temperature increases, to a single broad line. In an annealed sol-gel glass, the new signal has a shape of a single asymmetric hyperlorentzian line. Such spectra are attributed to superparamagnetic systems, an assembly of static single domain magnetic nanoparticles, randomly dispersed in a partially devitrified matrix.

Superparamagnetism is a specific type of magnetic behaviour observed with fine magnetically ordered particles dispersed in a diamagnetic matrix [1]. Understanding of this phenomenon has been a very important contribution to fundamental theories of magnetism and laid the foundation

J. Franse et al. (eds.), Smart Materials for Ranging Systems, 27–48.

for the development of new materials for high-density information storage. The corresponding magnetic resonance signal at $g_{eff} \approx 2.0$, usually much narrower than that arising from clusters of Fe^{3+} ions, is called superparamagnetic resonance (SPR) to distinguish it from the resonance of individual ions referred to as electron paramagnetic resonance (EPR). Initial superparamagnetic resonance studies have been carried out mostly on magnetic fluids (ferrofluids) or on thin films with deposited ultra fine magnetic particles, in view of various technical applications, e.g., as catalysts, or for magnetic recording.

Partially crystallised oxide glasses containing magnetic nanoparticles constitute a novel and much more complex class of superparamagnetic materials. Their properties are determined by morphology (size and shape distributions) and magnetic characteristics of the particles. The SPR, sensitive to all these parameters, is particularly well adapted to the study of the superparamagnetic nanoparticles. The Fe^{3+} ions dispersed in a glass matrix can be good probes of the devitrification process, particularly at its early stage when crystallised nuclei are arising and growing.

The superparamagnetic resonance, assisted by computer simulations, provides a new method of morphological analysis of magnetic nanoparticles.

2. Theoretical model

2.1 General expression of the SPR spectrum in a disordered system

Magnetic resonance in an assembly of ferromagnetic nanoparticles in ferrofluids, glasses and other systems has attracted much attention. But, up to now, only a few attempts have been made of quantitative, computer-simulation based analysis of SPR spectra of magnetic nanoparticles dispersed in glassy systems [2-11].

An assembly of single crystal ferromagnetic particles with size of the order of some nanometers (nanoparticles) isolated in a diamagnetic matrix, forms a superparamagnetic system. Nanoparticles of relatively greater sizes may incorporate several magnetic domains; however, below *ca.* 10 nm they are typically single-domain particles [12]. Such a nanoparticle (neglecting surface and disorder effects) represents a single giant magnetic moment formed by exchange interaction between all individual spins ; this magnetic moment is a product of the spontaneous magnetisation M (not necessarily the same as in the bulk material) and the particle volume V.

We consider a statistical assembly of *disordered* single-domain magnetic nanoparticles whose characteristics vary from one particle to another. In a particular (but most widespread in practice) case of *macroscopically isotropic* superparamagnetic systems (*e.g.*, powders or glasses), all orientations of the static magnetic field B of a spectrometer are equivalent. So, the joint distribution density of its polar and azimuthal angles with respect to the magnetic symmetry axes, ϑ_B and φ_B becomes $P(\vartheta_B, \varphi_B) = \sin\vartheta_B$. In the main axes frame each individual nanoparticle is described by a random vector ξ whose components are magnetic parameters and morphological (size and shape) characteristics. For such an assembly, no close analytical expression of the SPR spectrum can be obtained, but it can be computed numerically as follows [8]:

$$\mathsf{P}\ (B) = \int_0^{2\pi}\int_0^{\pi}\int_{\xi} f_V(\boldsymbol{\xi})\, F\left[B, B_{\mathrm{r}}, \Delta_{\mathrm{B}}\right] \sin\vartheta_B \,\mathrm{d}\boldsymbol{\xi}\,\mathrm{d}\vartheta_B\,\mathrm{d}\varphi_B \tag{1}$$

where the volume fraction of particles with a given $\boldsymbol{\xi}$, $f_V(\boldsymbol{\xi}) = P(\boldsymbol{\xi})V(\boldsymbol{\xi})$, $P(\boldsymbol{\xi})$ being a joint distribution density of the components of $\boldsymbol{\xi}$; B is the norm of $\boldsymbol{B}$, $B_{\mathrm{r}}(\boldsymbol{\xi}, \vartheta_B, \varphi_B)$ is the resonance magnetic field, $F(B, B_{\mathrm{r}}, \Delta_B)$ is the intrinsic lineshape for a particle with given characteristics and $\Delta_{\mathrm{B}}(\boldsymbol{\xi}, \vartheta_B, \varphi_B)$ is the individual linewidth for such a particle.

If we assume that the magnetic parameters M and K_1 (the first-order anisotropy constant) are not distributed (have fixed value for all the magnetic particles and throughout the volume of each particle), the components of the reduced random vector $\boldsymbol{\xi}$ in expression (16) will include only the geometrical characteristics of the nanoparticles. For particles with spherical shape, $\boldsymbol{\xi}$ reduces to the particle diameter D and $f_V(D) = P(D)\pi D^3/6$, $P(D)$ being the relative number of nanoparticles of a given diameter [11].

2.2 Resonance magnetic field

The magnetic moment of a particle in a magnetic field B evolves according to the equation of (damped) precession, and in the absence of damping the resonance occurs when the oscillating field frequency exactly equals the precession frequency. For simplicity, we assume the nanoparticles to have an ellipsoidal form with principal axes x, y and z. In this instance, the free energy density in a particle to the first order in the magnetic symmetry can be expressed as [13] :

$$E = -\boldsymbol{M}\cdot\boldsymbol{B} + K_1\Phi(\vartheta_M, \varphi_M) + \tfrac{1}{2}\mu_0\boldsymbol{M}\cdot\mathsf{N}\cdot\boldsymbol{M}\,. \tag{2}$$

The first term on the right-hand side of eq. (2) is the Zeeman energy. The second term is the magnetic anisotropy energy, ϑ_M and φ_M being the polar and azimuthal angles of $\boldsymbol{M}$ with respect to the magnetic symmetry axes. The third term is the magnetostatic energy, with N the demagnetising tensor whose eigenvalues, the demagnetising factors, are related by $N_x + N_y + N_z = 1$. The $\Phi(\vartheta_M, \varphi_M)$ function depends on the magnetic symmetry [13], namely, in cubic symmetry

$$\Phi(\vartheta_M, \varphi_M) = \tfrac{1}{4}\left(\sin^2 2\vartheta_M + \sin^4 \vartheta_M \sin^2 2\varphi_M\right) \tag{3}$$

and in axial symmetry

$$\Phi(\vartheta_M, \varphi_M) = \sin^2 \vartheta_M . \tag{4}$$

The equilibrium orientation of $\boldsymbol{M}$ minimises the value of E. In the absence of $\boldsymbol{B}$, in a spherical particle, $\boldsymbol{M}$ is oriented along an easy magnetisation axis. In typical SPR conditions (the strong external magnetic field case), the Zeeman coupling dominates over all other interactions, so that the direction of $\boldsymbol{M}$ is close to that of $\boldsymbol{B}$. If the deviation of $\boldsymbol{M}$ from its equilibrium orientation is small, $\boldsymbol{M}$ oscillates about its equilibrium position with a cyclic frequency at resonance ω_{r}. corresponding to the *applied* magnetic field $B_0 = \omega_{\mathrm{r}}/\gamma$ where. From the ω_{r} expression [13] one gets

$$\frac{\omega_{\mathrm{r}}}{\gamma} = \frac{1}{M \sin \vartheta_0}\left(\frac{\partial^2 E}{\partial \vartheta_M{}^2}\frac{\partial^2 E}{\partial \varphi_M{}^2} - \frac{\partial^2 E}{\partial \vartheta_M \partial \varphi_M}\right)^{1/2}\Bigg|_{\vartheta_M = \vartheta_0, \varphi_M = \varphi_0} \tag{5}$$

where γ is the gyromagnetic ratio and the derivatives are calculated for the equilibrium orientation of M given by $\vartheta_M = \vartheta_0$ and $\varphi_M = \varphi_0$. An iterative procedure of solving (5) yields the resonance magnetic field B_{r}. Different alternative procedures of calculating B_{r} have been reported [4, 12, 14-17]. If the Zeeman energy predominates over the magnetic anisotropy and magnetostatic energies, B_{r} is simplified to :

$$B_{\mathrm{r}} = B_0 + B_{\mathrm{a}} + B_{\mathrm{d}} . \tag{6}$$

The anisotropy field B_{a} and the demagnetising field B_{d} arise, respectively, from the last two terms of (2). These fields bring about an angular dependence of the spectra. In the cubic and axial symmetry,

$$B_{\mathrm{a}}^{\mathrm{cub}} = \frac{2K_1^{\mathrm{cub}}}{M_s}\left[5\left(\sin^2 \vartheta_B \cos^2 \vartheta_B + \sin^4 \vartheta_B \sin^2 \varphi_B \cos^2 \varphi_B\right) - 1\right] \tag{7}$$

and

$$B_{\mathrm{a}}^{\mathrm{ax}} = \frac{K_1^{\mathrm{ax}}}{M_s}\left(3\cos^2 \vartheta_B - 1\right) \tag{8}$$

where K_1^{cub} and K_1^{ax} are, respectively, the cubic and axial first-order anisotropy constants.

For the nanoparticle form of an ellipsoid of revolution, $N_x = N_y = N_{\parallel}$, $N_z = N_{\perp}$ and the demagnetising field is expressed as

$$B_d = \frac{\mu_0 M}{2}\left(N_{\parallel} - N_{\perp}\right)\left(3\cos^2\vartheta_B - 1\right). \tag{9}$$

2.3 Superparamagnetic narrowing

In a system of nanoparticles, *thermal fluctuations* of their magnetic moments severely reduce the anisotropy of the resonance magnetic field, resulting in superparamagnetic spectra narrowing. This reduction is the more pronounced the smaller is the particle size. Therefore, the SPR spectra of macroscopically isotropic nanoparticle systems characterised by a distribution in size usually maintain a distinct shape asymmetry characteristic of "powder patterns" of randomly oriented anisotropic particles. From an inspection of such spectra, one can conclude that the angular dependence of the resonance magnetic field of individual particles is *not* completely reduced.

A simple way to account for this particle volume-dependent narrowing of the SPR spectra, suggested by De Biasi and Devezas [18], consists in averaging the magnetisation and magnetic anisotropy over the thermal fluctuations of the magnetic moment as follows :

$$\langle M \rangle = M\langle \cos\vartheta \rangle \tag{10}$$

and

$$\langle K_1 \rangle = K_1 \langle P_n(\cos\vartheta) \rangle \tag{11}$$

where $\vartheta = \vartheta_M - \vartheta_B$ is the angle between $\boldsymbol{M}$ and $\boldsymbol{B}$, $P_n(\cos\vartheta)$ is the n-th order Legendre polynomial and the angular brackets denote averaged quantities. In typical magnetic resonance conditions, the total energy of a magnetic particle of a volume V can be approximated by $EV \approx -MVB\cos\vartheta$.

For an assembly of particles whose magnetic moments are much larger than the Bohr magneton, the partition function can be calculated as an integral over all possible values of ϑ. In this approximation, the averaging factor becomes in cubic symmetry ($n = 4$) :

$$\delta_{\text{cub}} = \frac{\langle P_4(\cos\vartheta) \rangle}{\langle \cos\vartheta \rangle} = \frac{1 + 35x^{-2}}{L(x)} - 10x^{-1} - 105x^{-3} \tag{12}$$

and in axial symmetry ($n = 2$) :

$$\delta_{\mathrm{ax}} = \frac{\langle P_2(\cos\vartheta)\rangle}{\langle\cos\vartheta\rangle} = \frac{1}{L(x)} - 3x^{-1} \tag{13}$$

where $\mathrm{L}(x) = \coth x - 1/x$ is the Langevin function of $x = MB_{\mathrm{r}}V/kT$.

Thus, in order to account for superparamagnetic narrowing of the SPR spectra, the cubic and axial anisotropy fields should be multiplied, respectively by δ_{cub} and δ_{ax}. The demagnetising field (9) should also be scaled by the latter factor (as shown by Raikher and Stepanov [19] which authors have noted an error in the De Biasi and Devezas paper [18]).

2.4 Distribution of particle diameters and demagnetising factors

The size of magnetic nanoparticles is of a primary importance in determining their physical properties. A number of experimental techniques, such as magnetic, optical and rheological measurements, X-ray diffraction, electron microscopy, atomic force microscopy, small-angle neutron scattering, as well as magnetic resonance, have been applied to evaluate the size of magnetic nanoparticles in superparamagnetic systems and, eventually, to get some insight into the size distribution functions [7-10, 12, 20-27]. Several authors have shown that if the nanoparticle shape is close to a sphere, the log-normal form of $f_V(D)$ and $P(D)$ is the most appropriate one, e.g., in the case of demixing processes of glasses during heat treatment [21, 28] or for fine magnetic particles dispersed in a fluid (ferrofluids) [22, 23, 25]. A number of interrelated expressions of $f_V(D)$ and $P(D)$ can be used [25], the most current one for the volume fraction of particles with diameter D being [6] :

$$f_V(D) = \frac{\exp(-\sigma^2/2)}{\sqrt{2\pi}\sigma D_{V\mathrm{m}}} \exp\left(-\frac{1}{2\sigma^2}\ln^2\frac{D}{D_{V\mathrm{m}}}\right), \tag{14}$$

where σ is the standard deviation of ln$D$ and $D_{V\mathrm{m}}$ corresponds to the maximum of $f_V(D)$. The expression of $P(D)$ is derived from (14) simply by replacing $D_{V\mathrm{m}}$ by the *most probable* diameter D_{m} corresponding to the maximum of $P(D)$, $D_{\mathrm{m}} = D_{V\mathrm{m}}\exp(-3\sigma^2)$.

In the majority of papers dealing with the analysis of the SPR spectra, a spherical particle shape has been assumed. Meanwhile, anisotropy of the particle shapes plays an important role in determining the magnetic birefringence of ferrofluids [23]. Non-sphericity of the magnetic nanoparticles must be still more pronounced in partially devitrified glasses, in which case considerable statistical distributions of the particle shapes are expected, as well. From the viewpoint of the SPR phenomenon, the basic manifestation of the particle shape anisotropy is the emergence of the demagnetising field, which contributes to the local magnetic field. The existence of this field has been overlooked in a number of SPR studies.

For an ellipsoidal particle, the ratios of the axes can be related to the demagnetising factors [13], so that it is more convenient to directly consider the distribution of the latter quantities. Here we remind only the principal steps of the procedure of obtaining the corresponding distribution function (see [8] for details). The demagnetising tensor is expressed as follows:

$$N = \tfrac{1}{3}1 + n \tag{15}$$

Here, *1* is a unit matrix of order 3, and *n*, a zero-trace tensor, is a sum of two tensors : a *fixed* one which describes the *average* distortions of particle shapes from that of a sphere (e.g., due to magnetostriction or interactions with the environment) and a *random* tensor which accounts for *departure from the average* of the individual particle shapes (manifestation of disorder in the system).

For simplicity, we take into consideration only axial distortions of the particle shapes, in which case $n_x = n_y = -\tfrac{1}{2}n_z$, so that the random vector ξ in (1) reduces to $\boldsymbol{\xi} = (D, n_z)$. In this case the distribution density of n_z can be derived from the non-central χ^2 distribution with five degrees of freedom [29-31]. Assuming that the deviations of the particle shape from that of a sphere are not large, the latter density can be then associated with the log-normal distribution density (14) in order to construct the joint distribution density of particle diameters and demagnetising factors, $P(D, n_z)$. The components of the N tensor may depend on the particle size, resulting in a correlation between D and n_z. Including for simplicity the correlation coefficient ρ in the " Gaussian " part of $P(D, n_z)$, *cf.* ref. [30], one gets [8] :

$$P(D, n_z) \propto n_z \, \mathrm{G}\left(\frac{n_z n_0}{\sigma_n^{\,2}}\right) e^{-\frac{1}{2(1-\rho^2)}\left[\frac{(n_z - n_0)^2}{\sigma_n^{\,2}} - 2\rho \frac{(n_z - n_0)\ln\frac{D}{D_\mathrm{m}}}{\sigma_n \sigma_D} + \frac{\ln^2 \frac{D}{D_\mathrm{m}}}{\sigma_D^{\,2}}\right]} \tag{16}$$

where n_0 is a mean value of n_z and $\mathrm{G}(y) = y - 1 + (y+1)\exp(-2y)$.

2.5 Intrinsic linewidth

In the literature concerning the SPR linewidth and, more generally, the magnetic resonance in powders, glasses or fluids, there has been much confusion between the *observed, e.g.*, peak-to-peak width of the spectral features and the *true* intrinsic linewidth [2, 4, 32]. In the approach based on eq. (1) the spectra " broadening " due to the spread of the resonance magnetic fields $B_{\mathrm{r}}(\boldsymbol{\xi}, \vartheta_B, \varphi_B)$ is explicitly taken into account in the dependence of B_{r} on the orientation and the size and shape parameters. The intrinsic linewidth Δ_B, in turn, accounts for the contribution to the SPR spectra from particles of a given size and shape and thus includes all remaining broadening sources. We should like to stress here that the corresponding broadening in most cases is *not* obvious from a visual inspection of resonance spectra and can only be deduced from accurate computer simulations.

Homogeneous broadening of the resonance lines arises from damping terms in the equation describing a precession of the magnetic moments. These terms may have different form, those most currently used yield the Bloch-Bloembergen, Gilbert, Landau-Lifshitz and Callen equations. We have carried out a comparative analysis of the resonance lineshapes resulting from these equations [33]. In the low damping approximation, valid for diluted paramagnets and usually applied in the paramagnetic resonance studies, all these equations result in a Lorentzian lineshape. In the strong damping case (ferromagnetic or low-temperature superparamagnetic resonance) the resonance conditions are considerably altered and the actual lineshape expression essentially depends on the form of the damping term. We have shown that the low-temperature behaviour of the SPR spectra is best described by the Landau-Lifshitz equation :

$$\dot{\boldsymbol{M}} = \gamma \boldsymbol{M} \wedge \boldsymbol{B} - \frac{\lambda}{|M|^2} \boldsymbol{M} \wedge (\boldsymbol{M} \wedge \boldsymbol{B}) \tag{17}$$

where $\lambda > 0$ is the damping factor. If the static magnetic field is sufficient to saturate the magnetisation, (17) leads to the normalised absorption lineshape including both the resonant and non-resonant terms[33]:

$$LL(B,B_r,\Delta_B)=\frac{\sqrt{3}\Delta_B}{2\pi}\left[\frac{1}{3\frac{B^2}{B_r^2}+\frac{(B-B_r)^2}{\Delta_B^2}}+\frac{1}{3\frac{B^2}{B_r^2}+\frac{(B+B_r)^2}{\Delta_B^2}}\right] \tag{18}$$

where the linewidth parameter is given by $\Delta_B = \lambda B_r / |\gamma M|$. In contrast to a Lorentzian or Gaussian, the lineshape (18) is characterised by a linewidth-dependent *apparent* resonance field $B_{apparent}$ (corresponding to the absorption maximum), namely, B_r decreases as Δ_B increases [33] :

$$B_{apparent} = \frac{B_r^2}{B_r^2+\Delta_B^2}\left[2B_r\left(B_r^2+\Delta_B^2\right)^{1/2} - B_r^2 - \Delta_B^2\right]^{1/2}. \tag{19}$$

Inside a single-domain nanoparticle, the exchange interaction is usually predominant. Several other interactions must also be taken into account, namely (i) the Zeeman interaction which tends to align the magnetic moment along the magnetic field, (ii) the magnetocrystalline anisotropy which tends to align the magnetic moment along one of the easy magnetisation axes and maintains an energy barrier between them, (iii) the magnetostatic interaction due to non-spherical particle shapes, (iv) the surface anisotropy and, finally, (v) the thermal excitation which results in fluctuations of the orientation of the magnetic moments [12]. In the X-band at room temperature for a particle diameter of 5 nm with typical magnetic parameter values, the Zeeman energy is more than an order of magnitude greater than the other energies.

Inhomogeneous broadening of the resonance lines can be related to distributions in the magnetic parameters M and K_1 The corresponding contribution to the linewidth is [8] :

$$\Delta_B^2 = \left(\frac{\partial B_r}{\partial M}\right)^2 \langle\Delta_M\rangle^2 + \left(\frac{\partial B_r}{\partial K_1}\right)^2 \langle\Delta_{K_1}\rangle^2 \tag{20}$$

where $\langle\Delta_M\rangle$ and $\langle\Delta_{K_1}\rangle$ are *dynamically averaged* distribution widths of the corresponding parameters. The site-to-site distributions of M contribute to the individual linewidth via distributions of both the anisotropy (7), (8) and demagnetising fields (9), while those of K_1 interfere only with the former one. As temperature increases, the distributions of M and K_1 are narrowing roughly in the same way as the corresponding mean values, eqs. (12), (13). In addition, an increase in temperature lowers the energy barrier between directions of different easy magnetisation axes and gradually removes the corresponding line broadening. Taking into account all these mechanisms, the following expression of the temperature dependence of the individual linewidth has been put forward :

$$\Delta_B = \Delta_T \left(\frac{K_1 V_s}{kT} \right) \mathrm{L} \left(\frac{MVB}{kT} \right) = \Delta_0 \delta \left(\frac{K_1 V_s}{kT} \right) \mathrm{L} \left(\frac{MVB}{kT} \right) \tag{21}$$

where Δ_0 is a saturation linewidth at 0 K for a particle of volume V_s (presumably the greatest volume in the statistical ensemble) and the function δ, depending on the symmetry, is chosen as δ_{cub}, eq.(12) or δ_{ax}, eq. (13) [9].

3. Results and discussion

3.1 Sample characterisation and experimental data

Two different superparamagnetic systems have been studied : a borate glass of molar composition 0.63 B_2O_3-0.37 Li_2O-0.75·10^{-3} Fe_2O_3 and a Fe_2O_3-doped sol-gel silica glass [11] with a molar ratio of Fe/Si of 2 %. The very low iron content has been chosen in order to avoid initial clustering.

The borate glass transition temperature T_g = 435 °C was determined by differential scanning calorimetry. The samples were annealed in air in an electric furnace by repeated stages for ½ h at increasing anneal temperatures T_a starting at T_g and ranging up to 550 °C and cooled down to room temperature after each anneal step. In the heat-treated glasses, X-ray diffraction has detected only two crystalline phases, $LiBO_2$ and $Li_2B_4O_7$. No iron-containing phases could be observed because of the very low iron content. We assume that the nanoparticles in the glasses have the same nature as those identified by X-ray diffraction as lithium ferrite, $LiFe_5O_8$ in annealed borate glasses with *high* iron oxide content [34]. Consequently, we have adopted the magnetic parameters of lithium ferrite, $M_s = 310$ kA m^{-1} and $K_1 = 8.0$ kJ m^{-3} (cubic symmetry).

Magnetic nanoparticles in the sol-gel silica glass were obtained by heat treatment at 1000°C in air during 6 h and identified by X-ray and Mössbauer spectroscopy as γ-Fe_2O_3 (maghemite) [11]. The magnetic parameters of maghemite used in computer simulations of the SPR spectra, are $M_s = 370$ kA m^{-1} and $K_1 = -4.64$ kJ m^{-3} (cubic symmetry) [35].

At room temperature the X-band EPR spectrum of a non-annealed borate glass exhibits an asymmetric $g_{eff} \approx 4.3$ signal, characteristic of isolated iron ions in the glassy matrix. As T_a increases from 460 to 550 °C, this line

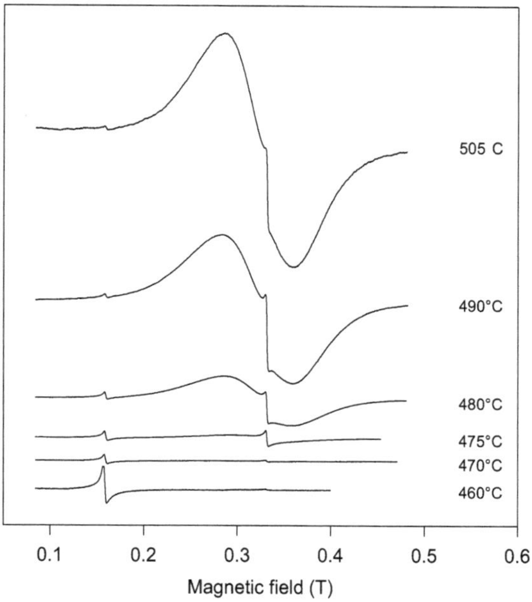

Figure 1. X-band EPR room temperature spectra of the glass 0.63 B_2O_3 - 0.37 Li_2O with $0.75 \cdot 10^{-3}$ Fe_2O_3, annealed at the indicated temperatures during ½ hour.

gradually decreases and finally disappears (figure 1). Simultaneously, a new resonance emerges at $g_{\text{eff}} \approx 2.0$, appearing as a narrow line superposed with a broader one. The narrow component predominates at lower anneal temperatures and it is progressively replaced by the broader one at higher temperatures. Such behaviour in a temperature range above T_g indicates the devitrification process, confirmed by X-ray diffraction results. The $g_{\text{eff}} \approx 2.0$ resonance is assigned to iron-containing crystalline particles.

Numerical integration of the experimental derivative-of-absorption spectra yields the resonance absorption shown in figure 2. One can see that in the course of the heat treatment the number of iron ions diluted in the glass matrix decreases while that in the magnetic nanoparticles increases. The total number of spins (in relative units) obtained by a second integration of the spectra, figure 3, indicates a continuous rapid growth of the total volume of the magnetic particles in the range of T_a from 475 to 550 °C.

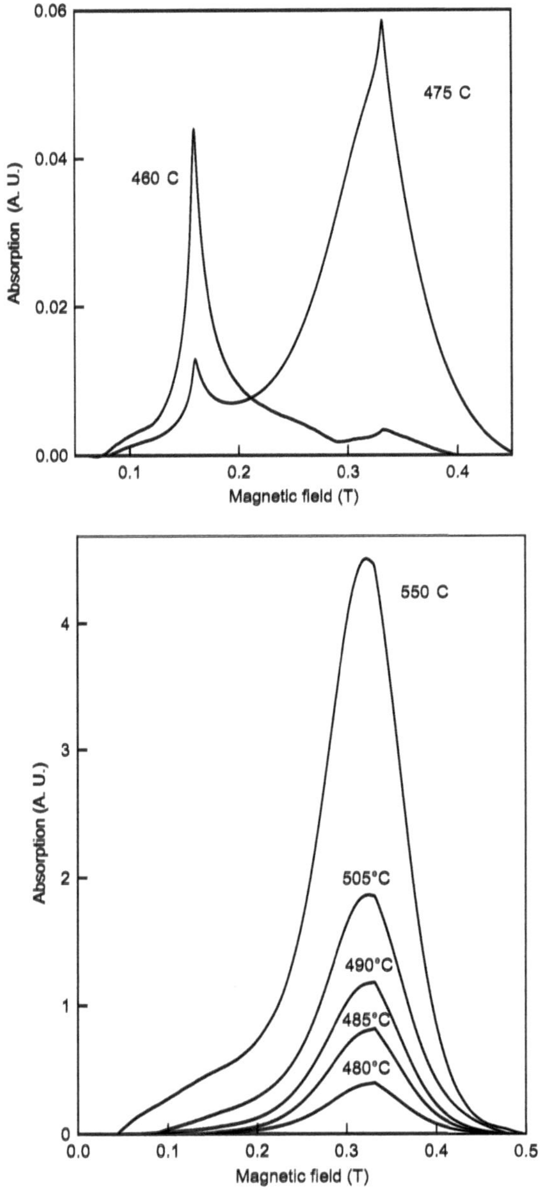

Figure 2. Absorption spectra at lower (top) and higher (bottom) anneal temperatures indicated alongside the curves.

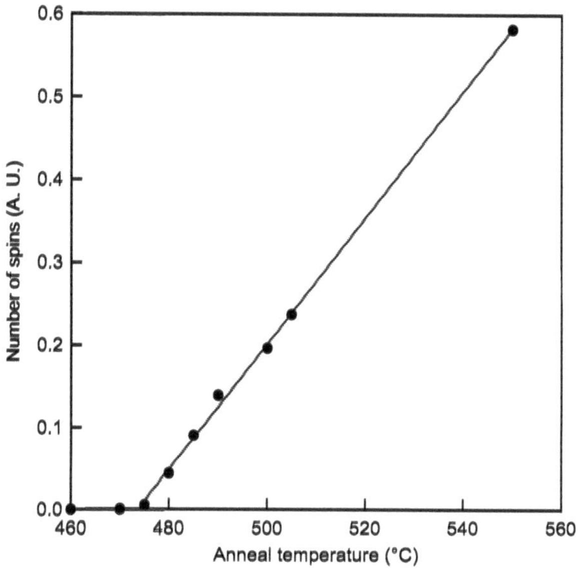

Figure 3. Number of spins (in relative units) in magnetic particles as a function of T_a. The straight line shows a linear dependence between 475 and 550°C.

3.2 Nanoparticles size and shape distribution

In order to evaluate the characteristics of the nanoparticle sizes and shapes, a series of computer simulations of the SPR spectra has been carried out using the approach based on the joint distribution density of the particle diameters and demagnetising factors $P(D,n_z)$. Figure 4 shows computer fits to the spectra of the borate and sol-gel silica glass as well the corresponding best-fit distribution densities. As the computer-generated spectra are very sensible to the variations of the simulation parameters, the good quality of the computer fits is quite significant, and the characteristics of the joint distribution density of diameters and demagnetising factors are determined from the simulations with sufficient accuracy and in a unique way.

It might seem surprising that an *unimodal* distribution of particle diameters could well reproduce the *two-line* pattern at $g_{\text{eff}} \approx 2.0$. Here we are dealing with a specific example of a very general characteristic of magnetic resonance spectra of disordered systems : singularities (sharp features) appear in such spectra if the resonance magnetic field is stationary with respect to the distributed magnetic parameters [36, 37]. This situation is similar to that of EPR spectra of diluted Fe^{3+} ions in oxide glasses: an unimodal distribution of the fine structure parameters, brought about by disorder inherent in the vitreous state, gives rise to two separate lines, at $g_{\text{eff}} \approx 4.3$ and $g_{\text{eff}} \approx 2.0$. In the present case, such singularity is the sharp feature with $g_{\text{eff}} \approx 2.0$, resulting from the smallest magnetic nanoparticles, for which the anisotropy and demagnetising magnetic fields are almost entirely averaged by thermal fluctuations of their magnetic moments.

Table 1. Parameters of the joint distribution density of diameters and demagnetising factors (16) and of the SPR linewidth[a)] for magnetic nanoparticles in annealed glasses.

Glass	Borate			Sol-gel silica
T_a (°C)	465	475	480	1000
D_m (nm)	3.4	3.9	4.7	3.0
σ_D	0.39	0.37	0.34	0.38
n_0	-0.022	-0.006	0.010	-0.027
σ_n	0.095	0.060	0.060	0.13
ρ	0.39	0.52	0.60	0.1
Δ_M (kA m^{-1})	45	45	36	60
Δ_{K_1} (kJ m^{-3})	2.0	2.0	1.0	0.5

Table 1 summarises the best-fit simulation parameters for the two glasses. From an inspection of these data, one can conclude that with the increase of the anneal temperature, the most probable diameter D_m of the magnetic nanoparticles formed in the borate glass increases while the standard deviation (the distribution width) σ_D slightly decreases, so that the assembly of nanoparticles becomes more ordered with respect to their size and shape distributions.

Concerning the shape of the magnetic nanoparticles in the borate glass, one can see that n_0 increases with the increase of the anneal temperature T_a, passing from negative to positive values. (We remind that negative and positive n_0 values indicate a preference for oblate and prolate spheroidal shapes, respectively.) Still more spectacular is the tendency towards ordering in the nanoparticle shapes, the distribution width σ_n being much larger at lower T_a values than at higher ones. Note that in all cases the absolute values of n_0 are small in comparison with the distribution widths of n_z, σ_n. It can be concluded that the statistical assembly of the magnetic nanoparticle shapes in partially devitrified glass is characterised by random distortions from the spherical shape, not necessarily involving any *average* distortion, rather than by a preference for an *average* prolate or oblate distortion. The

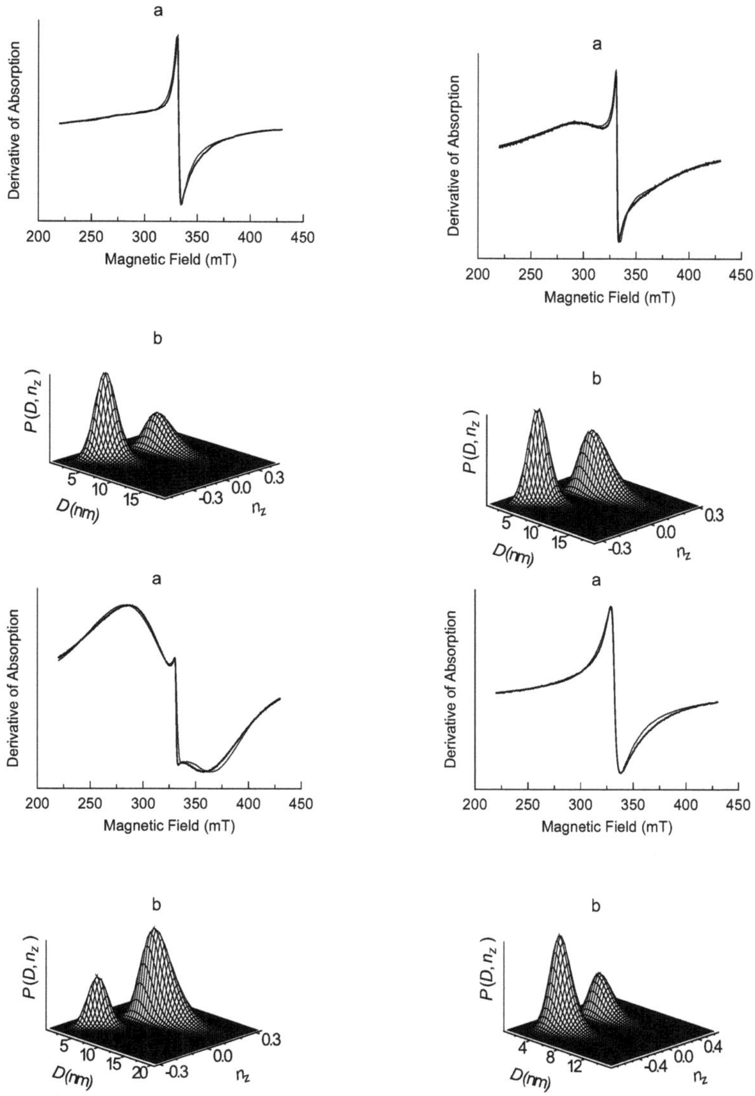

Figure 4. Computer fits to the room-temperature X-band SPR spectra with the joint distribution density of eq. (16). (a) Experimental spectra (full lines) and the best-fit computer-generated spectra (dashed lines); (b) reconstructions of the best-fit joint distribution density of diameters and demagnetising factors. Top, left : the borate glass annealed at 465 °C ; top, right : the borate glass annealed at 475 °C ; bottom, left : the borate glass annealed at 480 °C ; and bottom, right : the sol-gel silica glass. See table 5 for the simulation parameters.

correlation coefficient between D and n_z also increases with the anneal temperature. Moreover, in spite of the apparent general broadening of the SPR spectra, the intrinsic linewidth parameters Δ_M and Δ_{K_1} decrease at higher anneal temperatures. All these factors indicate a general tendency towards morphological ordering of the assembly of magnetic nanoparticles with the increase of the anneal temperature.

Relatively small particle sizes are obtained in the sol-gel silica glass, while the very large σ_n value found in this glass shows that the shapes of the magnetic nanoparticles are less regular in comparison with the borate glass, see table 5. This conclusion is corroborated by the very small correlation coefficient between D and n_z, indicating that the assembly of superparamagnetic nanoparticles in the sol-gel silica glass is more disordered that in the borate glass.

3.3 High-temperature behaviour

A series of SPR spectra of a borate glass sample annealed at 753 K are recorded at measurement temperatures T_m from 300 to 723 K, figure 5. The corresponding temperature dependence of the peak-to-peak linewidth are shown in figure 6. As T_m is increased, a drastic line narrowing occurs,

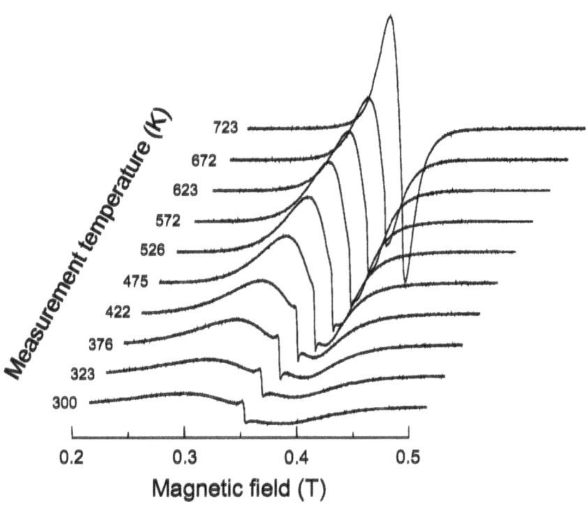

Figure 5. Electron magnetic resonance spectra of a borate glass sample annealed at 753 K and recorded at different temperatures T_m with a constant gain.

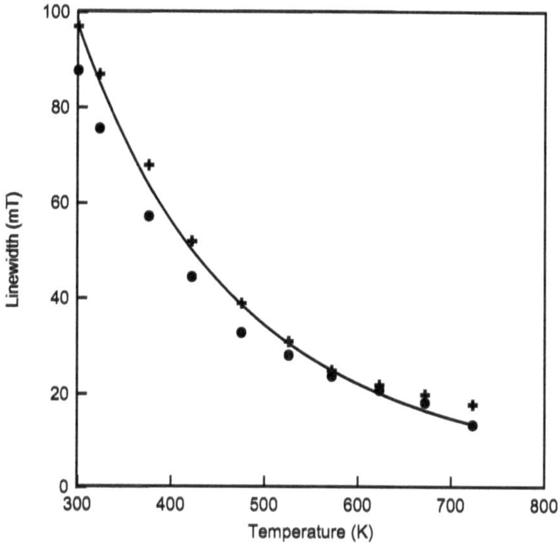

Figure 6. Experimental peak-to-peak width (•) and saturation linewidth $\Delta_0\delta\left(K_1V_s/kT\right)$ (✚) defined in eq. (21) vs. T_m. The curve is a fit calculated in accordance with eq. (21).

accompanied by an increase in the peak-to-peak amplitude. The two-component structure of the $g_{eff} \approx 2.0$ resonance is no longer resolved. These transformations are due to dynamic effects and not to a structural change, since they remain quite reversible for $T_m < T_a$. In contrast, at $T_m > T_a$, an irreversible broadening of the spectrum has been found after cooling the sample back to room temperature.

3.3 Low-temperature behaviour

Below room temperature the X-band SPR spectra of the borate glass broaden beyond the possibility of any meaningful study. On the other hand, the sol-gel glass, containing finer nanoparticles in comparison with the borate glass, *vide ultra*, exhibits much narrower lines at a given temperature. The experimental spectra of this glass at different temperatures are shown in figure 7.

Figure 8 shows the temperature dependence of the apparent resonance field $B_{apparent}$, cf. eq. (19), and the apparent peak-to-peak linewidth. The spectra broadening and shift towards lower fields with the decrease in temperature are clearly seen. (A weak resonance at $g_{eff} \approx 4.3$ is due to a small number of Fe^{3+} ions diluted in the glass matrix, as in the borate glass case.)

The results of computer simulations of the spectra recorded at different temperatures are shown in figure 7. The simulations have been carried out following eq. (1) and the individual lineshape has been chosen in the form of the first derivative of that of eq. (18). The following set of parameters provides the best fits in the whole temperature range of this study : Δ_0 = 0.2944 T, $D_{V\mathrm{m}} = 6.8$ nm, $\sigma = 0.40$, $V_\mathrm{s} = 6370$ nm^3, $N_\perp - N_\parallel = 0.33$.

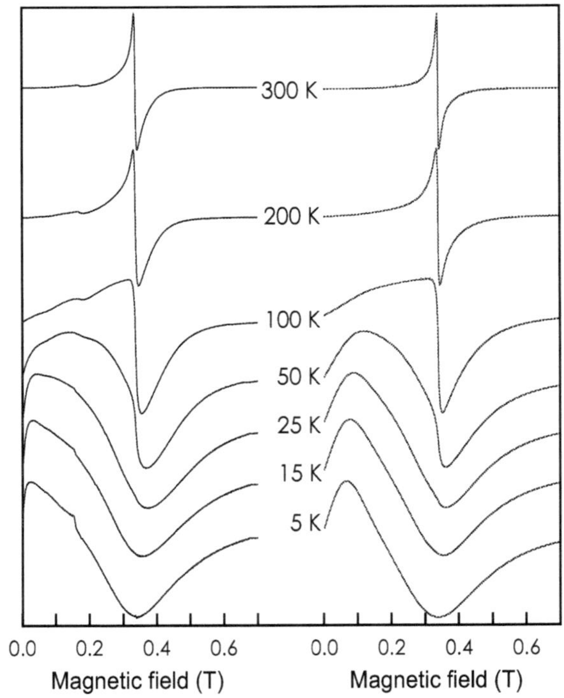

Figure 7. Experimental (left) and computer-generated (right) SPR spectra of the sol-gel glass at different temperatures.

The broadening and concomitant shift of the SPR spectra towards lower fields with the decrease in temperature have been observed in a number of systems, viz., silica supported nickel particles [2], fine grains of α-Fe_2O_3 in Al_2O_3 matrix [38], nanoscale FeOOH particles [39], dispersed ultrafine Mn-Zn ferrite particles [40], polydispersed ferrite particles in a magnetic fluid [41], Fe-based nanoparticles in a polyethylene matrix [42], a granular Cu-Co alloy [43] and maghemite (γ-Fe_2O_3) nanoparticles in ferrofluids [32]. No consistent explication of this behaviour had been put forward. On the other hand, in the framework of our approach, this behaviour is perfectly fitted to ; moreover, the whole shape of the SPR spectra at different temperatures is quite well reproduced. The minor discrepancies between experimental and computed spectra observed in the vicinity of $g_{eff} \approx 4.3$ are mainly due to the EPR signal of isolated Fe^{3+} ions.

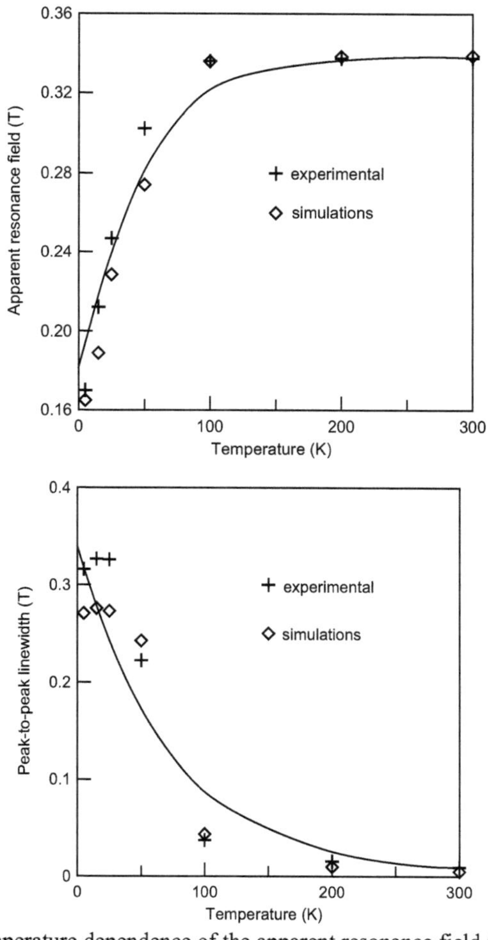

Figure 8. Temperature dependence of the apparent resonance field (top) and of the peak-to-peak linewidth (bottom) in the experimental and computer-simulated spectra. The full curve in the left part is the dependence of $B_{apparent}$ given by eq. (19), and that in the right part shows the theoretical $\Delta_T(T)$ dependence of eq. (21).

In figure 8, besides the corresponding experimental values, we have plotted the apparent resonance field and the "true" linewidth extracted from the best-fit computer-generated spectra for different temperatures. The apparent shift of the resonance field is well described assuming the Landau-Lifshitz line shape (18).

The temperature dependence of the individual linewidth, eq. (21), is rather complicated. At very low temperatures the Langevin function in this equation becomes saturated for the majority of the nanoparticles, so that the main mechanism of this dependence is the thermal modulation of the magnetic anisotropy energy. Indeed, one can see from figure 8 (bottom) that the Δ_T temperature dependence provides a good estimate of the experimental low-temperature linewidth. As the damping factor linearly depends on the linewidth, it follows the same temperature dependence [11].

4. Conclusion

In iron-doped borate and sol-gel silica glasses annealed above the glass transition temperature, superparamagnetic nanoparticle systems are formed, giving rise to characteristic resonance spectra. The computer-assisted superparamagnetic resonance technique is an effective tool for exploring the morphology of such nanoparticles. Convincing computer fits, based on the joint distribution density of diameters and demagnetising factors of dispersed ferromagnetic nanoparticles, show that the non-sphericity of the nanoparticles takes a prominent part in determining the shape of the experimental SPR spectra.

With the decrease in temperature, the SPR spectra of the sol-gel silica glasses undergo a shift towards lower magnetic fields with a marked broadening. Such behaviour is well described in the framework of a theoretical model based on the Landau-Lifshitz damped precession equation and an appropriate linewidth expression taking into account low-temperature freezing of the fluctuations of orientations of the magnetic moments.

References

[1] L. Néel, 1949, Ann. Geophys. **5** 99.
[2] V.K. Sharma and A. Baiker, 1981, J. Chem. Phys. **75** 5596.
[3] J. Dubowik and J. Baszynski, 1986, J. Magn. Magn. Mater. **59** 161.
[4] Yu. L. Raikher and V.I. Stepanov, 1995, J. Magn. Magn. Mater. **149** 34.
[5] R. Berger, J.-C. Bissey, J. Kliava and B. Soulard, 1997, J. Magn. Magn. Mater. **167** 129.
[6] R. Berger, J. Kliava, J.-C. Bissey and V. Baïetto, 1998, J. Phys. : Condens. Matter **10** 8559.

[7] R.D. Sánchez, M.A. López-Quintela, J. Rivas, A. González-Penedo, A.J. García-Bastida, C.A. Ramos, R.D. Zysler and S. Ribeiro Guevara, 1999, J. Phys : Condens. Matter **11** 5643.
[8] J. Kliava and R. Berger, 1999, J. Magn. Magn. Mat. **205** 328.
[9] R. Berger, J. Kliava, J-C. Bissey and V. Baïetto, 2000, J. Appl. Phys. **87** 7389.
[10] L.M. Lacava, B.M. Lacava, R.B. Azevedo, Z.G.M. Lacava, N. Buske, A.L. Tronconi and P.C. Morais, 2001, J. Magn. Magn. Mat. **225** 79.
[11] R. Berger, J-C. Bissey, J. Kliava, H. Daubric and C. Estournès, 2001, J. Magn. Magn. Mat. **234** 535.
[12] L. Dormann, D Fiorani and E. Tronc, 1997, Adv. Chem. Phys. **98** 283.
[13] G. V. Skrotskii and L. V. Kurbatov, 1966, Ferromagnetic Resonance, Ed. V Vonsovski, Oxford, Pergamon.
[14] M. Respaud, M. Goiran, F. Yang, J.M. Broto, T. Ould Ely, C. Amiens, B. Chaudret and S. Askenazy, 1998, Physica B **246-247** 580.
[15] M. Respaud, M. Goiran, J.M. Broto, F. H. Yang, T. Ould Ely, C. Amiens and B. Chaudret, 1999, Phys. Rev. **B 59** R3934.
[16] J. Smit and H. G. Beljers, 1955, Philips Res. Rep. **10** 113.
[17] L. Baselgia, M. Warden, F. Waldern, S. L. Hutton, J. E. Drumheller, Y. Q. He, P. E. Wigen, M. Marysko, 1988, Phys. Rev. **B 38** 2237.
[18] R.S. de Biasi and T.C. Devezas, 1978, J. Appl. Phys. **49** 2466.
[19] Yu.L. Raikher and V.I. Stepanov, 1992, Sov. Phys. JETP **75** 764.
[20] G. Ennas, A. Musinu, G. Piccaluga, D. Zedda, D. Gatteschi, C. Sangregorio, J. L. Stanger, G. Concas and G. Spano, 1998, Chem. Mater. **10** 495.
[21] J. Zarzycki, J. Appl. Crystallogr., 1974, **7** 200.
[22] J.-C. Bacri, F. Boué, V. Cabuil and R. Perzynski, 1993, Colloids and Surfaces **A 80** 11.
[23] J. Popplewell and L. Sakhnini, 1995, J. Magn. Magn. Mater. **149** 72.
[24] J.C. Bacri, R. Perzynski, D. Salin, V. Cabuil and R. Massart, 1986, J. Magn. Magn. Mater. **62** 36.
[25] R.V. Upadhyay, G.M. Sutariya and R.V. Mehta, 1993, J. Magn. Magn. Mater. **123** 262.
[26] C. Estournès, T. Lutz, J. Happich, P. Quaranta, P. Wissler and J.L.Guille, 1997, J. Magn. Magn. Mater. **173** 83.
[27] M. Jamet, V. Dupuis, P. Mélinon, G. Guiraud, A. Pérez, W. Wernsdorfer, A. Traverse and B. Baguenard, 2000, Phys. Rev. **B 62** 493.
[28] J. Zarzycki and F. Naudin, 1967, Phys. and Chem. Glasses **8** 11.
[29] G. Le Caër, R.A. Brand and K. Dehghan, 1985, J. Physique, Coll. **46** C8-169.
[30] M. Maurer, 1986, Phys. Rev. **B34** 8996.
[31] G. Le Caër, J.M. Cadogan, R.A. Brand, J.M. Dubois and H.J. Güntherodt, 1984, J. Phys. F : Met. Phys. **14** L73.
[32] F. Gazeau, J.C. Bacri, F. Gendron, R. Perzynski, Yu.L. Raikher, V.I. Stepanov and E. Dubois, 1998, J. Magn. Magn. Mater. **186** 175.
[33] R. Berger, J-C. Bissey and J. Kliava, 2000, J. Phys.: Condens. Matter. **12** 9347.
[34] E. Rezlescu, N. Rezlescu and M.L. Craus, 1997, J. Phys. IV, France **7** 553.
[35] E. Schmidbauer and R. Keller, 1996, J. Magn. Magn. Mater. **152** 99.
[36] J. Kliava, 1986, Phys. Status Sol.(b) **134** 411.
[37] J. Kliava, 1988, EPR Spectroscopy of Disordered Solids (EPR spektroskopia neuporiadochennyh tverdyh tel), Riga, "Zinàtne".
[38] R. Zysler, D. Fiorani, J.L. Dormann and A.M. Testa, 1994, J. Magn. Magn. Mater. **133** 71.
[39] M.M. Ibrahim, G. Edwards, M.S. Seerha, B. Ganguly and G. P. Huffman, 1994, J. Appl. Phys. **75** 5873.

[40] K. Nagata and A. Ishihara, 1992, J. Magn. Magn. Mater. **104-107** 1571.
[41] K. Parekh, R.V. Upadhyay and R.V. Mehta, 2000, J. Appl. Phys. **88** 2799.
[42] Yu. A. Koksharov, S.P. Gubin, I.D. Kosobudsky, G.Yu. Yurkov, D.A. Pankratov, L.A. Ponomarenko, M.G. Mikheev, M. Beltran, Y. Khodorkovsky and A.M. Tishin, 2000, Phys. Rev. **B 63** 12407.
[43] H. K. Lachowicz, A. Sienkiewicz, P. Gierlowski and A. Slawska-Waniewska, 2000, J. Appl. Phys. **88** 368.

EFFECT OF MAGNETIC FIELD ON THE MAGNETIC STATE OF COPPER METABORATE

G. Petrakovskii *, M. Popov, V. Zinenko
Institute of Physics SB RAS, 660036 Krasnoyarsk, Russia
* gap@iph.krasn.ru

B. Roessli, J. Schefer
Laboratory for Neutron Scattering, ETH Zurich & Paul Scherrer Institute, CH-5232 Villigen, Switzerland

M. Boehm
Laboratory for Neutron Scattering, ETH Zurich & Paul Scherrer Institute, CH-5232 Villigen Switzerland and Institut Laue-Langevin, 38042 Grenoble, France

U. Staub
Swiss Light Source, Paul Scherrer Institute, CH-5232 Villigen, Switzerland

J. Franse et al. (eds.), Smart Materials for Ranging Systems, 49–65.

Abstract: The influence of the external magnetic field on the magnetic state of copper metaborate CuB_2O_4 was investigated by means of magnetization, heat capacity and neutron diffraction measurements. It was found that the temperature of the magnetic phase transition from the commensurate to the incommensurate magnetic structure depends strongly on the value of the magnetic field. For temperatures below 10 K the magnetization of the crystal increases sharply at the critical magnetic field. A theory based on a symmetry analysis and using a phenomenological approach is developed to describe the field dependence of copper metaborate. The model uses two order parameters corresponding to ferromagnetic and antiferromagnetic moments directed in the tetragonal plane of the crystal. For the description of the incommensurate magnetic structure we introduced the modified Lifshits invariant for the case of two two-component order parameters. This theory gives a satisfactory description of the experimental data including the "magnetic field-temperature" phase diagram of copper metaborate.

Keywords: Copper metaborate, magnetic state, magnetic field, commensurate phase, incommensurate phase, phenomenological approach.

1. Introduction

Oxide compounds of copper attract the great interest of researchers due to a wide variety of magnetic and electric properties. In this class of compounds there were found a spin-Peierls singlet magnet [1], high-temperature superconductors [2], incommensurable magnetic structures [3], and magnets with cooperative Jahn-Teller effect [4]. Magnetic properties of such compounds are usually defined by exchange interactions of ions Cu^{2+}. As the spin of this ion $S = 1/2$, the analysis of a magnetic state and spin dynamics of copper oxides can be often fulfilled strictly enough.

Copper metaborate CuB_2O_4 represents the rare case where it is possible to investigate a surprising wealth of various manifestations in the formation of the magnetic ordering on the basis of a compound with a relatively simple crystal structure. Our previous investigations [5,6] have shown that the magnetic structure of CuB_2O_4 is the result of complex and mutual interactions of two magnetic subsystems. The first subsystem represents a 3D-network of copper spins. The second subsystem forms frustrated interpenetrating zigzag chains which remain in a non saturated magnetic state down to the lowest temperatures. The existence of the Dzyaloshinskii-Moria antisymmetric exchange interaction allows weak ferromagnetism over the temperature range 10÷20 K. A helix structure of "magnetic soliton lattice"-type occurs below 10 K as a result of competition between Lifshits invariant and magnetic crystallographic anisotropy. Moreover, at temperatures below 2 K there are additional magnetic transformations of the magnetic structure [6] which are still not understood. The low-dimensionality and the frustration of the zigzag chains are at the origin of the diffuse intensity observed by neutron scattering in an unusually wide temperature range 1.5÷30 K.

The magnetic field has a strong influence on the magnetic state of copper metaborate. In particular, it reduces significantly the transition temperature T_i from the incommensurate to the commensurate state [7]. In a previous work [8] we showed that a phenomenological theory could describe the main features of the magnetic state of copper metaborate in the absence of a magnetic field. In this paper we present results of experimental and theoretical investigations of the influence of the external magnetic field on the magnetic state of CuB_2O_4. The experimental results include an elastic neutron scattering study of the magnetic structure in the presence of a magnetic field together with measurements of the heat capacity in fields up to 80 kOe, and temperature and field dependencies of the magnetization at various temperatures. A phenomenological theory is developed to describe the experimental data.

2. Experimental data

In all our measurements we used crystals of copper metaborate prepared in the Institute of Physics SB RAS [5]. The crystalline structure of CuB_2O_4 is tetragonal with space group $I\bar{4}2d$ $\left(D_{2d}^{12}\right)$ and has lattice parameters a=11.528 Å and c=5.607 Å. The unit cell contains 12 formula units. The ions of copper Cu^{2+} occupy two non-equivalent positions: Cu(b) - position 4b with point symmetry S_4 (0,0,1/2), and Cu(d) - position 8d with point symmetry C_2 (0.0815,1/4,1/8). The ion Cu(b) is at the center of a square unit formed by four oxygen ions. The ion Cu(d) is surrounded by six oxygens located at the vertices of a distorted octahedron.

The crystalline structure does not undergo any phase transition between 1.5÷300 K [5]. Magnetic peaks measured by neutron diffraction in the temperature range 10÷20 K coincide with those of the lattice. Therefore, the magnetic unit cell of copper metaborate coincides with the chemical one. Hence, the phase transition at T = 20 K is described by the propagation vector at the center of the Brillouin zone, $\boldsymbol{q} = 0$. Upon cooling the sample below 10 K, we found that the magnetic reflections split into a set of two satellites, indicating that the magnetic structure becomes incommensurate along the tetragonal axis of the chemical structure [5].

The magnetic measurements were performed both with a SQUID-magnetometer and with a vibrating magnetometer. We find that the crystal magnetization is sharply anisotropic and shows peculiarities at temperatures 10 K and 20 K (Fig. 2). The dependencies of the magnetization on magnetic field for two orientations and different temperatures are shown in Figs. 3 and

4. In Fig. 5 the results of measurements of the heat capacity in magnetic fields up to 80 kOe are represented. These measurements were done using a PPMS-6000 from Quantum Design.

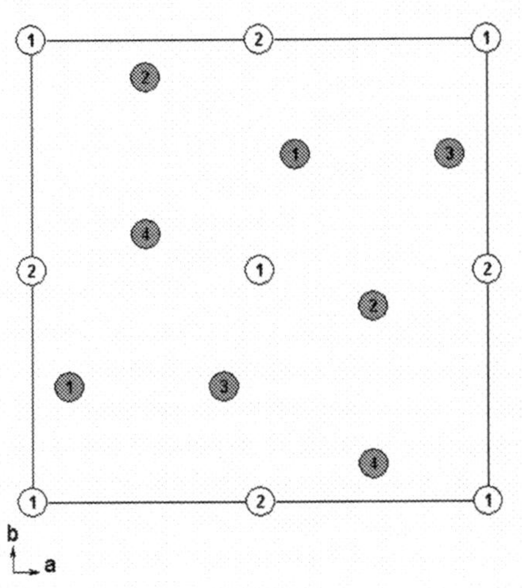

Figure 1. Arrangement and enumeration of the Cu-ions in the tetragonal unit cell: open circles – Cu(b) -ions, shaded circles – Cu(d) -ions.

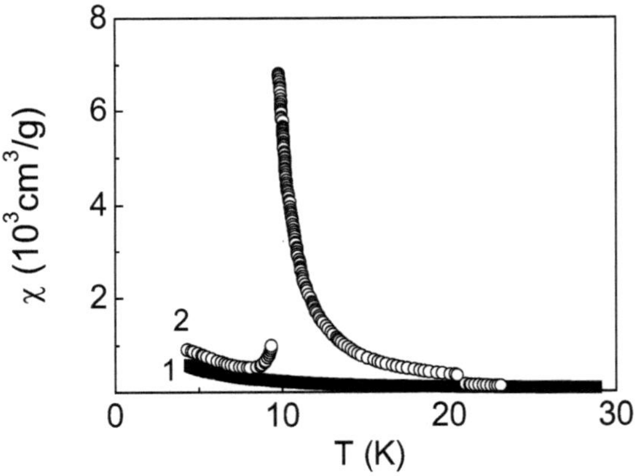

Figure 2. Temperature dependence of the magnetic susceptibility of single crystalline copper metaborate: (1) the magnetic field is parallel to the tetragonal crystal axis; (2) the magnetic field is parallel to the basal plane.

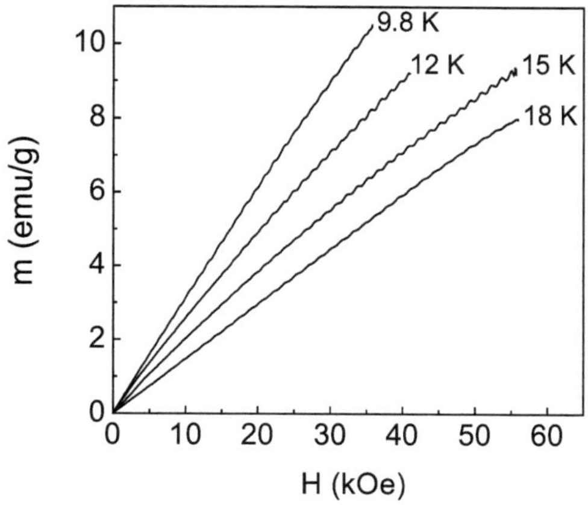

Figure 3. Magnetic field dependence of the magnetization of single crystalline copper metaborate with magnetic field parallel to the tetragonal crystal axis for different temperatures.

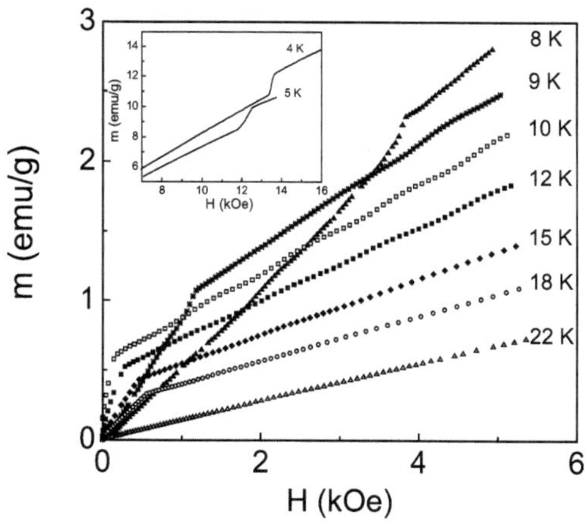

Figure 4. Temperature dependence of the magnetic moment of single crystalline copper metaborate for the magnetic field parallel to the tetragonal crystal axis.

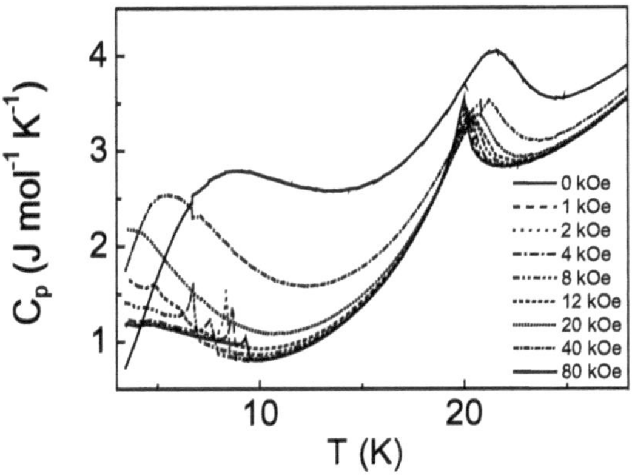

Figure 5. Temperature dependence of the heat capacity of single crystalline copper metaborate for different magnetic fields parallel to the basal plane.

We performed additional neutron scattering measurements to study the transformations of the magnetic structure of copper metaborate under application of an external magnetic field [7]. Increasing of the magnetic field applied in the tetragonal plane induces a transition from the incommensurate phase to a commensurate phase of weak ferromagnetic type. The above mentioned splitting of magnetic reflections abruptly vanishes (Fig. 6). This phase transition is also observed in the magnetization data (Fig. 3).

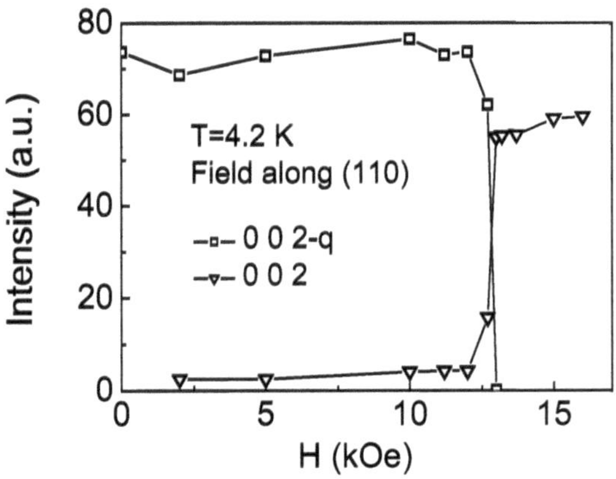

Figure 6. Magnetic field dependence of the intensity of the Bragg reflections (0,0,2) and (0,0,2-q) in copper metaborate at T = 4.2 K.

3. Phenomenological approach

For a phenomenological description of the magnetic properties of the crystal under consideration, it is necessary at first to carry out a symmetry analysis to define the magnetic modes of the irreducible representations, and to generate from them invariants which allow us to obtain a phenomenological potential.

The point group of copper metaborate crystal $\bar{4}2m$ contains eight elements of symmetry [9]: $1, \bar{4}_3^1, 4_3^2, \bar{4}_3^3, 4_1^2, 4_2^2, m_4, m_5$ and has five irreducible representations. Four of these ($\Gamma_1 - \Gamma_4$) are one-dimensional and one (Γ_5) is two-dimensional. The decompositions of the magnetic representations for the two non-equivalent sublattices of copper are:

$$\Gamma_{4b} = \Gamma_1 + \Gamma_2 + 5\Gamma_5 , \qquad \Gamma_{8d} = \Gamma_1 + 2\Gamma_2 + 2\Gamma_3 + \Gamma_4 + 3\Gamma_5 .$$

The magnetic modes can be converted into irreducible representations of the group $\bar{4}2m$ in the Cu(b) sublattice which yield:

Γ_1: $S_{b1z} - S_{b2z}$

Γ_2: $S_{b1z} + S_{b2z}$

$$\Gamma_5: \ (S_{b1x} + S_{b2x}, - S_{b1y} - S_{b2y}) \tag{1}$$

$$\Gamma_5: \ (S_{b1y} - S_{b2y}, S_{b1x} - S_{b2x}) \tag{2}$$

The modes corresponding to the representations Γ_1 and Γ_2 of the position 4b describe antiferromagnetic or ferromagnetic ordering along the tetragonal c-axis. Modes corresponding to the representation Γ_5 describe noncollinear magnetic structures in the tetragonal plane. Forthe Cu(d) -sublattice the magnetic modes are:

Γ_1: $S_{d1x} + S_{d2y} - S_{d3x} - S_{d4y}$

Γ_2: $S_{d1y} - S_{d2x} - S_{d3y} + S_{d4x}$

Γ_2: $S_{d1z} + S_{d2z} + S_{d3z} + S_{d4z}$

Γ_3: $S_{d1y} + S_{d2x} - S_{d3y} - S_{d4x}$

Γ_3: $S_{d1z} - S_{d2z} + S_{d3z} - S_{d4z}$

Γ_4: $S_{d1x} - S_{d2y} - S_{d3x} + S_{d4y}$

$$\Gamma_5: \ (S_{d1x} + S_{d2x} + S_{d3x} + S_{d4x}, - S_{d1y} - S_{d2y} - S_{d3y} - S_{d4y}) \tag{3}$$

$$\Gamma_5: \ (S_{d1x} - S_{d2x} + S_{d3x} - S_{d4x}, S_{d1y} - S_{d2y} + S_{d3y} - S_{d4y}) \tag{4}$$

Γ_5: $(S_{d2z} - S_{d4z}, S_{d1z} - S_{d3z})$.

For a description of an incommensurate state, it is necessary to take into account the linear derivatives to the coordinate z of these modes: for the position 4b

Γ_3: $(S_{b1z} - S_{b2z})'$

Γ_4: $(S_{b1z} + S_{b2z})'$

Γ_5: $(-S_{b1y} - S_{b2y}, S_{b1x} + S_{b2x})'$

Γ_5: $(S_{b1x} - S_{b2x}, S_{b1y} - S_{b2y})'$

and for the position 8d

Γ_1: $(S_{d1y} + S_{d2x} - S_{d3y} - S_{d4x})'$

Γ_1: $(S_{d1z} - S_{d2z} + S_{d3z} - S_{d4z})'$

Γ_2: $(S_{d1x} - S_{d2y} - S_{d3x} + S_{d4y})'$

Γ_3: $(S_{d1x} + S_{d2y} - S_{d3x} - S_{d4y})'$

Γ_4: $(S_{d1y} - S_{d2x} - S_{d3y} + S_{d4x})'$

Γ_4: $(S_{d1z} + S_{d2z} + S_{d3z} + S_{d4z})'$

Γ_5: $(-S_{d1y} - S_{d2y} - S_{d3y} - S_{d4y}, S_{d1x} + S_{d2x} + S_{d3x} + S_{d4x})'$

Γ_5: $(S_{d1y} - S_{d2y} + S_{d3y} - S_{d4y}, S_{d1x} - S_{d2x} + S_{d3x} - S_{d4x})'$

Γ_5: $(S_{d1z} - S_{d3z}, S_{d2z} - S_{d4z})'$.

where $f' \equiv df/dz$.

The analysis of the neutron diffraction pattern [5,6] using 25 magnetic peaks shows that in the commensurate phase ($T_i < T < T_N$) the spins of both Cu(b)- and Cu(d) -sublattices of copper metaborate form a noncollinear magnetic structure. In the Cu(b) -sublattice, the antiferromagnetic moment lying in the tetragonal plane dominates over the ferromagnetic moment directed in the plane perpendicular to it. The antiferromagnetic moment along the c-axis is rather small. In the Cu(d)-sublattice, the antiferromagnetic moment along the c-axis probably dominates, and the antiferromagnetic moment in the tetragonal plane is rather small. The ferromagnetic moment of this sublattice could not be detected experimentally. In this phase, the magnetic moment of copper ion located in the position 8d is much smaller than the value of the moment at position 4b.

In the incommensurate phase $T < T_i$ the magnetic structure forms a helix along the c-axis. For the theoretical analysis of the magnetic properties of copper metaborate based on a phenomenological thermodynamic potential, it is essential that the crystal symmetry has no center of inversion. The inversion operation enters only in combination with rotational displacement around the c-axis by 90°: $\overline{4}_3^1$ and $\overline{4}_3^3$. Therefore, in the thermodynamic potential the presence of invariants such as Lifshits-like invariants are possible. These invariants are bilinear on the two-component order parameters and their spatial derivatives, and they are essential for the occurrence of an incommensurate phase.

An account of all eight one-component and five two-component homogeneous magnetic modes at the phenomenological level results in an extremely complicated thermodynamic potential of the system. Therefore it is necessary to choose the modes permitting to describe the basic properties of copper metaborate obtained experimentally. For that, it is necessary to pay attention to the following: If one represents an arbitrary α-th homogeneous mode in the representation Γ_5 as $\eta_\alpha = (\eta_{\alpha 1}, \eta_{\alpha 2})$, then the first derivative on coordinate z of this mode in the same representation is equal to $\eta'_\alpha = (\eta_{\alpha 2}, \eta_{\alpha 1})'$. In this case the invariant $\eta_\alpha \cdot \eta'_\alpha = \eta_{\alpha 1} \cdot \eta'_{\alpha 2} + \eta_{\alpha 2} \cdot \eta'_{\alpha 1}$ is a complete derivative with respect to z and there is no Lifshits invariant for a single mode in the considered symmetry group. In terms of the theory of representations, the reason is that in this case the antisymmetric square $\{\Gamma_5^2\}$ is converted on the representation Γ_2, while a vector component longitudinal to the tetragonal axis is converted to the representation Γ_3 and vector components perpendicular to the tetragonal axis are converted to the representation Γ_5 [10].

Let us present now a thermodynamic potential of copper metaborate as a functional of two two-component order parameters, $\eta_1 = (\eta_{11}, \eta_{12})$ and $\eta_2 = (\eta_{21}, \eta_{22})$:

$$\Phi\{\eta\} = \int\{A_{11}\eta_1 \cdot \eta_1/2 + A_{22}\eta_2 \cdot \eta_2/2 + A_{12}\eta_1 \cdot \eta_2 + B_{22}(\eta_2 \cdot \eta_2)^2/4 + C_{12}(\eta_1 \cdot \eta'_2 - \eta'_1 \cdot \eta_2) + D_{11}\nabla\eta_1 \cdot \nabla\eta_1 + D_{22}\nabla\eta_2 \cdot \nabla\eta_2 + D_{12}\nabla\eta_1 \cdot \nabla\eta_2 - \eta_1 \cdot H_1\}dV, \quad (5)$$

where the integration is carried out over the volume of the whole crystal, $A_{11} = a_{11}(T-T_1)$, $A_{22} = a_{22}(T-T_2)$, $a_{11} > 0$, $a_{22} > 0$, $B_{22} > 0$, $D_{11} > 0$, $D_{11}D_{22} > D_{12}^2$, ∇ - nabla operator, $\nabla\eta_\alpha = (\nabla\eta_{\alpha 2}, \nabla\eta_{\alpha 1})$, H_1 - a field conjugate to the parameter η_1. In Eq. 5 the dependence of order parameters on coordinates is used implicitly.

The parameters η_1 and η_2 are linear combinations of the magnetic modes converted on the representation Γ_5. The fact that below T_N the magnetic subsystem of copper metaborate forms an easy-plane weak ferromagnet, twisted below T_i in a spiral, permits to compose η_1 as a combination of the ferromagnetic modes (1) and (3), and η_2 as a combination of the antiferromagnetic modes (2) and (4). Accordingly $H_1 = (H_{11}, H_{12}) = (H_x, -H_y)$. It is necessary to note that in the thermodynamic potential given by Eq. (5) the order parameter responsible for the transition at T_i is not chosen in an explicit form as it was done in our previous paper [8].

The invariant with coefficient C_{12} in Eq. (5) is a generalization of the well known shape of the Lifshits invariant [10] for the case of two two-component order parameters - instead of being formed by an antisymmetric product of components of one order parameter and their spatial derivatives, it is composed of antisymmetric products of different order parameters and their spatial derivatives.

Among invariants of fourth order, the invariant describing a temperature dependence of the antiferromagnetic order parameter is left only in (5). The invariants quadratic in the spatial derivatives of the order parameters are taken in the simple isotropic form. It will be shown below that the presence in Eq. (5) of other order parameters is not necessary for the evolution of the magnetic states in copper metaborate for temperatures decreasing from the paramagnetic phase above 20 K to the incommensurate phase below 10 K.

In absence of a Lifshits invariant, an equilibrium state of Eq. (5) is homogeneous due to the positive definiteness of the remaining nonuniform terms in the potential. A Lifshits invariant breaks the stability of such a state along the tetragonal axis. To find a new equilibrium state we make use of the fact that at zero external magnetic field the magnetic peaks of the higher-order harmonics in the neutron scattering spectra are visible only in a rather narrow temperature region close to transition into the incommensurate phase [5].

Therefore one can describe the state in an external magnetic field applied in the tetragonal plane as

$$\begin{aligned} \eta_1 &= (p_{11q}\cos(qz+\varphi_{11q}) + p_{11k}\cos(kz+\varphi_{11k}), \\ &\quad p_{12q}\cos(qz+\varphi_{12q}) + p_{12k}\cos(kz+\varphi_{12k})), \\ \eta_2 &= (p_{21q}\cos(qz+\varphi_{21q}) + p_{21k}\cos(kz+\varphi_{21k}), \\ &\quad p_{22q}\cos(qz+\varphi_{22q}) + p_{22k}\cos(kz+\varphi_{22k})), \\ H_1 &= (H\cos(kz), -H\sin(kz)), \end{aligned} \tag{6}$$

where the amplitudes $p_{\alpha\beta}$ and the phases $\varphi_{\alpha\beta}$ of the spatial waves do not depend on z. Without loss of generality by choosing $\varphi_{11} = 0$ we determine a reference point on the tetragonal axis. The case of a homogeneous external field ($k = +0$) can be obtained after the integration over a crystal volume.

The variational equations have two solutions for which the common relationships are

$$\begin{aligned} &p_{11q} = -(A_{12q}p_{21q} - C_{12q}p_{22q})/A_{11q}, && p_{12q} = -(A_{12q}p_{21q} + C_{12q}p_{22q})/A_{11q}, \\ &p_{11k} = (H - A_{12q}p_{21q})/A_{11}, && p_{12k} = (H - A_{12q}p_{22q})/A_{11}, \\ &\varphi_{11q} = \varphi_{21q} = \varphi_{11k} = \varphi_{21k} = 0, && \varphi_{12q} = \varphi_{22q} = \varphi_{12k} = \varphi_{22k} = \pi/2. \end{aligned} \tag{7}$$

One solution corresponds to the homogeneous state of the magnetic system:

$$p_{21q} = p_{22q} = 0, \qquad p_{21k} = p_{22k} = p_{2k}, \tag{8}$$

where p_{2k} is a solution of the equation

$$A_{11}B_{22}p_{2k}^3 + (A_{11}A_{22} - A_{12}^2)p_{2k} + A_{12}H = 0.$$

Another solution of the variational equations allows to describe a nonuniform state:

$$p_{21q} = p_{22q} = p_{2q}, \qquad p_{21k} = p_{22k} = p_{2k}, \qquad (9)$$

where p_{2q} and p_{2k} are determined from the combined equations

$$A_{11q}B_{22}p_{2q}^2 + A_{11q}(A_{22q} + 2B_{22}p_{2k}^2) = A_{12q}^2 + C_{12q}^2,$$
$$3A_{11}A_{11q}B_{22}p_{2k}^3 + [2A_{11}A_{11q}A_{22q} + A_{11q}A_{12}^2 - A_{11}A_{11q}A_{22} -$$
$$- 2A_{11}(A_{12q}^2 + C_{12q}^2)] = A_{11q}A_{12}H,$$

$A_{\alpha\beta q} = A_{\alpha\beta} + D_{\alpha\beta}q^2$, $C_{12q} = 2\,C_{12}q$. The propagation vector q is equal to zero or determined by the expression

$$q = \mathrm{Re}\{[(\{[4A_{11}D_{11}C_{12}^2 - (A_{11}D_{12} - A_{12}D_{11})^2]/[D_{11}D_{22}\text{-}D_{12}^2]\}^{1/2} -$$
$$- A_{11})/D_{11}]^{1/2}\} \qquad (10)$$

depending on the stability of the corresponding solutions in Eqs. (8) or (9). The stability of one or another solution depends on the relation between the parameters of the thermodynamic potential (5) and the value of the magnetic field.

According to Eqs. (7)-(9), in the incommensurate phase at $H = 0$ the order parameter η_1 forms an elliptic spiral along the tetragonal axis, the order parameter η_2 forms a circular spiral. An elliptic spiral implies a reorientation of the magnetic moments out of the tetragonal plane.

4. Discussion and conclusion

In difference from the analysis performed in Ref. [11] for a one-parametric potential, the peculiarity of the relations obtained here is the opportunity to distinguish between the ordering temperature of the magnetic system ($p_{21} \neq 0$ at $H = 0$) and the temperature of formation of the helicoidal structure ($q \neq 0$). The helicoidal magnetic structure exists in a temperature interval (T_{q1}, T_{q2}) given by:

$$T_{q1} = T_{q0} - \Delta T_q, \qquad T_{q2} = T_{q0} + \Delta T_q, \qquad (11)$$

where

$$T_{q0} = T_1 + (A_{12}D_{12} + 2C_{12}^2)/(a_{11}D_{22}),$$
$$\Delta T_q = [(A_{12}D_{12} + 2C_{12}^2)^2 - A_{12}^2D_{11}D_{22}]^{1/2}/(a_{11}D_{22}).$$

The magnetic ordering occurs at the temperature

$$T_p = T_{p0} + \Delta T_p, \qquad T_{p0} = [T_1 + T_2 - (D_{11}/a_{11} + D_{22}/a_{22})q^2]/2, \qquad (12)$$
$$\Delta T_p = \{[T_1 - T_2 - (D_{11}/a_{11} - D_{22}/a_{22})q^2]^2 + 4(A_{12}^2 + C_{12}^2)/(a_{11}a_{22})\}^{1/2}/2.$$

Such a distinction corresponds to the properties of copper metaborate as described above. Indeed, at particular values of parameters of the thermodynamic potential (5) it is possible to obtain $T_p > T_{q2}$ with a stable homogeneous state ($q = 0$) at intermediate temperatures. Then for copper metaborate the temperature T_N can be connected with T_p (12), and T_i - with T_{q2} (11).

In figures 7-13 the results of numerical calculation using formulae (6)-(12) are given with the following values of the coefficients of the incomplete thermodynamic potential (5):

$$a_{11} = a_{22} = 3.3 \cdot 10^2 \frac{G^2 g^2}{\text{erg cm}^3 \text{K}}, \qquad T_1 = 0.19\ \text{K}, \qquad T_2 = 20.4\ \text{K},$$

$$A_{12} = -1.8 \cdot 10^2 \frac{G^2 g^2}{\text{erg cm}^3}, \qquad B_{22} = 5.2 \cdot 10^1 \frac{G^4 g^4}{\text{erg}^3 \text{cm}^3},$$

$$C_{12}\left(\frac{2\pi}{c}\right) = -5.3 \cdot 10^3 \frac{G^2 g^2}{\text{erg cm}^3}, \qquad D_{11}\left(\frac{2\pi}{c}\right)^2 = 3.7 \cdot 10^4 \frac{G^2 g^2}{\text{erg cm}^3},$$

$$D_{22} = D11, \qquad D_{12} = 0.$$

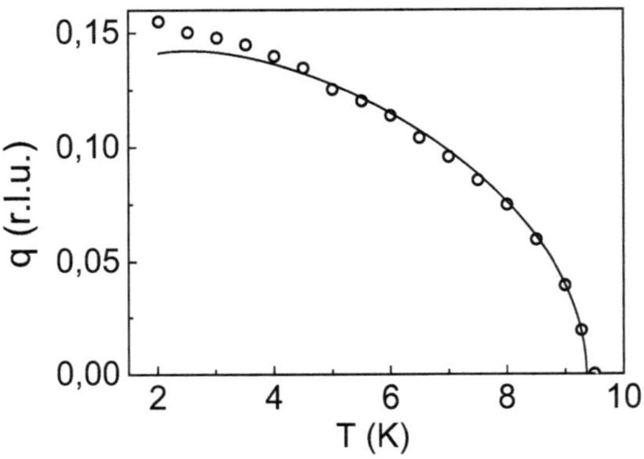

Figure 7. Simulation of the temperature dependence of the propagation vector q by Eq. (10) (solid line). Points are the experimental data [5].

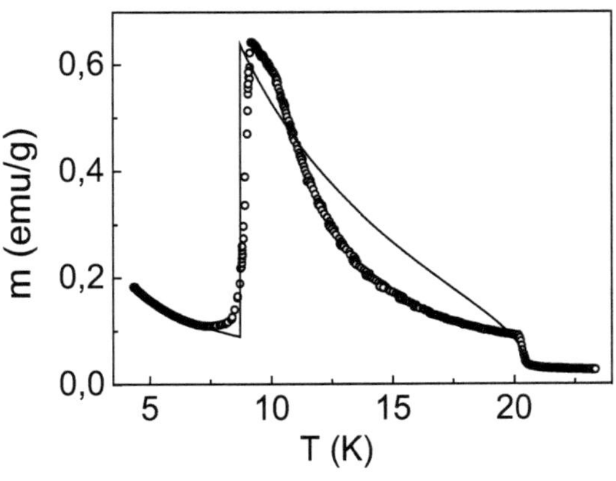

Figure 8. Temperature dependence of the magnetization of single crystalline copper metaborate for a value of magnetic field H = 0.25 K parallel to the basal plane: open circles – experimental data [7], solid line – calculated dependence.

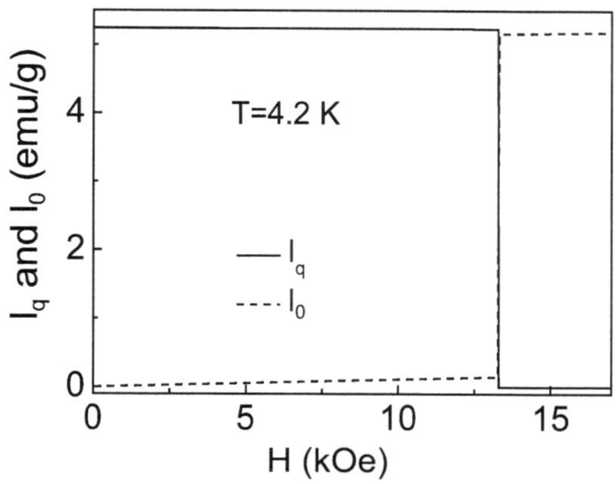

Figure 9. Calculated field dependence of the zero and first harmonics of the antiferromagnetic moment of copper metaborate at T=4.2 K (compare with Fig. 6).

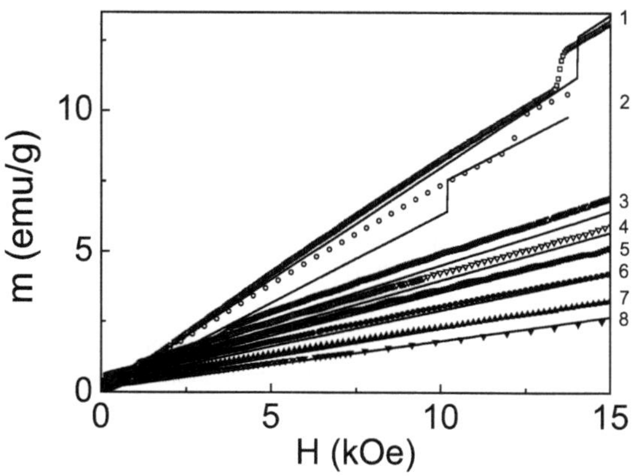

Figure 10. Magnetic field dependence of the magnetization of single crystalline copper metaborate with magnetic field parallel to the basal plane for different temperatures: 1 – 4.2 K, 2 – 5 K, 3 – 8 K, 4 – 9 K, 5 – 10 K, 6 – 12 K, 7 – 15 K, 8 – 18 K.

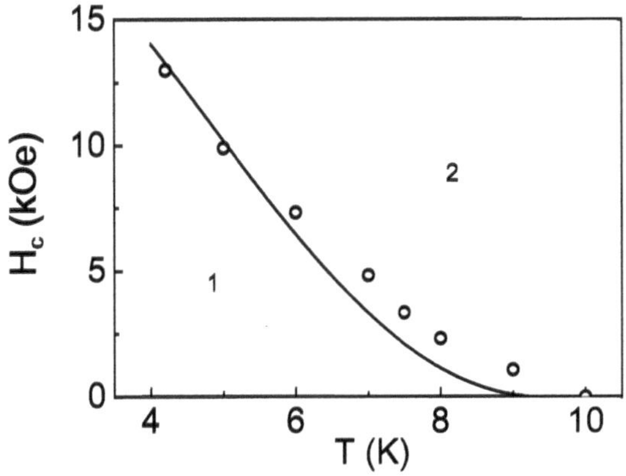

Figure 11. Temperature-magnetic field (parallel to the basal plane) phase diagram of single crystalline copper metaborate: 1 – incommensurate phase of helix magnet; 2 – commensurate phase of weak ferromagnet.

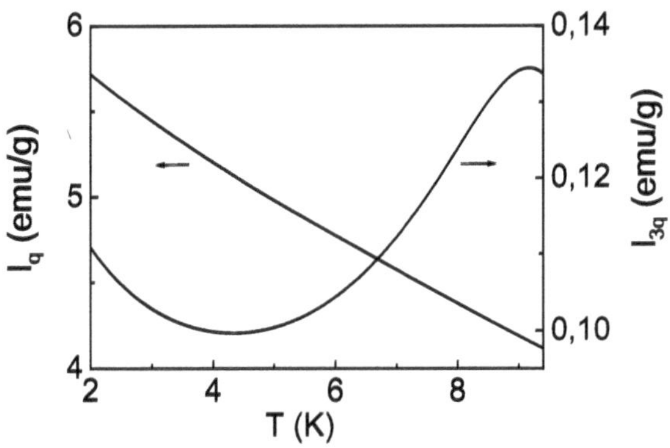

Figure 12. Simulation of the temperature dependence of the neutron intensities of the first and third harmonics of the incommensurate satellite reflection of copper metaborate at zero magnetic field.

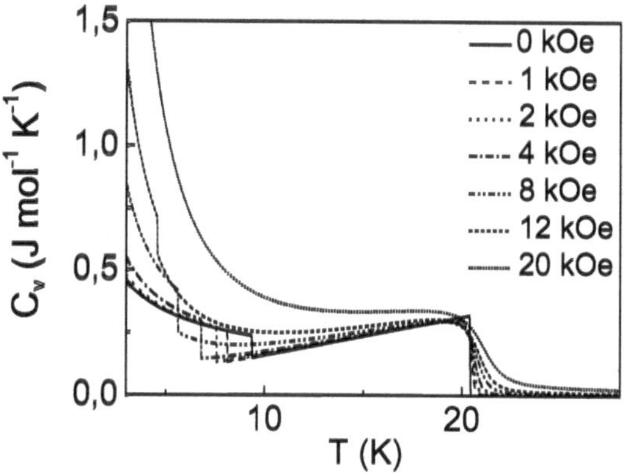

Figure 13. Simulation of the temperature dependence of the heat capacity of copper metaborate for different magnetic fields parallel to the basal plane.

As it can be seen on these figures, the calculated temperature and field dependencies of the copper metaborate magnetic parameters reproduce quite satisfactorily the experimental data.

Because of the helix formation in the incommensurate phase the spontaneous contribution to the magnetization of the whole crystal disappears, and there is only a magnetization because of the applied external magnetic field (Fig. 8). Under application of an increasing external field there is a first-order transition to the phase corresponding to the easy-plane weak ferromagnet. Namely, the component of the antiferromagnetic moment with $q \neq 0$ at a field $H_c(T)$ vanishes stepwise, while the component with a zero propagation vector increases stepwise (compare Fig. 9 to Fig. 6 [7]). Such a change of the magnetic subsystem state shows that the magnetization (Fig. 10) is bound with the antiferromagnetic moment through the Dzyaloshinskii invariant (with coefficient A_{12}). The experimental [7] and calculated temperature-field phase diagrams in the temperature interval 4÷10 K are given in Fig. 11.

The presence of the higher order harmonics of the magnetic helix in the incommensurate phase is characteristic for the temperature interval where the Lifshits invariant is comparable with an anisotropy invariant [11]. For magnetic systems with a one-parametric thermodynamic potential the propagation vector $\boldsymbol{q}$ is not equal to zero already at the temperature where the system orders, $T_i = T_p$. As an anisotropy invariant is proportional to $|\eta|^4$ for a crystal with tetragonal symmetry, then it becomes comparable with Lifshits invariant proportional to $q|\eta|^2$ much below T_i near the transition into a low-temperature commensurate phase. However, in copper metaborate $\boldsymbol{q}$ grows sharply from approximately zero at temperature $T_i < T_p$ (Fig. 7) [5].

So it is possible to compare these invariants only in a narrow interval close to T_i or in the range where q changes slowly. According to the experimental data [5], only the first one of the enumerated possibilities was observed.

The calculated temperature dependencies of neutron intensities of the first and third harmonics of incommensurate satellite reflection at zero external field are given in Fig. 12. To perform these calculation the anisotropy invariant of the form

$$E_{22}\int\eta_{21}^{2}\eta_{22}^{2}dV.$$

was added to the thermodynamic potential (5). The value E_{22} = 7 (G·g)^4/(erg·cm)^3 is fitted to the observed ratio of first and third harmonics l_q/l_{3q} at $T = 9.35$ K [5]. From the Fig. 12 it can be seen that with decreasing temperature the ratio l_q/l_{3q} rises. This is in good agreement with the neutron data detecting the higher harmonic satellite reflection only in the vicinity of the phase transition.

In Fig. 13 the result of the field dependent calculation of the heat capacity using Landau theory for fields within the tetragonal plane is shown. In agreement with the experiment (Fig. 5), the magnetic field broadens the anomaly observed at the Néel temperature T_N and lowers the transition point T_i of the commensurate to incommensurate phase transition. The increase of the calculated heat capacity while lowering the temperature below T_i is a consequence of the approach to the temperature T_i of the proper instability of the ferromagnetic subsystem. As such, the same increase is observed in the experimental data too and it is possible to assume that the broad maximum at low temperature is connected with the subsequent phase transition in the magnetic subsystem of copper metaborate.

Previously, sudden changes of amplitudes of magnetic satellites observable below 2 K [7] were interpreted in terms of lock-in transitions into phases with commensurate propagation vectors. Within the framework of the thermodynamic potential (5) it is impossible to describe a low-temperature transition from an incommensurate phase into a commensurate phase at zero external magnetic field.

In conclusion, field dependent single-crystal magnetization, specific-heat and neutron diffraction results are presented. They are compared with theoretical calculations based on the use of symmetry analysis and a phenomenological thermodynamic potential. For the description of the incommensurate magnetic structure of copper metaborate we introduced the modified Lifshits invariant for the case of two two-component order parameters. This invariant is the antisymmetric product of the different order parameters and their spatial derivatives. Our theory describes satisfactorily the main features of the behavior of the copper metaborate spin system under applied external magnetic field for the temperature range 2÷20 K. The definition of the nature of the low-temperature magnetic state anomalies observed at temperatures near 1.8 K and 1 K requires further consideration.

Acknowledgments

The work was financially supported by the Ministry of Education of Russian Federation (grant E02-3.4-227) and Russian Foundation for Basic Research (grant 03-02-16701).

References

[1] M. Nishi. J. Physics: Condens. Matter **6**, L19 (1994).
[2] R. Yoshizaki. Physica C: Superconductivity **173**, 89 (1991).
[3] Y. Fagot-Revurat, M. Horvatic, C. Berthier et al. Physical Review Letters **77**, 1861 (1996).
[4] H. Ohnishi, and T. Teranishi. J. Phys. Soc. Japan **16**, 35 (1961).
[5] B. Roessli, J. Schefer, G. Petrakovskii et al. Physical Review Letters **86**, 1885 (2001).
[6] M. Boehm, B. Roessli, J. Schefer et al. Physical Review B **68**, 024405 (2003).
[7] M. Boehm, B. Roessli, J. Schefer et al. Physica B **318**, 277 (2002).
[8] G. Petrakovskii, M. Popov, B. Roessli, and B. Ouladdiaf. JETP 93, 809 (2001).
[9] V. Kovalev. Irreducible representations of the space groups, Gordon and Breach, New York (1965).
[10] L.D. Landau and E.M. Lifshitz. Statistical physics, P. 1, Pergamon Press, Oxford (1980).
[11] I.E. Dzyaloshinskii. JETP **20**, 665 (1965).

INTRINSICALLY INHOMOGENEOUS MAGNETIC STATES IN ANTIFERROMAGNETS

V.V. Eremenko and V.A. Sirenko
Institute for Low Temperature Physics and Engineering,
47 Lenin ave., 61103, Kharkov, Ukraine
eremenko@ilt.kharkov.ua

Abstract: This review outlines inhomogeneous states which appear in antiferromagnets in the vicinity of phase transitions. It will be shown on the basis of a thermodynamic analysis of experimental data that these states are similar to intermediate and mixed states in superconductors or to those in non-ferromagnetic metals under conditions of de Haas-van Alphen effect, and are determined by the sign of the surface energy of the interface between co-existent phases, independently of its origin.

Keywords: Antiferromagnet, phase transition, magnetic properties, de Haas-van Alphen effect

1. Introduction

In the 70's of the passed century detailed studies were performed of the magnetic, magneto-resonance and magnetooptical properties of antiferromagnetic crystals in the vicinity of phase transitions, induced by a magnetic field [1]-[17]. At the same time the theory of the domain structure, which appears in antiferromagnets at first-order phase transitions, was constructed [18] - [21]. The results of these investigations, however, have not been summarized by now, neither as a monograph, nor as a review article. The present review is an attempt to fill this gap. The object of this review is to show that an inhomogeneous magnetic state arises in a magnet, independently of its nature (either being a superconductor, a normal diamagnetic metal, or an insulating antiferromagnet), if an external magnetic field induces a phase transition in it.

Besides the theory of domain structure in antiferromagnets [18] - [21], similar to that of the intermediate state in type 1 superconductors, the suggestion was made [9], [10], that in antiferromagnets a state may be

J. Franse et al. (eds.), Smart Materials for Ranging Systems, 67–99.

realized alike the mixed state arising in type 2 superconductors, when superconductivity is destroyed by a magnetic field. That is why we will start with reminding what the intermediate and mixed states are.

2. Intermediate and mixed states in superconductors

As early as the 1930's, in Kharkov, L.V. Shubnikov and co-workers [22] have shown that type 1 and type 2 superconductors exist, which differ qualitatively by their magnetization curves. The type 1 superconductors, being ideal diamagnets in the superconducting state, undergo a sharp transition to the normal state in an external field of magnitude H_c. The superconducting - normal state transition region of the type 2 superconductors covers a significant range of magnetic fields, restricted by the critical values H_{c1} and H_{c2}. For type 1 superconductors, in samples of finite size, the transition region ΔH_c is determined by the demagnetization. For type 2 superconductors the range $H_{c1} \div H_{c2}$ far exceeds the demagnetization factor effect.

L.D. Landau has shown theoretically that in type 1 superconductors a periodic structure with alternating domains of superconducting and normal phases should appear when $\mathrm{H_c}$ is reached [23].

L.V. Schubnikov and I.E. Nakhutin [24] have corroborated Landau's theory by electroresistivity measurements parallel and perpendicular to the direction of the external magnetic field on reaching the critical field intensity $\mathrm{H_c}$, and called such a domain structure "the intermediate state". A.G. Meshkovskyi and Yu.V. Scharvin [25] have demonstrated the existence of the intermediate state during its scanning by drawing a bismuth wire, well known for the remarkable sensitivity of its electroconductivity to magnetic field.

The magnetization process in type 2 superconductors was understood only when A.A. Abrikosov created the vortex structure theory of the magnetic-flux distribution in such superconductors [26]. At the same time N.V. Zavaritsky confirmed [27] Schubnikov's results experimentally on perfect thin-film samples.

From a thermodynamic standpoint the difference between type 1 and type 2 superconductors is that the surface energy of the interface between the superconducting and normal states is positive in the first case ($\sigma > 0$), and negative ($\sigma < 0$) in the second.

Destruction of superconductivity by a magnetic field is not the only example of phase transformations in metals. Condon [28], [29] has predicted theoretically and proved experimentally on diamagnetic beryllium and silver as the examples, that under condition of the de Haas - van

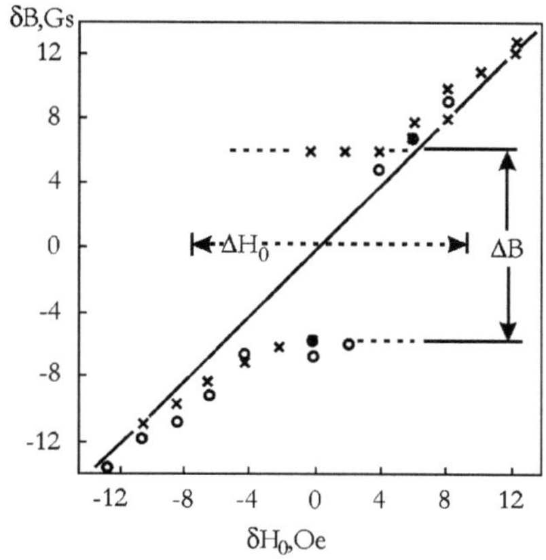

Figure 1. Dependence on applied magnetic field of the magnetic induction in the skin-layer. Silver plate in a field H, normal to its surface [29]. H = 90 kOe, $T <$ 2.5K.

Alphen (dHvA) effect every cycle of the magnetic oscillations is accompanied by a first-order phase transition, and by the appearance of a domain structure. Figure 1, taken from [29], illustrates the splitting of the NMR frequency of Ag^{109} in metallic silver under the condition of the dHvA effect, which unambiguously testifies to the occurrence of diamagnetic domains.

3. Magnetic phase transitions in antiferromagnets (phenomenological description)

Let us consider the simplest case of a two-sublattice antiferromagnet with uniaxial anisotropy. The relevant theoretical description is usually based on consideration of the thermodynamic potential [30], or the energy density, in the following form:

$$\Phi = \frac{A}{2} \cdot \vec{m}^2 + \frac{a}{2} \cdot m_z^2 + \frac{b}{2} \cdot l_z^2 - \vec{m} \cdot \vec{h}, \tag{1}$$

where $\vec{h} = \vec{H} \cdot M_0$,

$\vec{m} = (\vec{M}_1 + \vec{M}_2)/M_0$;

$\vec{l} = (\vec{M}_1 - \vec{M}_2)/M_0$;

$|\vec{M}_1| = |\vec{M}_2| = M_0$;

z is the direction along the principal crystallographic axis;

$\vec{M}_1$, $\vec{M}_2$ are the sublattice magnetic moments. Consideration is

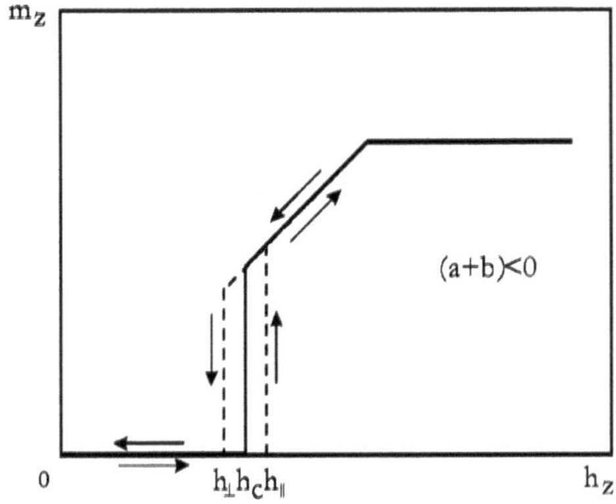

Figure 2. Dependence of the longitudinal component of the magnetization on the intensity of the magnetic field, parallel to the symmetry axis of an antiferromagnetic crystal with uniaxial anisotropy [30].

performed under assumption, that their modules are constant, which is true for the lowest temperatures $T \ll T_N$, where T_N is the Néel temperature, i.e. the temperature of the antiferromagnetic ordering.

$\vec{H}$ is the external field;

$\vec{m}$, $\vec{l}$ are the vectors of ferro- and antiferromagnetism.

The first term in expression (1) describes the exchange energy, the second and the third terms give the crystallographic anisotropy energy.

For antiferromagnets with "easy axis" anisotropy, the case of MnF_2 and $FeCO_3$, which are of interest for us, we have $b < 0$. If the condition $(a + b) < 0$ is also added, the result of minimization of (1) yields the dependence $m_z(h_z)$, presented in figure 2. This is the so called spin-flop transition. If, in addition, the absolute values of the anisotropy constants are high compared to the constant of exchange interaction A, then in an applied magnetic field h_z the antiferromagnet undergoes a transition directly to the saturated ferromagnetic state, omitting the spin-flop transition.

The first situation was observed by J.S. Jacobs for MnF_2 [31], the second one was registered by J.S. Jacobs [32], and later on by V.I. Ozhogin [33] for $FeCO_3$.

It should be taken into account that even while the spin-flop transition possesses the character of a 1-st order phase transition at a precise alignment of the magnetic field along the z-axis, the character of the 1-st order transition is lost when $\vec{h}$ tilts away from the z-axis by an angle ψ, larger than a certain critical value ψ_c. The critical angle value is deter-

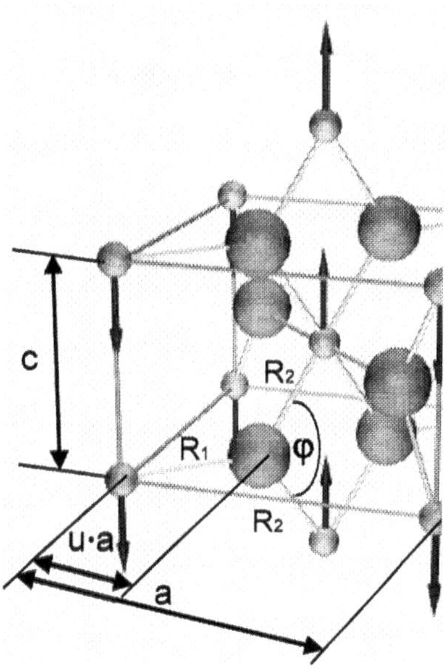

Figure 3. Unit cell of MnF_2 [1].

mined by the ratio of the magnetic anisotropy field to the exchange field, or differently speaking, by the ratio $|b|/A$ [34]. In the case of MnF_2 this angle is $\psi_c \approx 30'$ (thirty angular minutes).

4. Magnetic properties of antiferromagnetic MnF_2 in the vicinity of the spin-flop transition. Experimental results and discussion [1-3]

Following J.S. Jacobs [31], [32] we select MnF_2 ($T_N \approx 68K$) and $FeCO_3$ ($T_N \approx 35K$) for experimental study. The former has a tetragonal crystal structure (group $^{14}D_{4h}$), and the latter is a rhombohedral crystal (group $^{6}D_{3d}$). Their unit cells are shown in figures 3 and 4, respectively. The arrows indicate magnetic moments of the metal ions, which constitute antiferromagnetic sublattices.

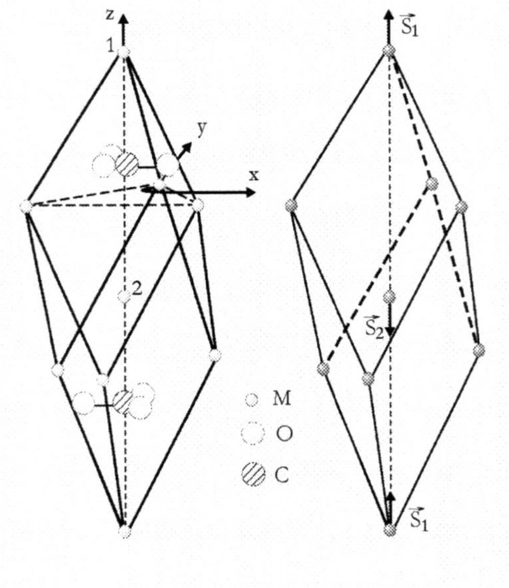

Figure 4. Crystal and magnetic unit cells of $FeCO_3$ [10].

4.1 Experimental technique and instrumentation [1,2]

Similar to the experiments of J.S. Jacobs [31], [32] and V.I. Ozhogin [33] a magnetic field was generated by pulse technique, and the measurements were performed using an inductive method. Still, the task of exploring the immediate vicinity of the phase transition, aimed at observation and study of the inhomogeneous magnetic states, particularly of the intermediate state in MnF_2, produces very strict requirements to the magnetic measurements. The major requirements may be summarized as follows:

1. For comparison with the theoretical results the properties under study should not differ significantly from those, expected for absolute zero temperature, i.e. the condition $T \ll T_N$ should be satisfied. Hence the samples were cooled by liquid helium ($T = 4.2K$).

2. In the vicinity of a phase transition a rapid change of magnetic properties with the field intensity occurs. Consequently, a highly uniform magnetic field, comparable to the experimental conditions in the studies of the dHvA effect in metals, is required.

3. In order to avoid the influence of relaxation processes in the sample, the rate of the magnetic-field sweep in the studied interval should not be too high. So the field was generated in a solenoid with a large inductance, which provided a pulse duration of $\approx 15 \cdot 10^{-3}$sec.

4. The spin-flop transition in MnF_2 is of first order at $\psi < \psi_c \approx 30'$ only. Extremely fine orientation of the magnetic field relative to the crystal axis should be achieved (as the permitted error should not exceed several angular minutes). For this reason the solenoid is mounted on a separate plate, and its slope is varied by 4 micrometer screws.

5. The inductive measuring technique should take account of the need to register all three components of the magnetization vector, which was achieved by using three coils wound on the sample (Fig. 5).

6. In order to elucidate the details of the magnetization distribution in the bulk of the sample, several pick-up coils were used. They were inductively coupled with different parts of the crystal (Fig. 6).

7. For separating the narrow region in the direct vicinity of the phase transition, a threshold scanning circuit was utilized, with the record (oscillography) beginning just on reaching the threshold field value H_{thr}, slightly lower than the critical field H_c. Then the scanning scale was magnified many times.

4.2 Peculiarities of the magnetic properties of MnF_2 in the vicinity of the spin-op transition. Experimental results. Evidences for appearence of the intermediate state [1-3]

By simultenous recording of the three components of the magnetization vector $\vec{M}$ the behavior of the absolute value M, as well as that of the angle Θ of its orientation relative to the crystal axes, was studied in the critical-field range.

1. At small angles of the field $\vec{H}$ tilt away from the fourfold crystal axis z ($\psi \ll 1$), though larger than the critical value $\psi_c \approx 30'$, a rapid increase of M_z is observed at the spin-flop transition, while $M_\perp$ passes through a maximum (Fig. 7). It means, that the increase of the magnetization M_z is concurrent with a turn of $\vec{M}$ to the angle $\Delta\Theta = \pi/2$. The shape

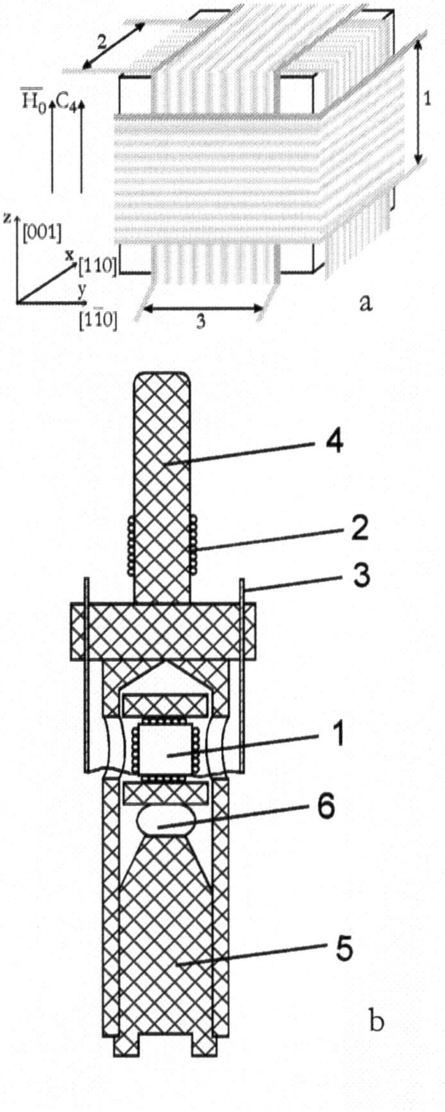

Figure 5. Disposition of pick-up coils on the sample. 1,2,3 - lead-ins of the coils aligned along [001], [110] and [1$\bar{1}$0] directions, respectively (a) and scheme of the sample mounting (b) [1,2].

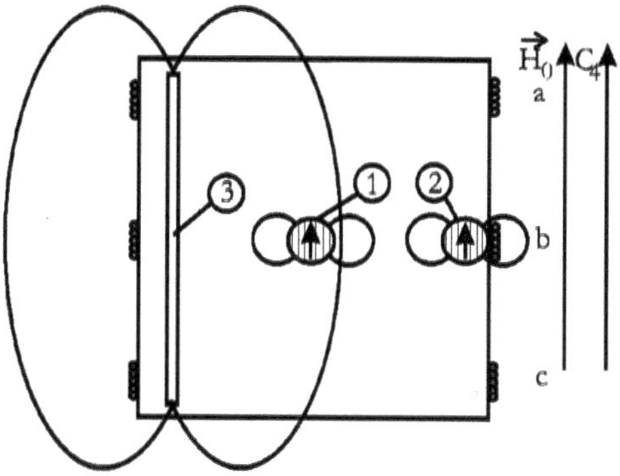

Figure 6. Cross section of the sample with three pick-up coils (a, b, c) aligned along [001]. Schematic presentation of magnetic flux coupling to the ddifferently shaped parts of the sample (1, 2, 3), together with the turns of the coil [1,2].

of the $M(\Theta)$ dependence is close to a semicircle (Fig. 8), which clearly demonstrates agreement with theoretical concepts about turning of the sublattice magnetization vectors.

When the angle ψ is decreased from $60'$ to $30'$ the shape of $M(\Theta)$ remains practically unchanged, in consistency with the properties of uniformly magnetized samples, expected for $\psi \ll 1$.

2. The dependence $M(\Theta)$ is examined in a narrow angle range $0 < \psi < \psi_c \approx 30'$ (Fig.9), where the theory predicts occurrence of the first-order phase transition. The angular dependence of the longitudinal component of the magnetic susceptibility χ_z and that of the transverse component of the magnetization $M_\perp$ ar edepicted in Fig. 10 and Fig. 11, respectively.

The observed change in the character of the $M(\Theta)$ dependence has unambiguously indicated that magnetic stratification of the sample into regions with different values of the magnetization vector $\vec{M}$ occurs at $\psi < \psi_c$. By measuring the integral sample magnetization $\vec{M}$ it was possible to find the values and directions of the local magnetization vectors in the co-existent phases (Fig. 9), as well as the turning angle $\Delta\Theta$ of this vector whithin the domain boundaries. $\Delta\Theta$ is strongly dependent on the angle ψ and is changed within the limits $\pi/2 > \Delta\Theta > 0$, in accordance with the theoretical model [18] - [21].

3. The absolute values of the differential magnetic susceptibilities were measured for samples of different shape in the critical range of magnetic fields and angles ψ. Under the conditions of magnetic stratification, the magnitudes of the susceptibilities appeared to be close to those expected

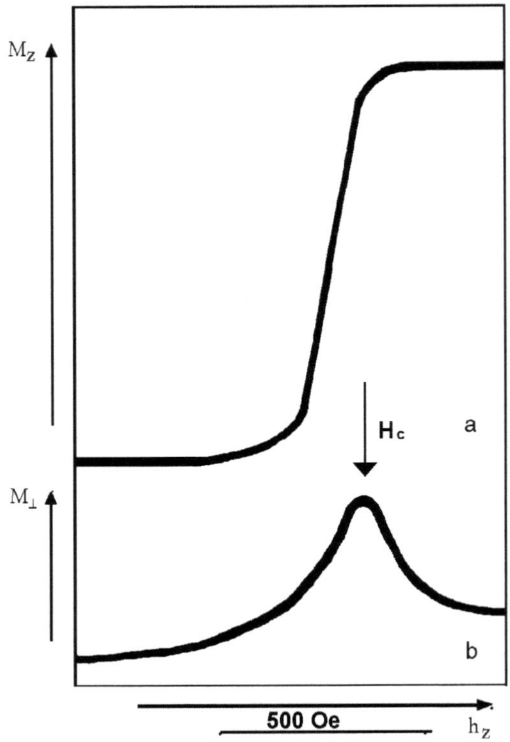

Figure 7. Oscillograms of the longitudinal (a) and the transverse (b) components of the magnetization vector in dependence on the magnetic-field intensity in the critical region. The enlarged portion of the magnetization curves near $H_c \approx 92$ kOe is shown. (MnF_2) [1,2].

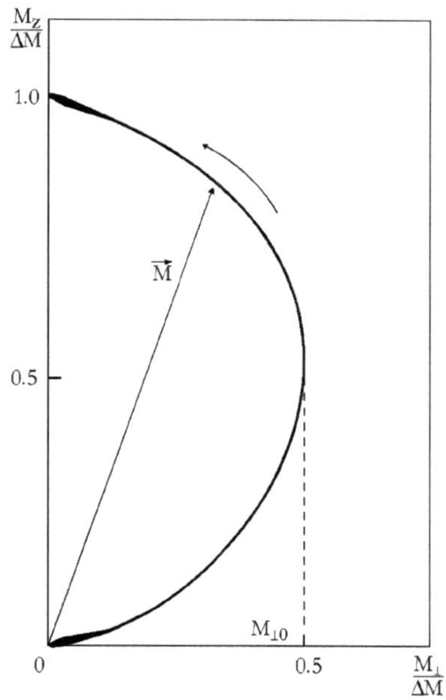

Figure 8. Oscillogram, reflecting growth and turn of the magnetization vector of the MnF_2 sample during sublattice flipping near $H_0 \approx 92$ kOe. The arrow shows the turn direction in increasing field ($\psi = 40'$) [1,2].

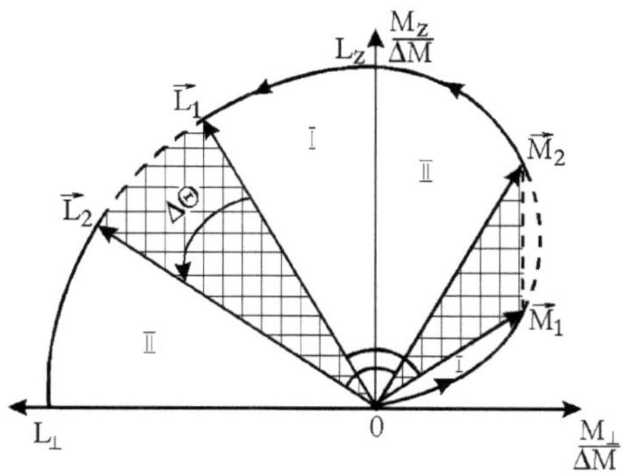

Figure 9. Diagram of the antiferromagnetism vector $\vec{L}$ (left-hand) and the magnetization $\vec{M}$ (right-hand) turn at sublattice flipping in tilted field ($\psi < \psi_c$). The arrows show the turn directions in the I and II phases. The dashed line encloses the angles, corresponding to the regions of phase instability and realization of domain boundaries [1,2].

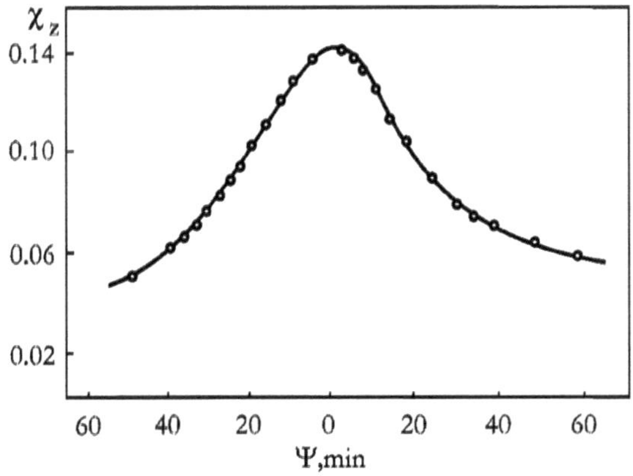

Figure 10. Angular dependence of the maximum value of the magnetic susceptibility $\chi_z = dM_z/dH_0$ in the region of sublattice flipping. ψ is the angle between the magnetic-field direction and the symmetry axis of the crystal [1,2].

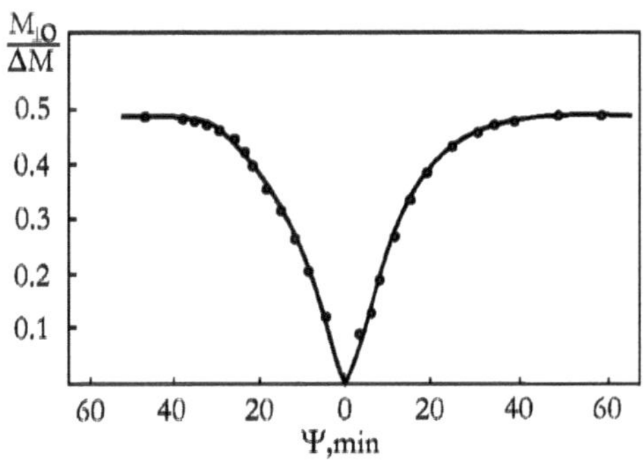

Figure 11. Angular dependence of the largest projection of the sample magnetization vector to the (001) plane [1,2].

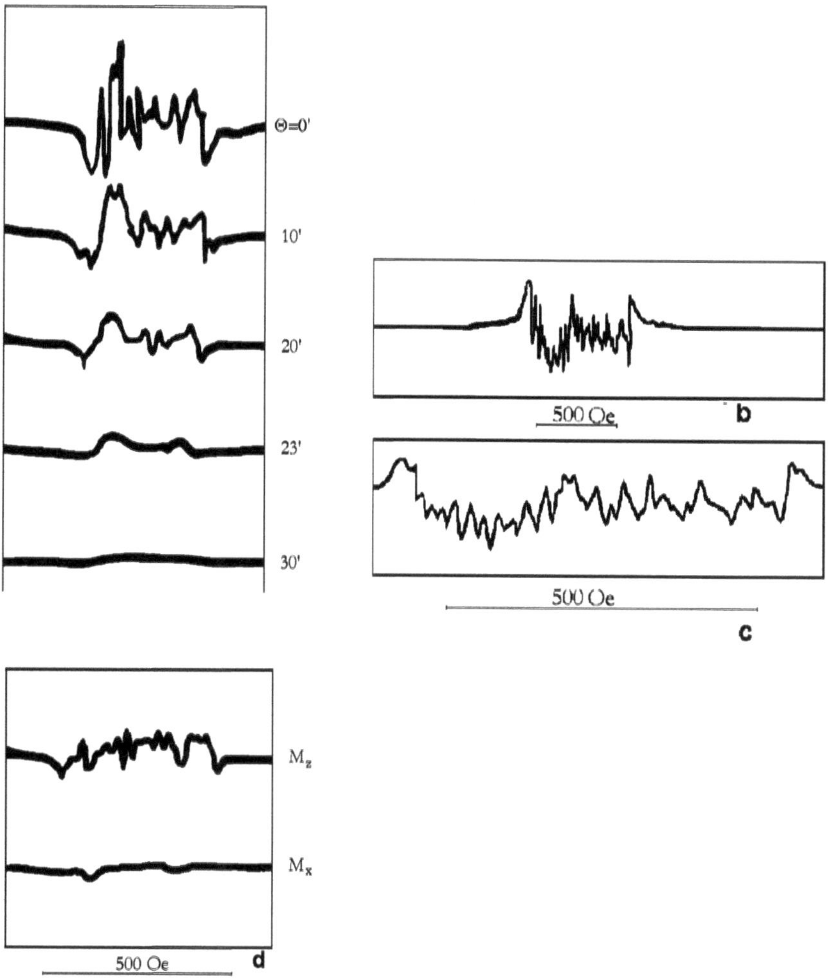

Figure 12. Discontinuous portion of the magnetization curve in the critical field range for different angles between the symmetry axis of the crystal and the external magnetic field direction [1,3].

in the presence of the domain structure, and dependent on the shape of the samples. This dependence establishes the relation of the magnetic stratification to the demagnetizing fields of the sample; actually they give rise to the domain structure.

4. The mgnetization process of MnF_2 at the spin-flop transition occurs discontinuously. Jumps with weak amplitudes, similar to Barkhausen jumps in ferromagnets, are observed in the critical ranges of field intensities ΔH and angles $\psi < \psi_c$ (Fig. 12). The different components of the vectorial quantities $\delta\vec{M}$ of these jumps were measured. They were used to estimate the vector difference between the specific magnetization of the phases co-existing in the domain structure, which appeared to be parallel to the principal crystal axis c_4. The volume of the regions with discontinous reversal of magnetization is about 10^{-5} cm^3; it gives an idea of the characteristic domain sizes. These regions are extended along the principal crystal axis c_4 to distances comparable with the sample dimensions in this direction.

Periodicity of the magnetization jumps reflects that of the domain structure.

4.3 Spectroscopic, magneto-optical, and magneto-resonance observations of the domain structure (intermediate state) in MnF_2

A number of works have shown that the change of the light absorption spectra in MnF_2 at the spin-flop transition correlates with the changes of the magnetic structure. (see, e.g., [35], [36].) At the spin-flop transition the frequency of one of the light absorption bands shifts significantly (for 52 cm^{-1}, while its half-width is 10 cm^{-1}). Measuring the absorption intensity at the frequency of the initial spectral position of this band ($H < H_c$) and at the frequency of its spectral position in the spin-flop phase ($H > H_c$) allowed the authors of [6], [7] to show that in the nearest neighbourhood of the phase transition absorption is revealed at both frequencies. With increasing field the absorption intensity decreases at the initial frequency and increases at the frequency corresponding to the spin-flop phase, which means that both phases co-exist. At the same time, at intermediate frequency, absorption does not appear in MnF_2 at any intensity of the magnetic field, if the latter is oriented strictly along the tetragonal axis of the crystal. These observations unambiguously testify to the discontinuous (jump-like) character of the transition, i.e. that it is a first order transition.

Even more direct evidence for the appearance of a domain structure at the spin-flop transition in MnF_2 was obtained by King and Paquette [5]: they succeeded in taking a photograph of this structure by transmitting monochromatic light through a MnF_2 crystal at the frequency corresponding to that of one of the phases (e.g., spin-flop).

The intermediate state in MnF_2 in the neighbourhood of the spin-flop transition was studied by means of nuclear magnetic resonance [5] and antiferromagnetic resonance [12], [14].

It was shown theoretically [21] that during the change of the external magnetic field, the internal magnetic field of the intermediate-state sample remains constant. In the case of MnF_2 the coupling between magnetic vibrations in the domains of antiferromagnetic and spin-flop phases cannot be strong due to the small value of the magnetic susceptibility. Therefore, it could be expected that in the intermediate state region the antiferromagnetic resonance frequency will remain unchanged when the intensity of the external magnetic field is varied, which was really observed in experiments [14], [17]: the resonances in antiferromagnetic and spin-flop phases occur independently of each other, and their intensity is proportional to the fraction of the substance in the corresponding phase.

Similar results were also obtained for a variety of intermediate-state antiferromagnets (e.g., for the antiferromagnetic crystal $CuCl_2 \cdot 4H_2O$ [15], [16]). For the low-frequency branch of a $CuCl_2{\cdot}4H_2O$ crystal a continuous change of frequency was observed [4], [8], which the authors of these works explained by strong coupling of low-frequency vibrations in the domains of antiferromagnetic and spin-flop phases.

It should be noted that the intermediate state was detected by a magnetic technique in $NiWO_4$ [13], and by a magnetooptical one in dysprosium orthoferrite [11].

5. Magnetic properties of $FeCO_3$ in the vicinity of the phase transition from antiferromagnetic to ferromagnetic [9, 10]

5.1 Experimental results [9, 10]

For fields orientated along the [111]-axis ($H \parallel c_3$), both longitudinal and transverse (relative to the field direction) components of the magnetization were measured. The rate of field sweep varied within $5 \div 30 \cdot 10^6$ Oe/sec. The angle ψ between the c_3-axis and $\vec{H}$ varied within $0 \div 10^o$. The samples were of natural origin, taken from different deposits. Samples of various geometries were examined, starting from a cylinder

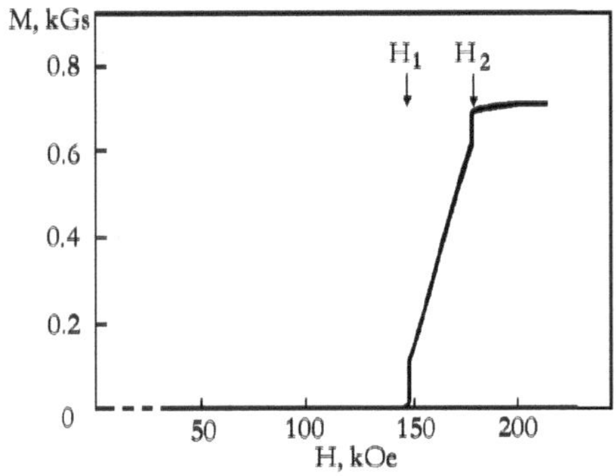

Figure 13. Dependence of the longitudinal magnetization of $FeCO_3$ on the intensity of the magnetic field, directed parallel to the symmetry axis of the crystal [9, 10].

with the axis along the [111] direction, to discs, with the face normal to the [111] direction.

Main results

1. The field dependence of the longitudinal magnetization for $\vec{H}$ parallel to the c_3 axis is shown in Fig.13. The change of magnetization starts in the field $H_1 = 148kOe$ and is finished in the field $H_2 = 176kOe$. The susceptibility averaged over the range $H_1 \div H_2$ amounts to $2.7 \cdot 10^{-2}$ CGSM.

2. The absolute value of the transverse magnetization component at any angle ψ $(0 < \psi < 10^o)$ is of such a value, that the total-magnetization vector coincides with the magnetic field vector $\vec{H}$. Hence, there are no magnetization components normal to $\vec{H}$ at any values of the intensity H and the angle ψ.

3. Near the critical fields H_1 and H_2 the magnetization changes more abruptly than in the $H_1 < H < H_2$ range. Figure 14 demonstrates in more detail the dependence of the magnetic susceptibility on the magnetic-field intensity near the critical fields. The maximum value of the susceptibility in the "critical" regions exceeds almost twice its value within the range $H_1 < H < H_2$. In cylindrical samples (with a ratio of height to diameter $\beta = 0.2$) the susceptibility amplitude is up to 40% larger than that in the disks ($\beta = 5$). The width of the magnetic field range $\Delta H = H_1 - H_2$ is practically independent of the shape of the samples.

The sharp change of the magnetic susceptibility (of the whole sample) near H_2 (Fig. 14b) allows one to estimate the upper limit of the

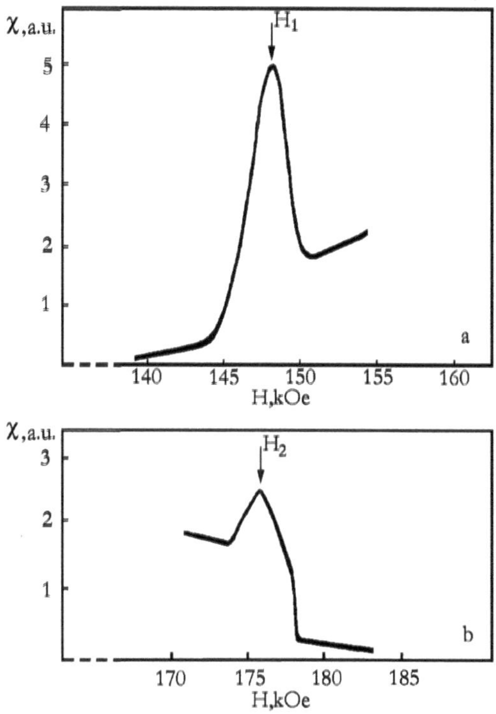

Figure 14. Differential magnetic susceptibility of $FeCO_3$ in the vicinity of the critical fields of the phase transition a) onset (H_1) and b) at end (H_2) [9,10].

magnetic-field inhomogeneity in the sample. It amounts to less than 1 kOe, that is much less than the widths of the characteristic features on the magnetization curve, which will be discussed further.

The presented results are the same for all the samples of natural origin, taken from different deposits.

5.2 Magnetic susceptibility in phenomenological consideration [9,10]

The phenomenological analysis of the magnetic properties of an antiferromagnet as usual implicates consideration of the thermodynamic potential, which contains the main interactions, associated with the occurrence of spontaneous magnetization of sublattices. The results of such an analysis for crystals, whose anisotropy is not very high [30], do not describe correctly the magnetization curve, measured for $FeCO_3$ crystals (Fig.13). Considering the pronounced anisotropy of the magnetic properties of $FeCO_3$, the experimental fact of the absence of transverse magnetic-moment components and the published data [37] - [39], we will include in the thermodynamic potential only interactions related to longitudinal components of the sublattice magnetization vectors. The crystal under study will be first regarded as a two-sublattice antiferromagnet, with the following denotations for the sublattice magnetizations:

$$\boldsymbol{m}_1 = \frac{\boldsymbol{M}_1}{M_0}, \qquad \boldsymbol{m}_2 = \frac{\boldsymbol{M}_2}{M_0}.$$

Here M_0 is the saturated value of the spontaneous magnetization, which is the same for both sublattices. The external magnetic field, denoted by $h = h_z = H_z/M_0$, will be directed along the symmetry axis of the crystal. Then the thermodynamic potential of the antiferromagnet may be presented in the following form:

$$\Phi = -\frac{1}{2}\delta\left(m_1^2 + m_2^2\right) - \gamma m_1 m_2 - (m_1 + m_2)\,h, \tag{2}$$

where the variables $m_1 = m_{1z}$, $m_2 = m_{2z}$, $m = m_1 + m_2$ change whithin the limits

$$-1 \leq m_1 \leq +1, \quad -1 \leq m_2 \leq +1, \quad -2 \leq m \leq +2.$$

Expression (2) is applicable for the description of homogeneous states (with inhomogeneities not exceeding the size of a unit cell), and also for inhomogeneous states; however in the latter case the quantities m_1 and

m_2 refer to sublattice magnetizations averaged over a region with a size larger than all possible inhomogeneities.

In the absence of a magnetic field the antiferromagnetic state corresponds to a minimum of (2) for the following relations of the constants:

$$\gamma < 0, \qquad |\gamma| > |\delta|$$

or

$$h_1 = -(\gamma - \delta) > 0, \quad h_2 = -(\gamma + \delta) > 0.$$

Minimizatation of (2) with respect to the variables m_1 and m_2, taking account of the limits of their variation, yields the following solutions for the sublattice magnetizations at different intensities of the magnetic field:

$$A) \qquad h < h_1: \; m_1 = +1, \; m_2 = -1, \; m = 0, \; \Phi_1 = -h_1;$$

$$B) \; h_1 < h < h_2: \begin{cases} m_1 = +1, \; m_2 = -\dfrac{h+\gamma}{\delta}, \\ m = -\dfrac{h + (\gamma - \delta)}{\delta}; \\ \Phi_2 = -h_1 - \dfrac{(h-h_1)^2}{h_2 - h_1} \end{cases} \qquad (\delta < 0);$$

$$C) \qquad h > h_2: \; m_1 = +1, \; m_2 = +1, \; m = +2, \; \Phi_3 = h_2 - 2h.$$

Solution A), which is realized in weak fields, corresponds to the antiferromagnetic state. In high fields solution B) describes the paramagnetic state. Transition from one state to another occurs near the critical field $h_{II} = {}^1\!/_2\,(h_1 + h_2) = -\gamma$. It is important to note that during the transition only the magnetization of one sublattice is changed, while the other sublattice remains magnetized up to saturation at all values of the magnetic field. The equations presented above show that the transition from antiferromagnetic to paramagnetic may be of two types, depending on the sign of the constant δ:

I) If $\delta > 0$, then $h_2 < h_1$, and the reorientation of the second sublattice occurs in a field $h = h_n$ by a first-order phase transition. The total magnetization in this field has a jump-like change $\Delta m = 2$. In the field interval $h_2 < h < h_1$ metastable states exist, and hysteresis is possible (Fig. 15a).

II) At $\delta < 0$ we have $h_2 > h_1$, and the transition occurs continuously in the field interval $\Delta h = h_2 - h_1$ (Fig.15b). In this interval solution B) is realized. In an increased magnetic field a second-order phase transition takes place at $h = h_1$ and the magnetic susceptibility reveals a jump-like

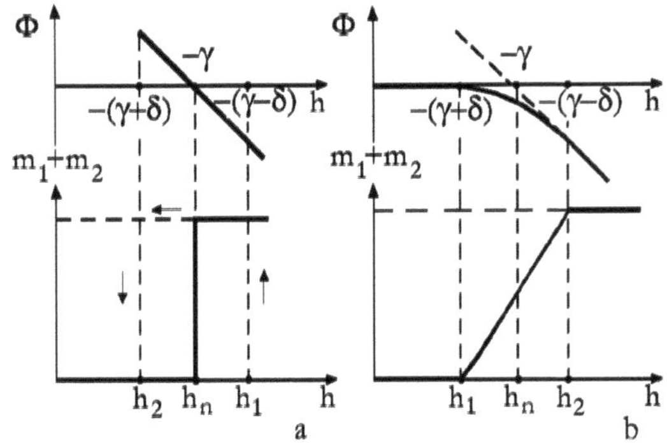

Figure 15. Dependence of the thermodynamic potential on the magnetic-field intensity in the vicinity of the transition from the antiferromagnetic state to the paramagnetic one: a) transition of the first type ($\delta > 0$), b) the second type transition ($\delta < 0$) [9,10].

increase $\Delta\chi = \dfrac{2M_0}{H_2 - H_1}$. The linear increase of magnetization ceases at $h = h_2$, where the second-order phase transition still occurs, and where the magnetic susceptibility falls again to zero.

The definition of the first term in the thermodynamic potential implies that the constant δ is the energy of the intrasublattice exchange interaction. Its sign is determined primarily by the sign of J_{11}, which is the interaction between the ion under consideration and the ions of the second coordination sphere. If the sign of J_{11} favours a ferromagnetic ordering inside the sublattice, the transition under study will be a first-order phase transition. This apparently takes place in the layered antiferromagnets $FeCl_2$, $FeBr_2$, etc. [40], [41]. If sign J_{11} = sign J_{12}, then the transition occurs smoothly, with second-order phase transitions at the beginning and at the end.

Comparison of the magnetization curve (Fig. 13), measured for $FeCO_3$, with the theoretical curves shows that the transition in this antiferromagnet follows type II . Accordingly, the intrasublattice interaction in antiferromagnetic $FeCO_3$ should have the same sign as that of the intersublattice interaction. The sign of the intrasublattice interaction in an antiferromagnet may also be estimated by the molecular-field approximation, comparing the Néel temperature (T_N) with the Weiss temperature (Θ). In accordance with experiment [42], the Weiss temperature of $FeCO_3$ is negative: $\Theta = -14$K, which does not contradict the assumption about the negative sign of the intrasublattice interaction.

The experimental curve in Fig.13 permits to determine the ratio of the intrasublattice interaction and the intersublattice interaction:

$$\frac{\delta}{\gamma} = \frac{h_2 - h_1}{h_2 + h_1} = +0.08.$$

Note that, if the presented consideration is really applicable to the magnetization curve of $FeCO_3$, its measurement is the most accurate way to determine the value of the intrasublattice exchange interaction among all of the known methods.

Therefore, the phenomenological thermodynamic potential (2), which includes interactions of longitudinal components of the sublattice magnetizations only, provides a classification of the transitions from antiferromagnetic to paramagnetic, determined by the sign of the constant δ of the intrasublattice exchange interaction. If the sign of δ is negative, it describes correctly the existence of two critical magnetic fields h_1 and h_2, and also explains a finite value of the magnetic susceptibility within the interval $h_1 < h < h_2$.

5.3 Transverse components of the magnetization and restrictions of the two-sublattice model [9,10]

The presented phenomenological analysis does not put any restrictions on the magnitudes of the transverse components of the magnetization. Although we have chosen a thermodynamic potential, assuming their absence, the solutions are formally true also for the case, when transverse components are present, but give no contribution to the interaction. That is why a discussion of their magnitudes requires additional assumptions.

The molecular-field approximation suggests the following relation for the field and temperature dependence of the sublattice magnetization:

$$M_1 = M_0\, B_s(y_1),$$

where $B_s(y_1)$ is the Brillouin function for spin S, and $y_i = H_i \mu_B g S/kT$. Therefore, at T = 0 the sublattice should be magnetized to saturation $M_i(0) = M_0$ in any non-zero field. In the range of magnetic fields $h_1 < h < h_2$ the longitudinal component of the magnetization of $FeCO_3$ varies continuously. If the modulus of the sublattice magnetization remains constant in this situation, then the vector $\vec{m}_2$ should turn continuously up to the angle π. The magnitude of the transverse magnetization component should approach M_0 in maximum.

The transverse magnetization components, however, are absent in the experiment. This feature persists under switching on the transverse components of the magnetic field, which could align the transverse magnetization components, if they existed and had different directions in the separate macroscopic parts of the crystal. Perhaps, it might be assumed that the transverse components of the sublattice magnetization are identically equal to zero, and that the modulus of the magnetization does change with magnetic field variation. This statement contradicts conclusions of the two-sublattice model.

A similar difficulty, which was already met in Néel's work [43], concerned with the analysis of the ferrimagnetic state, was avoided by Yafet and Kittel [44]. Their assumption, also referred to the case of negative intrasublattice interaction, is as follows. The considered sublattice actually divides up into two sublattices, moreover, each of the new sublattices is magnetized to saturation at $T = 0$. However, their total magnetization may acquire any intermediate value $-M_0 < M_i < M_0$, since they can be oriented at any angle to each other. Such interpretation may serve as a key to understanding the state of the sublattice in the field range $h_1 < h < h_2$.

Still this interpretation gets complicated by the fact that the magnetic Fe^{2+} ions in the crystal under study may, apparently, have only two orientations, namely parallel and antiparallel to the external magnetic field, and their transverse components should be absent, even in a one-ion approximation. Analysis of the cooperative spin-orbit coupling and crystal field effect on a Fe^{2+} ion with the spin 4/2 have shown, that the lowest one-ion state is a doublet, corresponding to two antiparallel spin orientations [45]. In this case an effective spin $S_{eff} = 1/2$ should be ascribed to the ion, and the Ising model should be used to describe the interaction between the ions. These assumptions are presently confirmed by numerous experiments [37] - [39]. Because of that, it could be accepted that in $FeCO_3$ the sublattices are oriented only along two fixed directions. This fact, on the one hand, is a ground for the thermodynamic potential (2). On the other hand, it is impossible to explain any intermediate value of sublattice magnetization by separating it into two.

It may be assumed that the sublattice under study is divide into a large enough number of sublattices. In the multisublattice model the successive changes in orientation of each sublattice moment are concurrent with a multistep character of the magnetization curve. However, as practically an unlimited number of sublattices may be introduced into consideration, the magnetization curve can look smooth enough for an accurate description of the experimental one.

The physical reason for a division in many sublattice is as follows. As it was shown in 4.2, if we take account of the interaction between the sublattices, the parallel spin orientation in each of them possesses the minimal energy. But, if one sublattice is considered as an isolated subsystem with $\delta < 0$, the energy minimum corresponds to the state of its splitting into several sublattices with different spin orientation in each of them.

The specifics of the current problem is that the external magnetic field h compensates the effective field of the intersublattice interaction $\gamma \cdot m_1$, which acts on one of the sublattices:

$$h_{2eff} = h + \gamma \cdot m_1 + \delta \cdot m_2.$$

So, the magnetic field removes the degeneracy of this sublattice. It splits into n components. Their field is presented as:

$$\sum_{j=1}^{n} \gamma_{ij} m_j = \delta + 2\sum_{j=1}^{k} \gamma_{ij} = -\delta - 2\sum_{j=k+1}^{n} \gamma_{ij}.$$

Here $m_j = M_j/M_{0j}$ is the reduced magnetic moment of the j-th sublattice; it may acquire the values $m_j = \pm 1$, $(\Delta m_j = \pm 2)$. The γ_{ij} are the relevant coefficients of the molecular field. The total value of this field may vary within the limits from $-\delta$ in the antiferromagnetic state (all $m_j = -1$) to $+\delta$ in the paramagnetic state ($m_j = +1$ for all j). In the right hand side of the equality a summing is performed only over k or $(n-k)$ sublattices with reversed or unreversed magnetization, respectively.

If the γ_{ij} are positive, the overturn of the j-th sublattice decreases the effective magnetic field acting on the other sublattice sites. It explains the fact that the alternation of their orientation does not occur simultaneously, but step by step (gradually), following an increase of the external magnetic field.

Therefore, the result derived in 4.2, describing the gradual change of the longitudinal magnetization, may be in agreement with the experimental result that transverse magnetization components are absent, just by assuming that one of the sublattices is split up into a large number of subsublattices, with the period much larger than that of the initially chosen unit cell. Therefore, the magnetic state of the crystal may be regarded as inhomogeneous in the field range $h_1 < h < h_2$; its periodicity, determined by that of the subsublattices, may change, if the magnetic field is varied.

5.4 Phase transitions and short-range order near the critical fields h_1 and h_2

It is impossible to calculate the total magnetization curve $M(H)$ in a multisublattice model, since the molecular-field coefficients γ_{ij} are unknown. Nevertheless it is possible to show that most of them are about zero. The latter circumstance permits to describe the magnetization curve and the magnetic structures in the vicinity of the critical fields h_1 and h_2.

It is known, that the exchange interaction rapidly decays in the crystal at distances r_0 of the order of several atomic spacings. In the multisublattice model, the sublattice period d is large enough ($d \gg r_0$) to make that the exchange field in a given point is either equal to zero, if all of the sublattice ions are far apart from it, or is induced by a single neighbor ion. The coefficients γ_{ij} are either equal to zero, or appear to be dependent on the distance r_{ij} of the nearest ion of the j-th sublattice to the i-th point. Hence, it is reasonable to pass in expression for total field acting on the site of i-th sublattice to summing over the neighbor ions.

Assume, that the ions on the i-th sublattice alternate their orientation in the external field h_i of such critical value, that the total effective field, acting on them, becomes zero:

$$h_i = -(\gamma - \delta) + 2\sum_{j=1}^{k} \gamma_{ij}(r_{ij}).$$

Here the summing is performed over the reversed ions located near the i-th site. If all γ_{ij} coefficients are positive, the magnetization process starts in field $h_1 = -(\gamma - \delta)$ with the change of spin orientation at individual sites, separated by the distance r_0.

For $r_0 \to \infty$, a second-order phase transition should occur at $h = h_1$ according to the results of the phenomenological analysis. Interaction between the reversed spins increases the total energy if $\gamma_{ij} > 0$, which is equivalent to their repulsion. Therefore, the ions with reversed spins have minimal energy, if they form a regular, close-packed structure[1] (Fig.16). Thin lines in Fig. 16 connect the points of equal exchange interaction, produced by the ion. The spacing of such a lattice should continuously decrease with increasing magnetic field.

The observed magnetization value may be used to calculate the average distance between the reversed spins. Therefore the $M(H)$ curve, measured in experiment, in principle allows one to obtain the dependence of the effective exchange field on the spacing between interacting spins $H_{eff}(r)$. The measured magnetization curve (Fig. 13), however,

Figure 16. Distribution of effective exchange fields, produced by ions with reversed spins in the (111) plane [9,10].

reveals a drastic ascent near H_1, and there is a drastic splash on the magnetic susceptibility curve (Fig. 14a) . In such cases it is difficult to guarantee that experimental errors do not affect the shape of the curve. For this reason we did not calculate $H_{eff}(r)$. Estimating the real experimental errors of the curve near H_1, we compare this portion with the model describing an infinite sudden jump, i.e. the first-order phase transition. This jump may be easily explained.

If the radius of interaction between the reversed spins r_0 is finite, then in the field H_1 a finite number of spins will reverse. They should form a regular lattice, in which the distances between them are equal to the radius of exchange interaction r_0. Here r_0 has the meaning of the minimal distance, at which this interaction does not yet affect the shape of the magnetization curve within the specified accuracy. From the experimental value of the magnetization change near H_1 it is possible to estimate that the concentration of reversed spins amounts to $\sim 12.5\%$, which corresponds to their average spacing in the (111) plane $r_0 = 9.4Å$. To discuss the magnetization jump, observed at H_2, the critical field at the site may be presented in the form

$$h_i = -(\gamma + \delta) - 2\sum_{j=k+1}^{n} \gamma_{ij}(r_{ij},$$

where summing is performed over the ions with unreversed spins. Analysis, similar to the one presented above, allows one to draw conclusions about the occurrence of a magnetization jump and about the magnetic structure near the critical field $h_2 = -(\gamma + \delta)$, similar to the structure near the field h_1.

The entity of the reversed spins in the crystal at $h = h_1$ may be presented in the form of individual threads (lines), parallel to the external magnetic field and carrying an additional quantum of magnetic flux.

$$\Delta\Phi_0 = 4\pi\mu_B \frac{4S}{d} \approx 0.2 \cdot 10^{-10}\,\mathrm{Oe} \cdot \mathrm{cm}^2,$$

where μ_B is the Bohr magneton, S is the spin of a Fe^{2+} ion, equal to $^4/_2$, d is the distance between the nearest overturned spins along the [111] direction, equal to $5 \cdot 10^{-8}$cm. This flux quantity appears to be 4 orders of magnitude smaller than the magnetic flux quantum in type 2 superconductors.

Thus, this calculation of the magnitude of the field, induced at the site by a single ion, and its dependence on distance explain the experimental observation of the jump-like magnetization changes at $h = h_1$ and $h = h_2$. As a result of the first phase transition, the reversed spins form a close-packed structure with a period $r_0 \sim 10^{-7}$cm.

5.5 Energy of the interface boundary and comparison with the transition of destroying type two superconductivity by a magnetic field

The transition described above from the antiferromagnetic state to a paramagnetic one has many common features with a transition of destroying superconductivity by a magnetic field. Classification of transitions in superconductors is based on a consideration of the surface energy of the interface between normal and superconducting phases. If the surface energy σ is positive (type one superconductors), in an infinite medium, the magnetization has a jump-like change at $h = h_c$, while the magnetic susceptibility is infinite, i.e a first-order phase transition takes place (Fig. 17a). If σ is negative (type two superconductor) the transition develops continuously, starting and finishing by second-order phase transitions. The field dependence of the magnetization and that of the magnetic susceptibility are shown for this case in Fig. 17b. In the field range $h_1 < h < h_2$ a mixed inhomogeneous state is realised in an infinite medium [46].

Magnetization curves for type 1 and type 2 superconductors (fig. 17a and 17b, respectively) are analogous to those, derived from the phe-

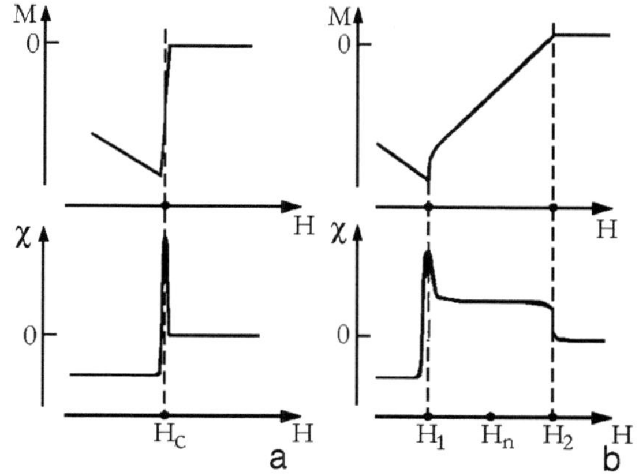

Figure 17. Magnetization and magnetic susceptibility at the transition from the superconducting to the normal state in a magnetic field: a) type 1 superconductor ($\sigma > 0$), b) type 2 superconductor ($\sigma < 0$) [9,10].

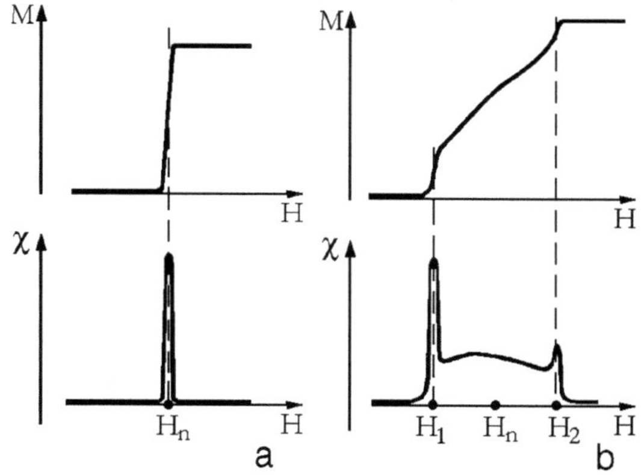

Figure 18. Magnetization and magnetic susceptibility at the transition from the antiferromagnetic state to the ferromagnetic state in a magnetic field, a) type 1 transition ($\delta > 0$), b) type 2 transition ($\delta < 0$) [9,10].

nomenological analysis of type one and type two transitions from the antiferromagnetic state to a paramagnetic one (fig. 18a,b). In order to enhance this analogy we will show that, in the framework of the model described above, the surface energy of the interface between antiferromagnetic and paramagnetic phases is positive for a transition of type 1, and is negative for a transition of type 2.

It was shown above that, apparently, individual spins in $FeCO_3$ may have only two opposite directions, intermediate orientations are forbidden. Therefore we will start directly with assuming a sharp phase boundary, in which the phase transition from one phase to another occurs over one interatomic distance. Fig.19 depicts the boundary in the perpendicular plane (111). At the left-hand side the crystal stays in the antiferromagnetic state, and at the right-hand side it is in the paramagnetic state; the external magnetic field $h = -\gamma$ is aligned along the [111] axis.

We suppose, for simplicity, that each ion interacts only with its neighbors in the first and the second coordination spheres. The Fe^{2+} ion in $FeCO_3$ has six nearest neighbors, located in the adjacent (111) layers with a spacing of 3.7Å. The interaction energy with each of them, at parallel spin orientation, will be denoted as J_{12}. In the second coordination sphere there are six ions in the plane of the same (111) layer, at distances 4.7 Å, the corresponding interaction energy being denoted as J_{11}. Then the interaction energy per magnetic ion in the antiferromagnetic phase, in the paramagnetic phase, and at the interface boundary, respectively, reads:

$$E_A = +6J_{12} - 6J_{11},$$

$$E_P = -6J_{12} - 6J_{11} - \mu_0 H,$$

$$E_B = -4J_{11} - \frac{\mu_0 H}{2}$$

In the absence of a magnetic field the antiferromagnetic phase with the accepted distribution of ions over the sublattices possesses the lowest energy, when $J_{12} < 0$ and $|J_{12}| > |J_{11}|$. If $h \gg -\gamma$ a paramagnetic phase is thermodynamically stable. In an external magnetic field $H = -12J_{12}/\mu_0$, the energies of both phases are equal to each other. The additional energy, associated with the interface boundary in a field H has the form:

$$E_{AP} = E_B - E_A = q2J_{11}$$

where q is the number of ions per unit cross section of the interface boundary. Consequently, in this model the energy of surface tension may have any sign depending on the sign of the energy of the intrasublattice exchange interaction.

Hence, the sign of the surface energy, unambiguously related to the sign of the intrasublattice exchange interaction, indicates directly the type of the transition to the paramagnetic state in an external field.

Usually, antiferromagnets are described in terms of magnetic sublattices. Still even for explaining the magnitude of jumps on the magneti-

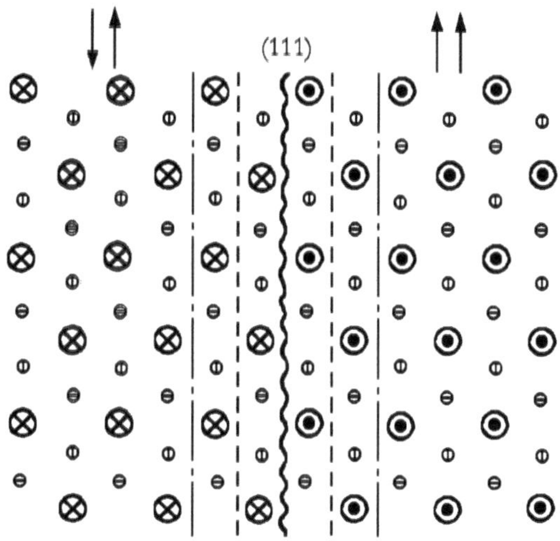

Figure 19. Model presentation of the interface boundary between antiferromagnetic (left-hand) and ferromagnetic (right-hand) phases. Large circles refer: ⊙ to reversed spins and ⊗ to unreversed spins in the same (111) plane, small circles refer to magnetic ions of two adjacent layers [9,10].

zation curve at H_1 and H_2, the number of sublattices should be much more than two. As shown above, a periodic inhomogeneous state forms in the critical range of magnetic fields in case the sign of the surface tension energy is negative. Similar to type 2 superconductors, a significant role in its formation belongs to the law of interaction decrease with distance, characterized by the correlation radius. But, in contrast to conventional type 2 superconductors, where this radius amounts to $\sim 10^{-5}$cm, in antiferromagnets it is $\sim 10^{-7}$cm. Because of this, the period of the inhomogeneous state in antiferromagnets is smaller by three orders of magnitude.

Notwithstanding these differences, the considered magnetization curves of $FeCO_3$ and those of type 2 superconductors have many common features. In both cases transition from one state to another occurs in a finite range of magnetic fields. The onset of the transition is accompanied by a magnetization jump and by a splash of magnetic susceptibility. And finally, the curves have appreciable hysteresis. These formal attributes initiated a comparison of transitions in two objects of different nature, considering the sign of the surface energy of the interface between two states of the crystal.

A certain similariry is also observed in the physical processes, which occur in the critical interval of magnetic fields. Let us consider these processes in the plane, perpendicular to the external magnetic field direction. Originally the transition to the new state occurs in separate points. At the moment of nucleation their spacing is infinite. A magnetic flux quantum is coupled to every point. Their repulsion results in the formation of a regular periodic structure by the points in the plane normal to the field. Their interaction decreases exponentially, so that the increasing field rapidly brings the points closer. The latter process leads to an infinite slope of the magnetization curve. If the interaction becomes zero at $r > r_0$, the magnetization initially has a jump-like change. As a result of the jump, an inhomogeneous structure with a period r_0 arises. Further bringing the points closer to each other in increasing field occurs more smoothly due to their essential interaction. Then the magnetization curve has a finite slope, and the magnetization increases until the whole sample undergoes a transition to the new homogeneous phase.

6. Summary

1. The domain structure, which appears in MnF_2 at the spin-flop transition illustrates a general thermodynamic law of intermediate state formation in the process of first-order phase transitions, induced by a magnetic field, and under the condition that the surface energy of the interface boundary ($\sigma > 0$) is positive.

A variety of substances, manifestating intermediate states, which was earlier represented by type 1 superconductors and weakly magnetic metals under conditions of the dHvA effect, is complemented by antiferromagnetic insulators.

2. The transition of $FeCO_3$ from the antiferromagnetic state to the paramagnetic one is similar to the transition of destroying type 2 superconductivity by a magnetic field. The surface energy of the interface between antiferromagnetic and paramagnetic states is negative. The transition occurs in a finite range of magnetic fields, where according to the model an inhomogeneous periodic magnetic structure is realized, alike the mixed state in type 2 superconductors.

Notes

1. This structure, similar to the mixed state of superconductivity, in the (111) plane, looks like a triangular lattice

References

[1] K.L. Dudko, (1971) Experimental evidence for existence of domain structure at the first order phase transition in antiferromagnetic crystals. - Thesis, Institute for Low Temperature Physics, Kharkov, 1971.

[2] K.L. Dudko, V.V. Eremenko, and V.M. Fridman, Magnetic stratification during flipping of antiferromagnetic manganese fluoride sublattices. Sov. Phys.-JETP, **34**, No. 2, 362-367 (1971)

[3] K.L. Dudko, V.V. Eremenko, V.M. Fridman, Magnetization jumps and domain structure of manganise fluoride at sublattice spin-flopping. Sov. Phys.-JETP, **61**, No 4, 1553-1563 (1971)

[4] A.A. Galkin, S.N. Kovner, V.A. Popov, Antiferromagnetic resonance in $CuCl_2 \cdot 2H_2O$ in inclined magnetic field near intermediate state. Phys. Stat. Sol. (b), **57**, 485-495 (1973)

[5] A.R. King, D. Paquette, Spin-flop domains in MnF_2. Phys. Rev. Lett., **30**, No. 14, 662-666 (1973)

[6] A.A. Mil'ner, Yu.A. Popkov, V.V. Eremenko, Spectroscopic investigation of the intermediate state in antiferromagnetic MnF_2. JETP Letters, **18**, No 1, 20-22 (1973)

[7] A.A. Mil'ner, Yu.A. Popkov, V.V. Eremenko, Spectroscopic study of intermediate state in antiferromagnetic MnF_2. Sov. Phys.-JETP, **18**, No 1, 39-42 (1973)

[8] V.G. Bar'yachtar, A.A. Galkin, V.T. Telepa, Intermediate state in the region of sublattice spin-flopping in antiferromagnetic single crystal $CuCl_2 \cdot 2H_2O$. Fizika Nizkich Temperatur, **1**, No 4, 483-485 (1975)

[9] K.L. Dudko, V.V. Eremenko, and V.M. Fridman, Relaxation and hysteresis of magnetization of antiferromagnetic $FeCO_3$. Sov. Phys.-JETP, **41**, No. 6 1157-1160 (1975);
K.L. Dudko, V.V. Eremenko, V.M. Fridman, Investigation of antiferromagnetic - pparamagnetic state transition in FeO_3 under strong magnetic field. Sov. Phys.-JETP, **68**, No 2, 659-671 (1975).

[10] V.M. Fridman, Investigation of antiferromagnetic-ferromagnetic state transition in $FeCO_3$, induced by a magnetic field. Thesis, Institute for Low Temperature Physics, Kharkov, 1971.

[11] N.F. Kharchenko, H. Szymczak, V.V. Eremenko, S.L. Gnatchenko, R. Szymczak, Magnetic intermediate state in disprosium orthferrite. Sov. Phys.-JETP Letters, **25**, No 5, 258-262 (1977)

[12] S.M. Rezende, A.R. King, R.M. White, J.P. Timble, Stability limit at the antiferromagnetic phase near spin-flop boundary in MnF_2. Phys. Rev. B, **16**, No 3, 1126-1131 (1977)

[13] N.V. Gapon, K.L. Dudko, Orientation phase transition and intermediate state in monoklinic antiferromagnet $NiWO_4$ in longitudinal external field. . Sov. Phys.-JETP, **77**, No 4(10), 1528-1543 (1979)

[14] V.V. Eremenko, A.V. Klochko, V.M. Naumenko, Antiferromagnetic resonance in the intermediate state of MnF_2. Sov. Phys.-JETP Letters, **35**, No 11, 479-481 (1982)

[15] V.V. Eremenko, A.V. Klochko, V.M. Naumenko, V.V. Pishko, Antiferromagnetic resonance in the intermediate state of $CuCl_2 \cdot 2H_2O$. Sov. Phys.-JETP Letters, **40**, No 6, 219-221 (1984)

[16] V.V. Eremenko, A.V. Klochko, V.M. Naumenko, V.V. Pishko, Magnetic resonance in the intermediate state of $CuCl_2 \cdot 2H_2O$. Fizika Nizkich Temperatur, **11**, No 3, 327-331 (1985)

[17] V.V. Eremenko, A.V. Klochko, V.M. Naumenko, Investigation of intermediate state in MnF_2 using antiferromagnetic resonance and Faradey effect. Sov. Phys.-JETP, **88**, No 9, 1002-1017 (1985)

[18] V.G. Bar'yachtar, A.E. Borovik, V.A. Popov, Theory of intermediate state in antiferromagnets at the first order phase transitions in external magnetic field. Sov. Phys.-JETP Letters, **9**, 634-637 (1969)

[19] V.G. Bar'yachtar, A.E. Borovik, V.A. Popov, , E.L. Stefanovsky, About antiferromagnet domain structure, arising with the change of magnetic anisotropy. Sov. Phys.-JETP, **59**, No 10, 1299-1306 (1970)

[20] V.G. Bar'yachtar, A.E. Borovik, V.A. Popov, , E.L. Stefanovsky, Domain wall vibrations in antiferromagnets. Sov. Phys. - Solid State physics, **12**, No 11, 3289-3297 (1970)

[21] V.G. Bar'yachtar, A.E. Borovik, V.A. Popov, Theory of intermediate state in antiferromagnets. Sov. Phys.-JETP, **62**, No 3, 2233-2241 (1972)

[22] L.V. Shubnikov, V.I. Khotkevich, Yu.D. Shepelev, Yu.N. Ryabinin, Dependence of magnetic induction of the stregth of external magnetic field for Pb_{95}-Tl_5 alloy at different temperatures. Sov. JETPh, **77**, 221-237 (1937); L.W. Schubnikov, W.I. Chotkewitsch, J.D. Schepelew, J.N. Rjabinin, Magnetische eigenschaften supraleitender Mrtalle und Legierungen. Phys.Z.Sow., **10**, No. 2 , s.165-169 (1936)

[23] L.D. Landau, Theory of intermediate state. JETPh, **7**, 371 (1937)

[24] L.V. Shubnikov, I.E. Nakhutin, Electroconductivity of superconducting sphere in intermediate state. JETPh, **7**, No. 4, 566 (1937); L.V. Schubnikov, I.E. Nakhutin, Superconductivity in intermediate state. JETPh, **8**, No. 6, 713-716 (1938)

[25] A.G. Meshkovskyj, A.I. Schalnikov, Scanning of intermediate state by Bi-wire JETPh, **17**, 851-861 (1947)

[26] A.A. Abrikosov, About magnetic properties of type 2 superconductors. JETPh, **32**, 1442 (1957); A.A. Abrikosov, Dokladi Akademii Nauk SSSR, **86**, 489-493 (1952)

[27] N.V. Zavaritskij, Investigation of superconducting properties of Tl- and Sn films, condensed at low temperatures. Dokladi Akademii Nauk SSSR, **86**, 501-505 (1952)

[28] J.H. Condon, Nonlinear de Haas-van Alphen effect and Magnetic Domains in Beryllium, Phys.Rev. **145**, 526-535 (1966)

[29] J.H. Condon, R.E. Walstedt, Direct evidence for magnetic domains in silver. Phys. Rev.Lett., **21**, No. 9, 612-614 (1968)

[30] E.A. Turov, Physical properties of magnetically ordered crystals. Edition of The Academy of Sciences of USSR, Moscow, 1963.

[31] J.S. Jacobs, Spin-flopping in MnF_2 by high magnetic fields. J. Appl. Phys., **32**, 61S-62S (1961)

[32] J.S. Jacobs, Metamagnetism of Siderite ($FeCO_3$). J. Appl. Phys., **34**, 1106-1107 (1963).

[33] V.I. Ozhogin, Antiferromagnets $CoCO_3$, CoF_2, and $FeCO_3$ in high fields. JETPh, **45**, 1687-1693 (1963)

[34] M.I. Kaganov, G.K. Chepurnikh Peculiarities of magnetization of uniaxial antiferromagnet near field of sublaatice flipping. **11**, 911-917 (1969); M.I. Kaganov, G.K. Chepurnikh About phase diagram of uniaxial antiferromagnet. **12**, 2988-2992 (1970)

[35] V.V. Eremenko, Introduction into optical spectroscopy of magnets. Kiev: Naukova Dumka, 1975, 469 p.

[36] V.V. Eremenko, N.F. Kharchenko, Yu.G. Litvinenko, V.M. Naumenko, Magneto-optics and spectroscopy of antiferromagnets. Springer-Verlag, 1992, 276 p.

[37] R.A. Alikhanov. Neutron study of antiferromagnetism in the carbonates of Mn and Fe. JETPh, **36**, 1690-1696 (1959)

[38] D.W. Forester, N.C. Koon, J. Appl. Phys., **40**, 1316 (1969)

[39] Hang Nam Ok, Relaxation effects in antiferromagnetic ferrous carbonate. Phys. Rev., **185**, 472-476 (1969)

[40] J.S. Jacobs, P.E. Lawrence, Metamagnetic phase transitions and Hysteresis in $FeCl_2$. Phys. Rev. **164**, 866-878 (1967)

[41] J.S. Jacobs, P.E. Lawrence, Metamagnetism and exchange in Ferrous Bromide. J. Appl. Phys., **35**, 996-997 (1964)

[42] H. Bizette, J. Phys. Radium, **12**, 161 (1951)

[43] L. Neel, Ann. de Phys. **3**, 137 (1948)

[44] Y. Yafet, C. Kittel, Antiferromagnetic arrangements in ferrites. Phys. Rev., **87**, 290-294 (1952)

[45] J. Kanamori, Prog. Theor. Phys., **20**, 890 (1958)

[46] A.A. Abrikosov, Modern state of the problem of superconductivity. Uspekhi physicheskih nauk, **87**, 125-142 (1965)

NMR STUDY OF THE SUCCESSIVE PHASE TRANSITIONS IN CuB_2O_4

M. Chiba, H. Nakamura, Y. Fujii and H. Kikuchi
Department of Applied Physics, University of Fukui, Fukui 910-8507, Japan
chiba@apphy.fukui-u.ac.jp

Y. Yamamoto and H. Hori
Physics Department, School of Materials Science, JAIST, Ishikawa 923-1292, Japan

G. Petrakovskii, M. Popov and L. Bezmaternikh
Institute of Physics, SB RAS, 660036, Krasnoyarsk, Russia

Abstract: The magnetic compound CuB_2O_4 exhibits successive phase transitions. With decreasing temperature the phase transition occurs at $T_N = 21$ K from the paramagnetic to the commensurate weakly ferromagnetic phase. Further, the second phase transition occurs at around $T^* \sim 10$ K to the incommensurate helix phase. Just below T^*, the magnetic soliton lattice has been found in a neutron diffraction experiment. In order to study the spin dynamics in such successive phase transitions an NMR experiment was performed on ^{11}B in a CuB_2O_4 single crystal. The NMR results on the spin structure in the commensurate phase are discussed. Also a phenomenological understanding of the commensurate to incommensurate transition is discussed on the basis of the molecular-field approximation.

Keywords: NMR, Successive phase transitions, CuB_2O_4

1. Introduction

The magnetic oxide compound CuB_2O_4 hass attracted much attention because of its nature, such as successive phase transitions, the soliton lattice, etc. The properties have been studied by several experimental methods, involving the magnetic susceptibility [1], the specific heat [2], ESR [3, 4], μSR [5], neutron diffraction [6, 7], nonlinear optics [8] etc. In these investigations, it has been found that the material undergoes

J. Franse et al. (eds.), Smart Materials for Ranging Systems, 101–113.

successive magnetic phase transitions. At $T_N = 21$ K,a transition occurs from the paramagnetic phase to a commensurate weakly ferromagnetic phase. Then, at $T^* \sim 10$ K the second magnetic transition produces an ncommensurate helical phase. At a temperature just below T^* higher order magnetic satellites have been observed, which is discussed in terms of the magnetic soliton lattice [6]. When the temperature is decreased below $T = 1.8$ K, an additional magnetic phase transition has been suggested by the neutron scattering and the μSR experiments [5]. Such complex phase transitions may be due to the fact that the association of the Dzyaloshinskii-Moriya interaction and the anisotropy plays an important role. In order to study the behavior of the spin system in this material we performed a nuclear magnetic resonance (NMR) experiment on ^{11}B in CuB_2O_4.

2. Crystal structure of CuB_2O_4

The crystal structure of CuB_2O_4 is shown in Fig. 1. The crystal belongs to the tetragonal space group $I\bar{4}2d(D_{2d}^{12})$ with lattice constants $a = 11.528$ Å, and $c = 5.607$ Å. The unit cell contains 12 formula units. The Cu^{2+} ions occupy two inequivalent position at Cu(A) = (site $4b$, point symmetry $S_4, 0\ 0\ \frac{1}{2}$) and Cu(B) = (site $8d$, point symmetry

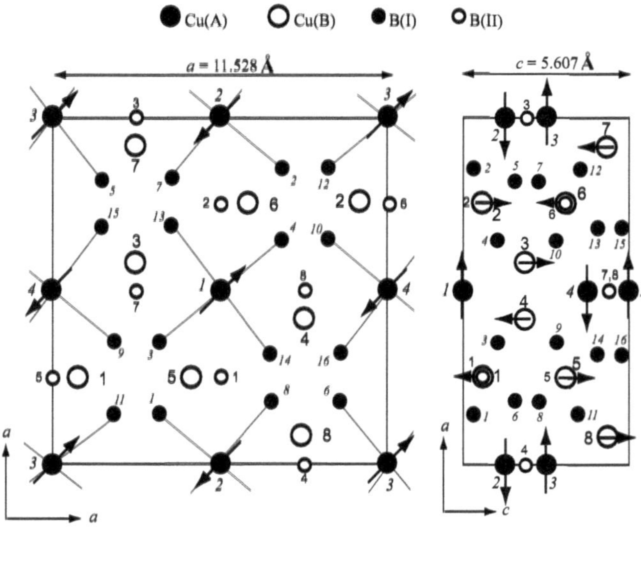

Figure 1. Crystal structure of CuB_2O_4. Positions of Cu(A), Cu(B), B(I) and B(II) sites are indicated. Arrows in the figure denote the magnetic structure in the commensurate phase reported in ref. [7].

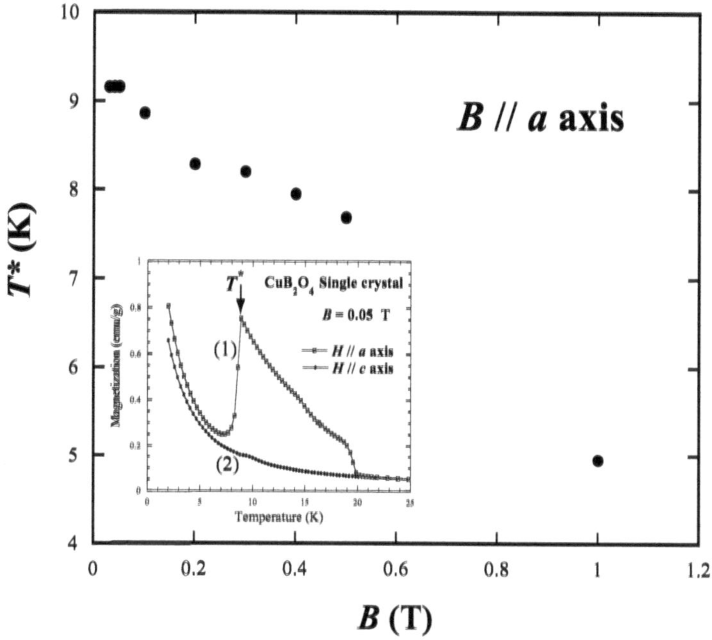

Figure 2. Applied-field dependence of T^* determined from the anomaly in magnetization. Inset is the magnetization observed at 0.05 T: (1) B // a-axis, (2) B // c-axis.

$C_2, x\ \frac{1}{4}\ \frac{1}{8}, x = 0.0815$) [7]. In this figure O^{2-} ions are not indicated in order to clarify the positional relations between Cu^{2+} ions and the ^{11}B nuclei for the convenience of the analysis of the NMR experiment. Single crystals of good quality were grown at the Institute of Physics (RAS) in Krasnoyarsk.

3. Magnetization

The magnetization of CuB_2O_4 was measured in applied fields both parallel to the a-axis and parallel to the c-axis of the tetragonal crystal. It was consistent with the results reported in ref. [1]. Anisotropy in the magnetization was observed in the temperature range $T^* < T < T_N$, as is indicated in the inset of Fig. 2. The Néel temperature T_N is almost independent of the applied field, while T^* decreases with increasing applied field. The field dependence of T^* obtained by the magnetization measurement is shown in Fig. 2. A qualitative discussion of the field dependence of T^* will be given in Section 7 on the basis of the molecular-field approximation.

4. Experimental results

The temperature dependence of the NMR spectra of ^{11}B in CuB_2O_4 are shown in Fig. 3 ~ 5. The experiments were made for three crystal orientations, namely, B // [001] (B // c-axis), B // [110] and B // [100] (B // a-axis). The operating frequency was 6.80 MHz and the applied magnetic field was around 0.5 T. The spectra show a characteristic structure depending on the magnetic phase.

In each crystal orientation the spectrum at the paramagnetic phase consists of two lines with a small splitting whose intensity ratio is 1/2. The splitting of the resonance curve in the paramagnetic phase is caused by the presence of boron ions at two nonequivalent positions of the crystallographic unit cell: B(I) = (site 16e, 0.1838 0.1484 0.6992) and B(II) = (site 8d, -0.0011 1/4 5/8). Since the nuclear spin of ^{11}B is $I = \frac{3}{2}$, usually the nuclear quadrupole splitting appears in the NMR spectrum. However, such a splitting was not observed. The quadrupole splitting is considered to be very small compared with the resonance line width, which means that the electric field gradient at both sites of B(I) and B(II) is small.

In the commensurate phase ($T^* \leq T \leq T_N$) the spectral line splits into several lines, and in the incommensurate phase below T^* the spectral

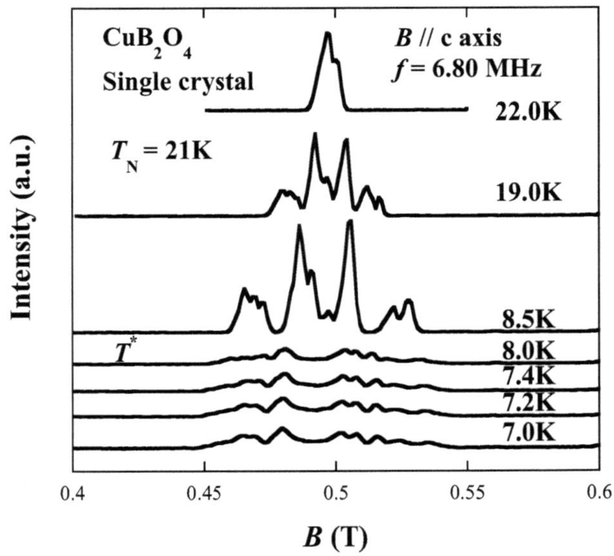

Figure 3. Temperature dependence of the ^{11}B-NMR spin echo spectrum. B // c-axis.

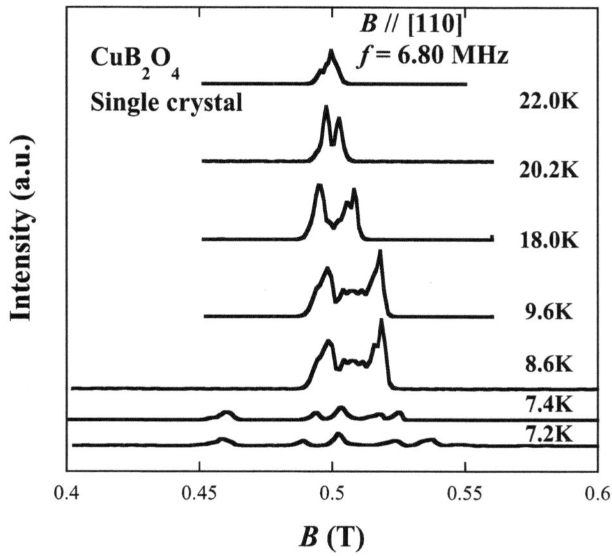

Figure 4. Temperature dependence of the ^{11}B-NMR spin echo spectrum. B // [110].

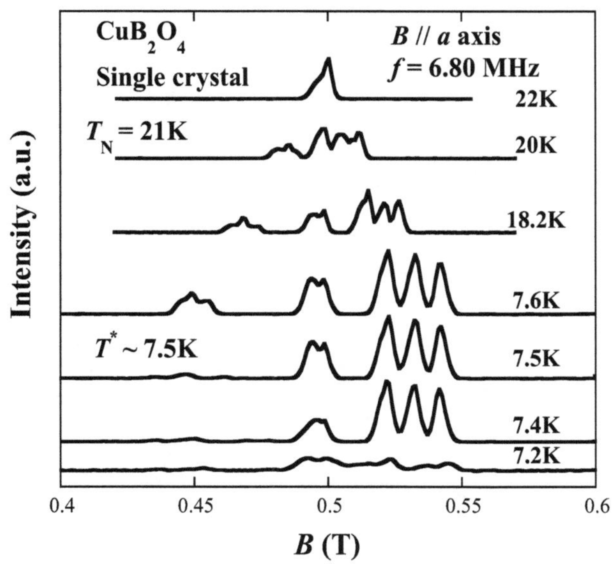

Figure 5. Temperature dependence of the ^{11}B-NMR spin echo spectrum. B // *a*-axis.

lines are broadened, suggesting that the spin fluctuations become large. In Fig. 5, for the configuration $B//a$-axis, we find the following characteristics in the spectrum. When the temperature is decreased down to near 7.5 K, the lower field resonance line broadened first. As the temperature is decreased further, the higher field resonance lines become weak. The temperature dependence of the resonance spectrum is attributed to the fact that T^* decreases with increasing applied field as is indicated in Fig. 2. In other words, for example, at T = 7.5 K the spin system is in the incommensurate phase at B = 0.45 T, while it is still in the commensurate phase in magnetic fields $B \geq 0.5$ T.

An example of the angular dependence of the NMR spectrum in the commensurate phase is given in Fig. 6. The experiment was performed at T = 16 K by keeping the direction of the applied magnetic field perpendicular to the crystal c-axis. The angles 0° and 90° correspond to the crystal a-axes in the tetragonal crystal of CuB_2O_4. The NMR spectrum for the configuration $B//a$-axis in the commensurate phase appears in both Fig. 5 and Fig. 6 They should be identical, but there is some difference between them. It may be due to a little misorientation of the single crystal sample in the experiment presented in Fig. 5.

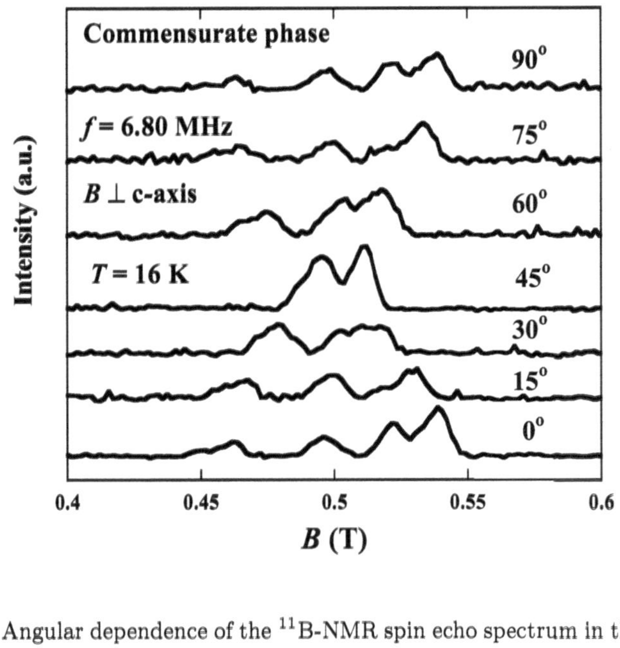

Figure 6. Angular dependence of the ^{11}B-NMR spin echo spectrum in the commensurate phase. $T = 16$ K. $B \perp c$-axis. The angle is measured from the crystal a-axis in the tetragonal crystal.

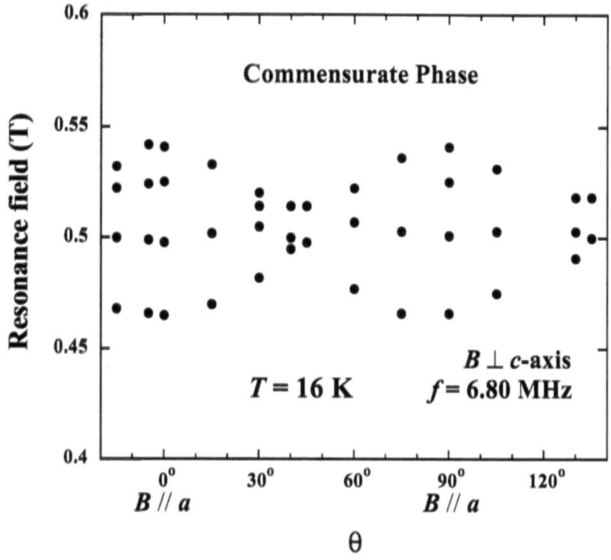

Figure 7. Angular dependence of the resonance field of the ^{11}B-NMR spin echo spectrum in the commensurate phase. $B \perp$ c-axis. θ is the angle between the crystal a-axis and the applied-field direction.

5. NMR spectrum in the commensurate phase

Here we discuss the spectrum in the commensurate phase. Usually ,in NMR experiments on magnetic materials the local field at the nucleus originated from the magnetic moment contributes to the shift of the resonance line. As for the local field two types of origins are known, namely classical and quantum origins. The classical origin is the dipole field from the magnetic ions. The quantum origin is the hyperfine field due to the covalency effect directly from the magnetic ions or the nonmagnetic ions in the path of the superexchange interaction. In the present results the amount of the spectral line shift is comparable with the amount of the classical dipole field. The contribution to the local field caused by the quantum origin is considered to be negligible. Hereafter, in the analysis of the NMR results, the local field at the site of the ^{11}B nucleus is regarded as originating from the point dipole moments at the Cu^{2+} ion sites.

The magnetic structure of CuB_2O_4 in zero magnetic field is reported through the analysis of the neutron diffraction experiment [7]. According to their result, the spins at the Cu(A) site exhibit an antiferromagnetic alignment in the commensurate phase at $T = 12$ K. In general, in the antiferromagnetic spin configuration of a two-sublattice system, the nuclei at the crystallographically equivalent sites feel two types of local fields with opposite signs. As a result the pattern of the NMR spectrum is symmetric. Since both spin configurations at Cu(A) and at Cu(B) sites are reported to be almost antiferromagnetic in the commensurate phase [7], the pattern of the NMR spectrum should be symmetric. As can be seen in Fig.3 and 4 both the spectral patterns for the crystal orientations B // $c-$axis and B // [110] are roughly symmetric. On the other hand the spectral pattern in the configuration of $B//a$-axis shown in Fig. 5 is not symmetric. Although with respect to the positions of the resonance lines the observed spectrum is almost symmetric, the intensity of the spectral line of the resonance signal in the lower applied field is always weaker than that in the higher field. The reason is not clear so far.

6. Angular dependence

In order to analyze the results of the angular dependence as shown in Fig. 7, we calculate the classical dipole field from the Cu^{2+} ions at 24 ^{11}B sites in the unit cell. We assume the spin structure in the com-

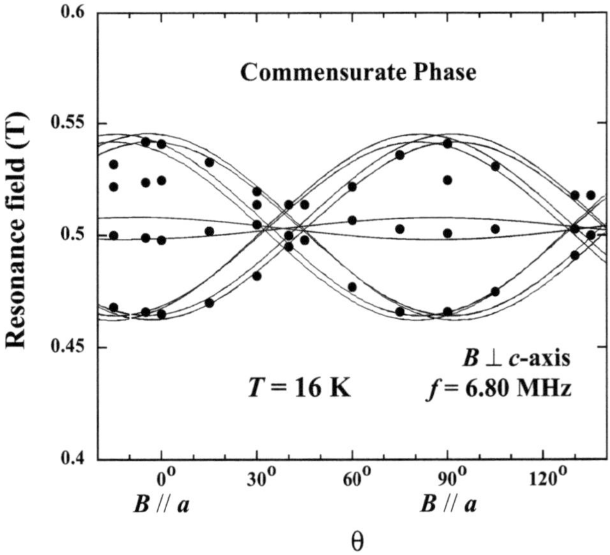

Figure 8. Calculated angular dependence of the resonance field. θ is the angle between the crystal a-axis and the applied-field direction. Solid circles represent the experimental data shown in Fig. 7.

mensurate phase on the basis of the structure reported in ref. [7], with a slight modification. The following model is considered for the present analysis. As is in ref. [7], Cu(A) moments ordersantiferromagnetically. The magnetic moment lies almost in the c-plane but is inclined about 10° from the c-plane. As for the amount of the moment in the c-plane we used 0.45 μ_B in order to fit the experimental data, instead of 0.86 μ_B as is reported. The magnetic moment with antiferromagnetic configuration is assumed to align perpendicular to the direction of the applied field, since the anisotropy field in the c-plane of CuB_2O_4 of the tetragonal crystal might be smaller than the applied field. In order to reproduce the weak ferromagnetism we added the moment of 0.05 μ_B along the applied field direction. As for Cu(B) sites, we assume that the magnetic moment is absent, because the spin at Cu(B) site has been reported to have a paramagnetic nature from the experiment of nonlinear optics [8]. The calculated result is shown in Fig. 8. Because the fitting is rather good, we can conclude that the magnetic moment at the Cu(A) site is 0.45 μ_B instead of 0.86 μ_B as reported from the neutron diffraction experiment.

Two possibilities are considered for the discrepancy in the amount of the magnetic moment. One is as follows: since the observing time of the NMR signal is longer than that of the neutron diffraction, it is considered that rapid spin fluctuations do not contribute to the static moment in NMR experiment. The other possibility is ascribed to a field induced phase transition. In ref. [4] it is reported from the ESR experiment that a phase transition occurs around the applied field of 0.05 T in the temperature region between T_N and T^*. Since the NMR experiment was performed in an applied field of 0.5 T, the magnetic phase is different from that of the neutron diffraction experiment that was carried at zero magnetic field.

7. Commensurate-incommensurate transition

With regard to the commensurate-incommensurate phase transition in CuB_2O_4, a phenomenological model based on the thermodynamic potential [9] has explained well the experimental results of the magnetization and the neutron diffraction experiments. Here, instead, we give a brief understanding on the basis of the molecular-field approximation. At a glance it is peculiar that the static spin structure in the commensurate phase existing at high temperatures is disturbed at low temperatures. We consider the behavior of the spin system on the basis of the following model as is indicated in Fig. 9.

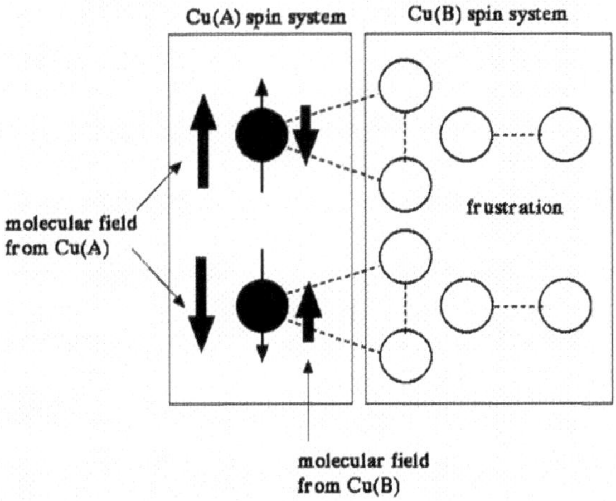

Figure 9. Model for understanding the commensurate-incommensurate phase transition. The Cu(A) system is ordered. The Cu(B) system is disordered because of the frustration caused by the geometrically competing interaction. The molecular field from the Cu(B) system disturbs the order in the Cu(A) system.

The material CuB_2O_4 has two inequivalent Cu^{2+} sites, namely, Cu(A) and Cu(B). There are 8 Cu(A)'s and 16 Cu(B)'s in the unit cell as is shown in Fig. 1. The spin system Cu(A) orders into the antiferromagnetic state by a cooperative phenomenon. On the other hand the Cu(B) spins form an orthogonal dimer system in 3-dimensions, which is a geometrically competing system. Now we assume that the magnetic couplings within Cu(A) and also within the Cu(B) systems are strong, but that the coupling between the Cu(A) and the Cu(B) systems is weak. When the temperature decreases, the Cu(A) system orders but the Cu(B) system is still in disorder because of the frustration due to the geometrical competing interaction. This phase corresponds to the commensurate phase.

Now we recall the discussion on the magnetic order in a ferromagnetic spin system through the molecular-field approximation as is described in the usual textbook [10]. In the spin system of $S = \frac{1}{2}$, the magnetization M at the temperature T is described as

$$M = N\mu \tanh \frac{\mu B}{kT}, \tag{1}$$

where N, μ and k are spin number, magnetic moment and Boltzman constant, respectively. Here B is the effective magnetic field acting on the spin, namely

$$B = B_0 + \lambda M \tag{2}$$

where B_0 is the applied magnetic field and λM is the molecular field. If $B_0 = 0$,

$$\frac{ktx}{N\mu^2\lambda} = \tanh x, \qquad \text{where } x = \frac{N\mu^2\lambda M}{kT}. \tag{3}$$

By putting both sides of the equation equal to y,

$$\begin{aligned} y &= \frac{kT}{N\mu^2\lambda}x, \\ y &= \tanh x. \end{aligned} \tag{4}$$

we find the solution from the intercept of two curves by plotting these equations.

Now we discuss our problem in a similar manner. For simplicity,the Cu(A) spins are regarded as forming a ferromagnetic spin system under the applied magnetic field B_0. Although the actual system is antiferromagnetic, the sublattice magnetization is similar to that of a ferromag-

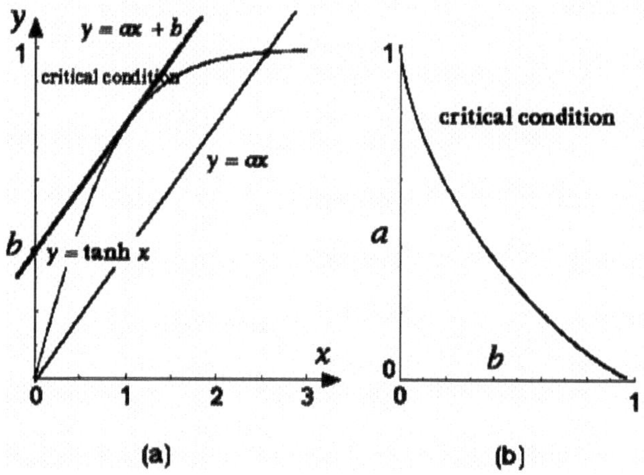

Figure 10. (a) Graphic solution. Critical condition means the condition that the straight line $y = ax + b$ is the tangent line of the curve $y = \tanh x$. (b) Relation between a and b at the critical condition.

net. In this case eq. (4) is modified to be

$$\begin{aligned} y &= \frac{kT}{N\mu^2\lambda}x - \frac{B_0}{N\mu\lambda}, \\ y &= \tanh x. \end{aligned} \tag{5}$$

Here B_0 contains not only the external applied field B_{external} but also the molecular field from the Cu(B) spins $B_{\text{mol(B)}}$,

$$B_0 = B_{\text{external}} - B_{\text{mol(B)}}. \tag{6}$$

The external magnetic field stabilizes the spin system Cu(A), while the molecular field from the Cu(B) spins makes the system unstable. Rewriting eq. (5) we obtain the following equations.

$$\begin{aligned} y &= ax + b, \\ y &= \tanh x. \end{aligned} \tag{7}$$

Here, we plot these curves in Fig. 10(a). Since

$$\text{slope:} \qquad a = \left(\frac{k}{N\mu^2\lambda}\right) T, \tag{8}$$

$$\text{intercept} \qquad b = -\left(\frac{1}{N\mu\lambda}\right) B_0, \tag{9}$$

a is proportional to T and b is proportional to B_0. The relation between a and b in the critical condition is plotted in Fig. 10(b), where the critical condition means that the straight line $y = ax + b$ is the tangent line of the curve $y = \tanh x$. In the critical condition, a corresponds to T^*. The result in Fig. 10(b) shows that if B_0 increases then T^* decreases. From eq. (6),

$$B_{\text{external}} = B_0 - B_{\text{mol(B)}}. \tag{10}$$

the molecular field from the Cu(B) system, $B_{\text{mol(B)}}$, may increase with decreasing temperature. Since both B_0 and $B_{\text{mol(B)}}$ in eq. (10) are decreasing functions of the temperature, then B_{external} becomes a decreasing function of the temperature. Thus, T^* decreases with increasing applied field, in accordance with the magnetization measurements shown in Fig. 2. The incommensurate helical spin structure is not discussed here, because the Dzyaloshinskii-Moriya interaction is not taken into account in this model.

8. Conclusion

The NMR experiment was performed on ^{11}B in CuB_2O_4. The spectrum in the commensurate phase is discussed. The magnetic moment at the Cu(A) site is estimated to be 0.45 μ_B, which is almost 50% of the moment derived from the neutron diffraction experiment. The magnetic moment at the Cu(B) sites is absent in the commensurate phase. The asymmetric nature of the spectral pattern is not understood so far. A phenomenological understanding of the commensurate to incommensurate transition is discussed on the basis of the molecular-field approximation. More precise discussion of the commensurate phase and also of the incommensurate phase will be presented in the near future.

Acknowledgments

This work is partially supported by Grant-in-Aid for Scientific Research on Priority Areas (B) (No. 13130204) and by the Grant-in-Aid for Young Scientists (B) (No. 14740207), from the Ministry of Educations, Culture, Sports, Science and Technology, Japan.

References

[1] G. A. Petrakovskii, A. D. Balaev, A. M. Vorotynov, Phys. Solid State **42** 321 (2000).

[2] G. A. Petrakovskii, D. Velikanov, A. Vorotinov, A. Balaev, K. Sablina, A. Amato, B. Roessli, J. Schefer, U. Staub, J. Magn. Magn. Mater. **205** 105 (1999).

[3] A. I. Pankrats, G. A. Petrakovskii, N. V. Volkov, Phys. Solid State **42** 96 (2000).

[4] A. I. Pankrats, G. A. Petrakovskii, M. A. Popov, K. A. Sablina, L. A. Prozorova, S. S. Sosin, G. Szimczak, M. Baran, JETP Lett. **78** 569 (2003).

[5] M. Boehm, B. Roessli, J. Schefer, B. Ouladdiaf, A. Amato, C. Baines, U. Staub, G. A. Petrakovskii, Physica B **318** 277 (2002).

[6] B. Roessli, J. Schefer, G. A. Petrakovskii, B. Ouladdiaf, M. Boehm, U. Staub, A. Vorotinov, L. Bezmaternikh, Phys. Rev. Lett. **86** 1885 (2001).

[7] M. Boehm, B. Roessli, J. Schefer, A. S. Wills, B. Ouladdiaf. E. Lelievre-Berna, U. Staub, G. A. Petrakovskii, Phys. Rev. B **68** 024405 (2003).

[8] R. V. Pisarev, I. Sänger, G. A. Petrakovskii, M. Fiebig, Phys. Rev. Lett. **93** 037204 (2004).

[9] M. A. Popov, G. A. Petrakovskii, V. I. Zinenko, Phys. Solid State **46** 491 (2004).

[10] For example, see C. Kittel, *Introduction to Solid State Physics 7th ed.*. John Wiley & Sons, Inc. 1996.

MAGNETIC PROPERTIES OF CUPRATE PEROVSKITES

A. Sherman
Institute of Physics, University of Tartu, Riia 142, 51014 Tartu, Estonia
alexei@fi.tartu.ee

Abstract: The magnetic susceptibility of underdoped yttrium and lanthanum cuprates is interpreted based on the self-consistent solution of the t-J model of a Cu-O plane. The calculations reproduce correctly the frequency dependencies of the susceptibility in $YBa_2Cu_3O_{7-y}$ and $La_{2-x}Sr_xCuO_4$ attributing their dissimilarity to the difference in the damping of spin excitations. In $YBa_2Cu_3O_{7-y}$ these excitations are well defined at the antiferromagnetic wave vector $\mathbf{Q} = (\pi, \pi)$ even in the normal state which manifests itself in a pronounced maximum – the resonance peak – in the susceptibility. In $La_{2-x}Sr_xCuO_4$ the spin excitations are overdamped which leads to a broad low-frequency feature in the susceptibility. The low-frequency incommensurability in the magnetic response is attributed to a dip in the magnon damping at $\mathbf{Q}$. The calculated concentration and temperature dependencies of the incommensurability parameter conform with experimental observations. Generally the incommensurate magnetic response is not accompanied with an inhomogeneity of the carrier density.

Keywords: High-T_c superconductors, magnetic properties, the t-J model.

1. Introduction

Among the results obtained with inelastic neutron scattering is the detailed information on the magnetic susceptibility in $YBa_2Cu_3O_{7-y}$ and lanthanum cuprates which reveals a considerable difference in their magnetic properties. A sharp magnetic collective mode – the so-called resonance peak – was observed at frequencies $\omega_r = 20 - 40$ meV in $YBa_2Cu_3O_{7-y}$ and some other cuprates, in underdoped case both below and above T_c [1]. In the momentum space the mode is strongly peaked at the antiferromagnetic wave vector $\mathbf{Q} = (\pi, \pi)$. Contrastingly, no resonance peak was observed in lanthanum cuprates. Instead for low temperatures a broad feature was detected at $\omega_f \approx 7$ meV and for

J. Franse et al. (eds.), Smart Materials for Ranging Systems, 115–128.

this and lower frequencies the susceptibility was found to be peaked at incommensurate momenta $(\pi \pm 2\pi\delta, \pi)$ and $(\pi, \pi \pm 2\pi\delta)$ [2]. Recently the low-frequency incommensurate magnetic response was also observed in $YBa_2Cu_3O_{7-y}$ [3]. Presently the fundamental difference of the susceptibility frequency dependencies in these two groups of cuprates is actively debated, as well as the nature of the resonance peak and the low-frequency magnetic incommensurability.

The aim of this work is to demonstrate that the above-mentioned unusual properties of cuprates can be interpreted in the framework of the t-J model of a Cu-O plane which is a common structure element of these crystals. The model was shown to describe correctly the low-energy part of the spectrum of the realistic extended Hubbard model [4]. To take proper account of strong electron correlations inherent in moderately doped cuprate perovskites the description in terms of Hubbard operators and Mori's projection operator technique [5] are used. The self-energy equations for hole and spin Green's functions obtained in this approach are self-consistently solved for the ranges of hole concentrations $0 \leq x \lesssim 0.16$ and temperatures 2 K$\leq T \leq$1200 K. Lattices with 20×20 sites and larger are used.

The calculations reproduce correctly the frequency and momentum dependencies of the resonance peak in $YBa_2Cu_3O_{7-y}$, and its variation with doping and temperature in normal and superconducting states. The peak is connected with the excitation branch of localized Cu spins and the peak frequency is close to the frequency of these excitations at $\mathbf{Q}$. The absence of the resonance peak in lanthanum cuprates is related to an increased damping of spin excitations which is possibly connected with a large hole damping in these crystals. For low frequencies and temperatures the susceptibility is peaked at incommensurate wave vectors $(\pi \pm 2\pi\delta, \pi)$ and $(\pi, \pi \pm 2\pi\delta)$. The incommensurability is connected with a dip in the magnon damping at $\mathbf{Q}$. In agreement with experiment the incommensurability parameter δ is approximately proportional to x for $0.02 \lesssim x \lesssim 0.12$ and saturates for larger concentrations. The incommensurability disappears with increasing temperature. Generally the incommensurate magnetic response is not accompanied by an inhomogeneity of the carrier density.

2. Main formulas

The Hamiltonian of the two-dimensional t-J model reads

$$H = \sum_{\mathbf{nm}\sigma} t_{\mathbf{nm}} a^\dagger_{\mathbf{n}\sigma} a_{\mathbf{m}\sigma} + \frac{1}{2} \sum_{\mathbf{nm}} J_{\mathbf{nm}} \left(s^z_{\mathbf{n}} s^z_{\mathbf{m}} + s^{+1}_{\mathbf{n}} s^{-1}_{\mathbf{m}} \right), \qquad (1)$$

where $a_{\mathbf{n}\sigma} = |\mathbf{n}\sigma\rangle\langle\mathbf{n}0|$ is the hole annihilation operator, $\mathbf{n}$ and $\mathbf{m}$ label sites of the square lattice, $\sigma = \pm 1$ is the spin projection, $|\mathbf{n}\sigma\rangle$ and $|\mathbf{n}0\rangle$ are site states corresponding to the absence and presence of a hole on the site. These states may be considered as linear combinations of the products of the $3d_{x^2-y^2}$ copper and $2p_\sigma$ oxygen orbitals of the extended Hubbard model [6]. We take into account nearest neighbor interactions only, $t_{\mathbf{nm}} = -t\sum_{\mathbf{a}} \delta_{\mathbf{n},\mathbf{m}+\mathbf{a}}$ and $J_{\mathbf{nm}} = J\sum_{\mathbf{a}} \delta_{\mathbf{n},\mathbf{m}+\mathbf{a}}$ where the four vectors $\mathbf{a}$ connect nearest neighbor sites. The spin-$\frac{1}{2}$ operators can be written as $s^z_{\mathbf{n}} = \frac{1}{2}\sum_\sigma \sigma|\mathbf{n}\sigma\rangle\langle\mathbf{n}\sigma|$ and $s^\sigma_{\mathbf{n}} = |\mathbf{n}\sigma\rangle\langle\mathbf{n},-\sigma|$.

Properties of the model are determined from the hole and spin retarded Green's functions

$$G(\mathbf{k}t) = -i\theta(t)\langle\{a_{\mathbf{k}\sigma}(t), a^\dagger_{\mathbf{k}\sigma}\}\rangle, \quad D(\mathbf{k}t) = -i\theta(t)\langle[s^z_{\mathbf{k}}(t), s^z_{-\mathbf{k}}]\rangle, \tag{2}$$

where $a_{\mathbf{k}\sigma}$ and $s^z_{\mathbf{k}}$ are the Fourier transforms of the respective site operators, operator time dependencies and averaging are defined with the Hamiltonian $\mathcal{H} = H - \mu\sum_{\mathbf{n}} a^\dagger_{\mathbf{n}\sigma}a_{\mathbf{n}\sigma}$ with the chemical potential μ. As mentioned, to obtain self-energy equations for these functions we used Mori's projection operator technique [5, 7]. In this approach the Fourier transform of Green's function $\langle\langle A_0|A_0^\dagger\rangle\rangle$ is represented by the continued fraction

$$\langle\langle A_0|A_0^\dagger\rangle\rangle = \cfrac{|A_0 \cdot A_0^\dagger|}{\omega - E_0 - \cfrac{V_0}{\omega - E_1 - \cfrac{V_1}{\ddots}}}. \tag{3}$$

The elements of the fraction E_i and V_i are determined from the recursive procedure

$$\begin{aligned} &[A_n, H] = E_n A_n + A_{n+1} + V_{n-1}A_{n-1}, \\ &E_n = |[A_n, H]\cdot A_n^\dagger|\,|A_n \cdot A_n^\dagger|^{-1}, \\ &V_{n-1} = |A_n \cdot A_n^\dagger|\,|A_{n-1}\cdot A_{n-1}^\dagger|^{-1}, \quad V_{-1} = 0, \quad n = 0,1,2,\ldots \end{aligned} \tag{4}$$

The operators A_i constructed in the course of this procedure form an orthogonal set, $|A_i \cdot A_j^\dagger| \propto \delta_{ij}$. In Eqs. (3) and (4) the definition of the inner product $|A_i \cdot A_j^\dagger|$ depends on the type of the considered Green's function. For example, for functions (2) these are $\langle\{A_i, A_j^\dagger\}\rangle$ and $\langle[A_i, A_j^\dagger]\rangle$, respectively. The method described by Eqs. (3) and (4) can be straightforwardly generalized to the case of many-component operators which is necessary, for example, to consider Green's functions for Nambu spinors in the superconducting state [7].

The residual term of fraction (3) is the Fourier transform of the quantity

$$\mathcal{T} = |A_{nt} \cdot A_n^\dagger| \, |A_{n-1} \cdot A_{n-1}^\dagger|^{-1}, \tag{5}$$

where the time evolution of the operator A_n is determined by the equation

$$i\frac{d}{dt}A_{nt} = \prod_{k=0}^{n-1}(1 - P_k)[A_{nt}, \mathcal{H}], \quad A_{n,t=0} = A_n \tag{6}$$

with the projection operators P_n defined as $P_n Q = |Q \cdot A_n^\dagger| \, |A_n \cdot A_n^\dagger|^{-1} A_n$. The residual term $\mathcal{T}$ is a many-particle Green's function which can be estimated by the decoupling. The decoupling is also used in calculating the elements of the continued fractions (3). Following Ref. [8] this approximation is improved by introducing the vertex correction α which is determined from the constraint of zero site magnetization

$$\langle s_{\mathbf{n}}^z \rangle = \frac{1}{2}(1 - x) - \langle s_{\mathbf{n}}^{-1} s_{\mathbf{n}}^{+1} \rangle = 0. \tag{7}$$

Since the statistical averaging includes samples with different ordering orientations, this condition is fulfilled also in magnetically ordered states.

Equations for Green's functions obtained in this way for the normal state read [7]

$$\begin{aligned}
G(\mathbf{k}\omega) &= \frac{\phi}{\omega - \varepsilon_{\mathbf{k}} + \mu - \Sigma(\mathbf{k}\omega)}, \; D(\mathbf{k}\omega) = \frac{4(1-\gamma_{\mathbf{k}})(J|C_1| + tF_1)}{\omega^2 - \omega\Pi(\mathbf{k}\omega) - \omega_{\mathbf{k}}^2}, \\
\operatorname{Im}\Sigma(\mathbf{k}\omega) &= \frac{16\pi t^2}{N\phi}\sum_{\mathbf{k}'}\int_{-\infty}^{\infty} d\omega' \Bigg[\gamma_{\mathbf{k}-\mathbf{k}'} + \gamma_{\mathbf{k}} + \operatorname{sgn}(\omega')(\gamma_{\mathbf{k}-\mathbf{k}'} - \gamma_{\mathbf{k}}) \\
&\times \sqrt{\frac{1+\gamma_{\mathbf{k}'}}{1-\gamma_{\mathbf{k}'}}}\Bigg]^2 [n_B(-\omega') + n_F(\omega - \omega')] \\
&\times A(\mathbf{k}-\mathbf{k}', \omega - \omega')B(\mathbf{k}'\omega'), \\
\operatorname{Im}\Pi(\mathbf{k}\omega) &= \frac{9\pi t^2 J^2(1-x)}{2N(1-\gamma_{\mathbf{k}})(J|C_1| + tF_1)}\sum_{\mathbf{k}'}(\gamma_{\mathbf{k}+\mathbf{k}'} - \gamma_{\mathbf{k}'})^2 \\
&\times \int_{-\infty}^{\infty} d\omega' A(\mathbf{k}'\omega')A(\mathbf{k}+\mathbf{k}', \omega + \omega')\frac{n_F(\omega + \omega') - n_F(\omega')}{\omega},
\end{aligned} \tag{8}$$

where $\phi = \frac{1}{2}(1 + x)$, $\gamma_{\mathbf{k}} = \frac{1}{2}[\cos(k_x) + \cos(k_y)]$, $C_1 = \langle s_{\mathbf{n}}^{+1} s_{\mathbf{n+a}}^{-1} \rangle$ and $F_1 = \langle a_{\mathbf{n}}^\dagger a_{\mathbf{n+a}} \rangle$ are the spin and hole correlations on neighboring sites which, as well as the hole concentration x, can be calculated from Green's functions,

$$\omega_{\mathbf{k}}^2 = 16J^2\alpha|C_1|(1-\gamma_{\mathbf{k}})(\Delta+1+\gamma_{\mathbf{k}}), \; \varepsilon_{\mathbf{k}} = -(4\phi t + 6C_1\phi^{-1}t + 3F_1\phi^{-1}J)\gamma_{\mathbf{k}} \tag{9}$$

are the energies of spin excitations and holes with the parameter Δ describing the spin gap at $\mathbf{Q}$, $A(\mathbf{k}\omega) = -\pi^{-1}\text{Im}\, G(\mathbf{k}\omega)$ and $B(\mathbf{k}\omega) = -\pi^{-1}\text{Im}\, D(\mathbf{k}\omega)$ are the hole and spin spectral functions, N is the number of sites, $n_F(\omega) = [\exp(\omega/T) + 1]^{-1}$, $n_B(\omega) = [\exp(\omega/T) - 1]^{-1}$ with the temperature T. Real parts of the self-energies are calculated from their imaginary parts and the Kramers-Kronig relation. The self-energy equations for the superconducting state can be found in Ref. [7].

Equations (7)–(9) form a closed set which can be solved by iteration for fixed chemical potential μ and temperature T. We carried out such calculations for the parameters of the model $J = 0.1$ eV, $t = 0.5$ eV which correspond to hole-doped cuprates [9]. To stabilize the iteration procedure an artificial damping $i\eta$, $\eta = (0.015 - 0.05)t$, was added to the frequency in the hole Green's function.

The results of such calculations can be directly compared with the imaginary part of the magnetic susceptibility χ'' derived from the neutron scattering experiments, since this quantity is connected with $B(\mathbf{k}\omega)$,

$$\chi''(\mathbf{k}\omega) = 4\pi\mu_B^2 B(\mathbf{k}\omega) = -\frac{16\mu_B^2(1-\gamma_{\mathbf{k}})(J|C_1| + tF_1)\omega\text{Im}\,\Pi}{(\omega^2 - \omega\text{Re}\,\Pi - \omega_{\mathbf{k}}^2)^2 + (\omega\text{Im}\,\Pi)^2}, \quad (10)$$

where μ_B is the Bohr magneton. Equation (10) is the most general form for the magnetic susceptibility which follows from the Mori formalism. With the spin excitation damping $\Pi(\mathbf{k}\omega)$ described by the fermion bubble (8) this equation acquires some similarity with the susceptibility derived in the random phase approximation [10] which is frequently used for the discussion of the magnetic properties of cuprates. For this similarity the term $\omega_{\mathbf{k}}^2$ in the denominator of Eq. (10) has to be negligibly small in comparison with $\omega\text{Re}\Pi(\mathbf{k}\omega)$. For the hole spectrum self-consistently calculated here this can happen only in the overdoped region $x \gtrsim 0.3$ when the spin correlation C_1 which determines the magnitude of $\omega_{\mathbf{k}}$ [see Eq. (9)] tends to zero [7, 11].

3. The spectrum of spin excitations

As follows from Eq. (10), the frequencies of spin excitations satisfy the equation

$$\omega^2 - \omega\text{Re}\,\Pi(\mathbf{k}\omega) - \omega_{\mathbf{k}}^2 = 0. \quad (11)$$

As seen from Eq. (8), $\text{Im}\,\Pi(\mathbf{k}\omega)$ is finite for $\mathbf{k} \to 0$, whereas $\omega_{\mathbf{k}}$ vanishes in this limit. Therefore the spin Green's function has a purely imaginary, diffusive pole near the Γ point, in compliance with the result of the hydrodynamic theory [12].

The dispersion of spin excitations near $\mathbf{Q}$ is of special interest, because this region gives the main contribution into the neutron scattering. This

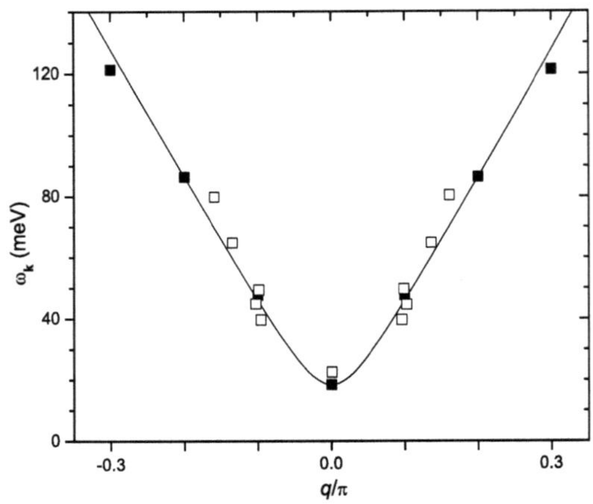

Figure 1. The dispersion of spin excitations near $\mathbf{Q}$ along the edge of the Brillouin zone. The dispersion was calculated in a 20×20 lattice for $x = 0.06$ and $T = 17$ K (filled squares, the solid line is the fit with $\omega_{\mathbf{k}} = [\omega_{\mathbf{Q}}^2 + c^2(\mathbf{k} - \mathbf{Q})^2]^{1/2}$). Open squares are the experimental dispersion [1] of the maximum in the frequency dependence of the odd $\chi''(\mathbf{q}\omega)$, $\mathbf{q} = \mathbf{k} - \mathbf{Q}$, in $YBa_2Cu_3O_{6.5}$ ($x \approx 0.075$ [13]) at $T = 5$ K.

dispersion is shown in Fig. 1. As seen from this figure, in contrast to the usual spin-wave theory the frequency of spin excitations at the antiferromagnetic wave vector $\mathbf{Q}$ is finite. This frequency $\omega_{\mathbf{Q}} = 4J(2\alpha|C_1|\Delta)^{1/2}$ is directly connected with the spin correlation length ξ. Indeed, using Eq. (8) and taking into account that the region near $\mathbf{Q}$ gives the main contribution to the summation over $\mathbf{k}$, we find for large distances and low temperatures

$$\langle s_{\mathbf{l}}^z s_{\mathbf{0}}^z \rangle = N^{-1} \sum_{\mathbf{k}} \mathrm{e}^{i\mathbf{kl}} \int_0^\infty d\omega \coth\left(\frac{\omega}{2T}\right) B(\mathbf{k}\omega) \propto \mathrm{e}^{i\mathbf{Ql}} (\xi/|\mathbf{l}|)^{1/2} \mathrm{e}^{-|\mathbf{l}|/\xi}, \tag{12}$$

where $\xi = \frac{1}{2} a \Delta^{-1/2}$ and a is the distance between in-plane Cu sites. As follows from the calculations, for low concentrations and temperatures $\Delta \approx 0.2x$ and consequently $\xi \approx a x^{-1/2}$. An analogous relation between the spin correlation length and the concentration has been derived from experimental data in $La_{2-x}Sr_xCuO_4$ [14].

In accord with the Hohenberg-Mermin-Wagner theorem [15] the considered two-dimensional system is in the paramagnetic state for $T > 0$. This result can also be obtained using Eq. (8). Also it can be shown [7] that in the infinite crystal at $T = 0$ the long-range antiferromagnetic order is destroyed for the hole concentrations $x > x_c \approx 0.02$.

4. The resonance peak

The imaginary part of the spin susceptibility at the antiferromagnetic wave vector, obtained in our calculations, is shown in Fig. 2. The

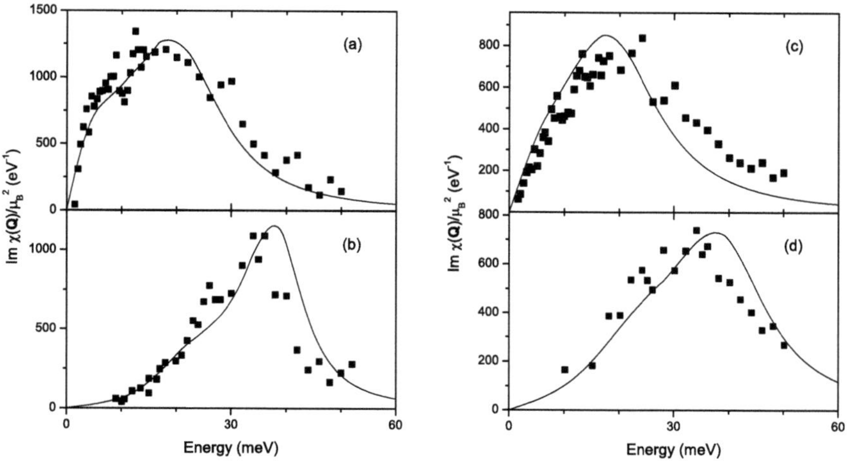

Figure 2. The imaginary part of the spin susceptibility at $\mathbf{Q}$ in the d-wave superconducting (a,b) and normal (c,d) states. Curves show the results of our calculations for $\eta = 0.015t$, $T = 17$ K (a,b) and $T = 116$ K (c,d), $x = 0.06$ (a,c) and $x = 0.12$ (b,d). Squares are the odd susceptibility measured [1] in $YBa_2Cu_3O_{6.5}$ ($T_c = 45$ K, $x \approx 0.075$, a and c) and in $YBa_2Cu_3O_{6.83}$ ($T_c = 85$ K, $x \approx 0.14$ [13], b and d) at $T = 5$ K (a,b) and at $T = 100$ K (c,d). To depict the experimental data in the same scale they were multiplied by the factor 1.5.

experimental data on χ'' in $YBa_2Cu_3O_{7-y}$ [1] are also depicted here. $YBa_2Cu_3O_{7-y}$ is a bilayer crystal and the symmetry allows one to divide the susceptibility into odd and even parts. The odd part, which for the antiferromagnetic intrabilayer coupling can be compared with our single-layer calculations, is shown in the figure.

The maximum in Fig. 2 is the resonance peak. As seen from this figure, the model reproduces satisfactorily the evolution of the peak position and shape with doping and temperature. In the t-J model the peak is connected with the coherent excitations of localized Cu spins near the antiferromagnetic wave vector $\mathbf{Q}$ and the peak frequency for $\mathbf{k} = \mathbf{Q}$ is close to the frequency $\omega_{\mathbf{Q}}$ in Fig. 1. It is also seen in this figure that the experimental dispersion of the peak is close to the dispersion of the mentioned spin excitations. Thus, we came to conclusion that the resonance peak in $YBa_2Cu_3O_{7-y}$ and apparently in other cuprate perovskites where it was observed is a manifestation of the magnon branch modified in the short-range antiferromagnetic order.

The above results are related to the underdoped case where the resonance peak is observed both in the normal and superconducting states. In the overdoped case in $YBa_2Cu_3O_{7-y}$ the peak is observed in the superconducting state only [1]. Calculations for the t-J model show that in the normal state on approaching the overdoped region the maximum in χ'' rapidly loses its intensity, broadens and shifts to higher frequencies [16]. This result correlates with the broad feature observed experimentally in these conditions. The mechanism of the peak reappearance in the superconducting state in the overdoped region was considered in Ref. [17]. The opening of the d-wave superconducting gap with the magnitude $2\Delta^s > \omega_{\mathbf{Q}}$ decreases considerably the spin excitation damping near $\mathbf{Q}$ which restores the peak in χ''.

As seen from Fig. 2, the resonance peak has a low-frequency shoulder which is more pronounced for low concentrations and temperatures. The shoulder stems from the frequency dependence of the magnon damping $\operatorname{Im}\Pi(\mathbf{Q}\omega)$ which in its turn is a consequence of the hole Fermi surface nesting existing in the t-J model for moderate doping [7]. An analogous nesting is expected in the two-layer $YBa_2Cu_3O_{7-y}$ between the parts of the Fermi surface belonging to the bonding and antibonding bends [10].

It is worth noting that for all four curves in Fig. 2 the value of $|\Pi(\mathbf{Q},\omega_{\mathbf{Q}})|/2$ is smaller than $\omega_{\mathbf{Q}}$. The satisfactory agreement between the experimental and calculated results allows us to conclude that near $\mathbf{Q}$ the spin excitations are not overdamped in underdoped $YBa_2Cu_3O_{7-y}$. In calculating the curves in Fig. 2 the artificial broadening η in the hole Green's function was set equal to $0.015t$. This broadening simulates contributions to the hole damping from mechanisms differing from the spin excitation scattering. The magnitude of these interactions can essentially vary in different crystals. The example of the magnetic susceptibility calculated with the larger hole damping $\eta = 0.05t$ is shown in Fig. 3. As seen from this figure, the spin excitation damping is very sensitive to the hole damping. The increase of this latter damping leads to a marked growth of the spin excitation damping which results in the overdamping of these excitations. Instead of the pronounced peak at the frequency $\omega_{\mathbf{Q}}$ a broad low-frequency feature is observed in the susceptibility (see Fig. 3b). This shape resembles that observed in $La_{2-x}Sr_xCuO_4$ [2]. The comparison was carried out in Fig. 3a. It should be noted that the use of a comparatively small 20×20 lattice did not allow us to describe the incommensurability of the magnetic response – χ'' is peaked at $\mathbf{Q}$ and our calculated data belong to this momentum (the use of a larger lattice is too time-consuming for the self-consistent calculations). In $La_{1.86}Sr_{0.14}CuO_4$ the low-frequency susceptibility is peaked at incommensurate momenta $\mathbf{k} = (\pi \pm 2\pi\delta, \pi), (\pi, \pi \pm 2\pi\delta)$ and the ex-

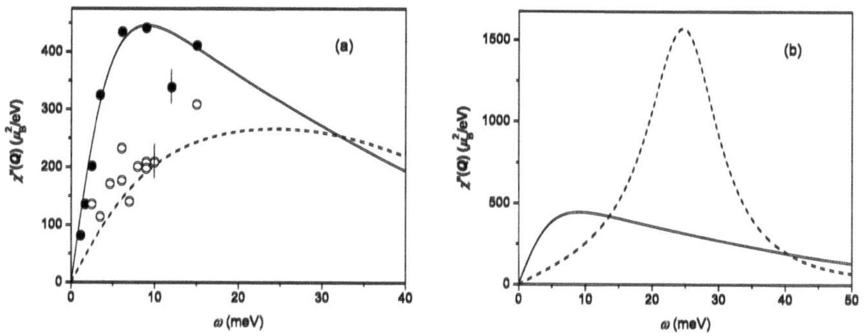

Figure 3. The frequency dependence of the imaginary part of the susceptibility. (a) The solid line corresponds to $T = 29$ K, $\eta = 0.05t$ and $x \approx 0.1$. The dashed line is for $T = 116$ K and the same other parameters. Symbols are experimental results [2] in $La_{1.86}Sr_{0.14}CuO_4$ at $T = 35$ K (filled circles) and $T = 80$ K (open circles). Vertical bars show experimental errors. The calculated susceptibility is given for $\mathbf{k} = \mathbf{Q}$, the experimental data are for a wave vector of the incommensurate peak. (b) The solid line is the same as in part (a). The dashed line is calculated for $\eta = 0.015t$, other parameters are the same as for the solid line.

perimental data [2] shown in Fig. 3a belong to one of these momenta. These data were increased by approximately 2.5 times to show them in the scale with the calculated results. The need for scaling is apparently connected with the splitting of the commensurate peak into the four incommensurate maxima. Apart from this difference in the momentum dependencies, the calculated frequency and temperature dependencies are in good agreement with the experimental results. Thus, it can be concluded that the dissimilarity of the frequency dependencies of the susceptibility in $YBa_2Cu_3O_{7-y}$ and $La_{2-x}Sr_xCuO_4$ is connected with different values of the damping of spin excitations, which are well defined at the antiferromagnetic wave vector in the former crystal and overdamped in the latter. This property of spin excitations in $La_{2-x}Sr_xCuO_4$ is not changed in the superconducting state due to the small superconducting gap $2\Delta^s \approx 9\text{meV} \lessdot \omega_{\mathbf{Q}}$ in this crystal [17].

5. Incommensurability in the magnetic response

For low frequencies susceptibility (10) is essentially simplified,

$$\chi''(\mathbf{k}\omega) \propto \frac{\omega \text{Im}\Pi(\mathbf{k}\omega)}{\omega_{\mathbf{k}}^4}. \tag{13}$$

As seen from Fig. 1 and Eq. (9), $\omega_{\mathbf{k}}$ is the increasing function of the difference $\mathbf{k} - \mathbf{Q}$ and therefore the denominator of the fraction (13) favors the appearance of a commensurate peak centered at the antiferromag-

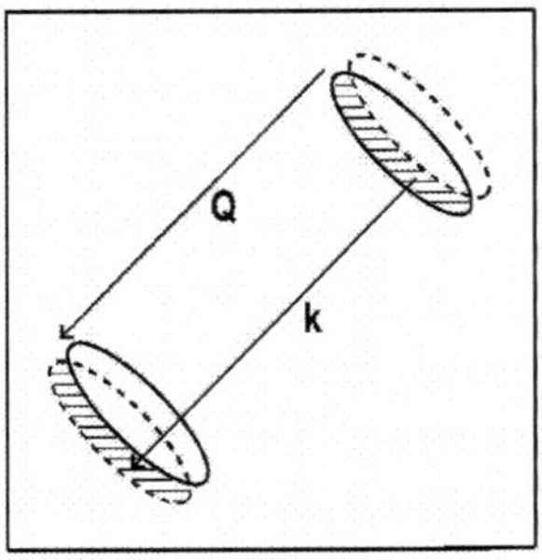

Figure 4. The Brillouin zone of the square lattice. Solid curves are two of four ellipses forming the Fermi surface at small x. Dashed lines are the Fermi surface contours shifted by $\pm(\mathbf{k}-\mathbf{Q})$. Regions of $\mathbf{k}'$ and $\mathbf{k}+\mathbf{k}'$ contributing to the damping (14) are shaded.

netic momentum in the susceptibility. However, if the spin excitation damping has a pronounced dip at $\mathbf{k}=\mathbf{Q}$ the peak splits into several peaks shifted from $\mathbf{Q}$. To make sure that the damping may really have such a behavior let us consider the case of low hole concentrations and temperatures. In this case the hole Fermi surface consists of four ellipses centered at $(\pm\pi/2, \pm\pi/2)$ [7]. Two of them are shown in Fig. 4. The spin excitation damping described by the fermion bubble (8) is simplified to

$$\begin{aligned} \operatorname{Im}\Pi(\mathbf{k}\omega) \propto & \sum_{\mathbf{k}'}(\gamma_{\mathbf{k}'}-\gamma_{\mathbf{k}'+\mathbf{k}})^2\Big[n_F(\varepsilon_{\mathbf{k}'+\mathbf{k}}-\mu)-n_F(\varepsilon_{\mathbf{k}'}-\mu)\Big] \\ & \times\ \delta(\omega+\varepsilon_{\mathbf{k}'}-\varepsilon_{\mathbf{k}'+\mathbf{k}}) \end{aligned} \tag{14}$$

in the considered case. Here $\varepsilon_{\mathbf{k}}$ is the hole dispersion. In the process described by Eq. (14) a fermion picks up an energy and momentum of a defunct spin excitation and is transferred from a region below the Fermi level to a region above it. However, for $\mathbf{k}=\mathbf{Q}$ this process is impossible, because for initial states $\mathbf{k}'$ interior to an ellipse in Fig. 4 all final states with momenta $\mathbf{k}+\mathbf{k}'$ will be inside another ellipse, i.e. below the Fermi level. Thus, in this simplified picture the damping vanishes for $\mathbf{k}=\mathbf{Q}$ and grows with increasing the difference $\mathbf{k}-\mathbf{Q}$ approximately proportional to the shaded area in Fig. 4 [18].

With increasing the hole concentration the Fermi surface of the t-J model is transformed to a rhombus centered at $\mathbf{Q}$ [16]. This result is in agreement with the Fermi surface observed in $La_{2-x}Sr_xCuO_4$ [19] [however, to reproduce the experimental Fermi surface terms describing the hole transfer to more distant coordination shells have to be taken into account in the kinetic term of Hamiltonian (1)]. For such x another mechanism of the dip formation in the damping comes into effect. The interaction constant $\gamma_{\mathbf{k}'} - \gamma_{\mathbf{k}'+\mathbf{k}}$ in Eq. (14) vanishes in the so-called "hot spots" – the crossing points of the Fermi surface with the boundary of the magnetic Brillouin zone. For small frequencies and $\mathbf{k} = \mathbf{Q}$ the nearest vicinity of these points contributes to the damping. Consequently, the damping has a dip at $\mathbf{Q}$. With the inclusion of the hole transfer to more distant coordination shells the interaction constant is changed, however the conclusion about the dip at the antiferromagnetic momentum in the damping remains unchanged.

The dip disappears when the hole damping η exceeds the frequency $\omega > 0$. Since the main contribution to the damping is made by states with energies $-\omega \leq \varepsilon_{\mathbf{k}'} - \mu \leq 0$ and $0 \leq \varepsilon_{\mathbf{k}'+\mathbf{k}} - \mu \leq \omega$, this means that the above consideration is valid when there exist well defined quasiparticle excitations near the Fermi surface.

Results of our calculations in a 120×120 lattice and experimental data are shown in Fig. 5. In agreement with experiment [20, 22] in

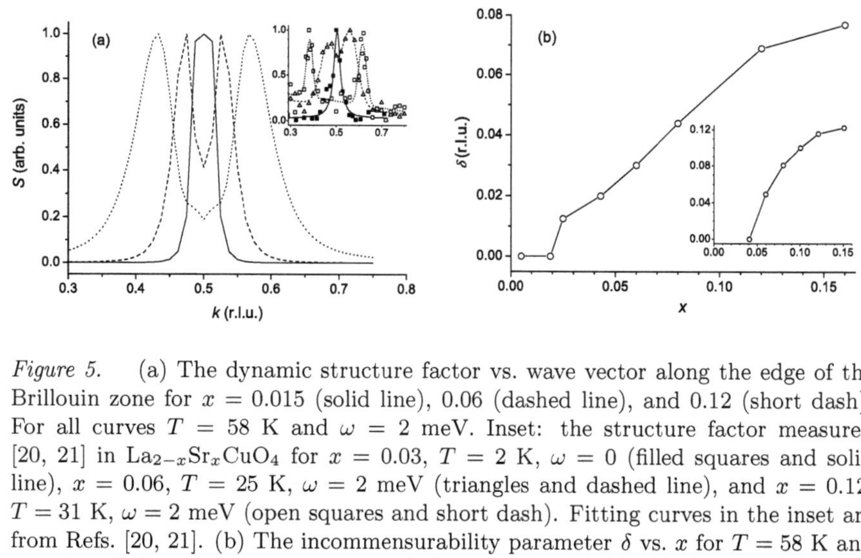

Figure 5. (a) The dynamic structure factor vs. wave vector along the edge of the Brillouin zone for $x = 0.015$ (solid line), 0.06 (dashed line), and 0.12 (short dash). For all curves $T = 58$ K and $\omega = 2$ meV. Inset: the structure factor measured [20, 21] in $La_{2-x}Sr_xCuO_4$ for $x = 0.03$, $T = 2$ K, $\omega = 0$ (filled squares and solid line), $x = 0.06$, $T = 25$ K, $\omega = 2$ meV (triangles and dashed line), and $x = 0.12$, $T = 31$ K, $\omega = 2$ meV (open squares and short dash). Fitting curves in the inset are from Refs. [20, 21]. (b) The incommensurability parameter δ vs. x for $T = 58$ K and $\omega = 2$ meV. Inset: experimental data [20] for $La_{2-x}Sr_xCuO_4$. The wave vector and the incommensurability parameter are given in the reciprocal lattice units $2\pi/a$.

our calculations peaks in the dynamic structure factor $S(\mathbf{k}\omega) = [1 - \exp(-\omega/T)]^{-1}\chi''(\mathbf{k}\omega)$ along the edge of the Brillouin zone [the direction $(\pi,\pi) - (0,\pi)$] are more intensive than those along the diagonal [the direction $(\pi,\pi) - (0,0)$]. As seen from Fig. 5, the calculations reproduce correctly the main known features of the incommensurate response except that the experimental δ is somewhat larger than the calculated one. In experiment with increasing T the incommensurability decreases and finally disappears. The calculations reproduce this peculiarity also. In the calculations the concentration dependence of δ is mainly determined by the spin gap parameter Δ in $\omega_{\mathbf{k}}$ in Eq. (9). For $x < 0.12$ it grows with x and saturates at larger concentrations [7].

With increasing ω the dip in the spin excitation damping is shallowed and finally disappears. Besides, on approaching $\omega_{\mathbf{Q}}$ the denominator in Eq. (10) will favor the appearance of the commensurate peak at $\mathbf{Q}$. Thus, the low-frequency incommensurate maxima of χ'' converge to the commensurate peak at $\omega \approx \omega_{\mathbf{Q}}$. The dispersion of the maxima in χ'' above $\omega_{\mathbf{Q}}$ is determined by the denominator in Eq. (10) and is close to that shown in Fig. 1. Consequently, the dispersion of the maxima in χ'' resembles two parabolas converging near the point $(\mathbf{Q}, \omega_{\mathbf{Q}})$. The upper parabola with branches pointed up reflects the dispersion of spin excitations, while the lower parabola with branches pointed down stems from the momentum dependence of the spin excitation damping. Such kind of the dispersion is indeed observed in cuprates [3, 23].

Generally the incommensurate magnetic response is not accompanied by an inhomogeneity of the carrier density. In works based on the stripe mechanism [23] the magnetic incommensurability is connected with the appearance of a charge density wave. Notice however, that the magnetic incommensurability is observed in all lanthanum cuprates, while the charge density wave is detected in neutron scattering in those cuprates which are in the low-temperature tetragonal or less-orthorhombic phases ($La_{2-x}Ba_xCuO_4$, $La_{2-y-x}Nd_ySr_xCuO_4$) [24]. In these phases the charge density wave is stabilized by the corrugated pattern of the in-plane lattice potential. It can be supposed that the magnetic incommensurability may be a precursor rather than a consequence of the charge density wave.

6. Concluding remarks

In this paper the projection operator technique was used for investigating the magnetic properties of the t-J model of cuprate high-T_c superconductors. It was demonstrated that the calculations reproduce correctly the frequency and momentum dependencies of the experimental magnetic susceptibility and its variation with doping and temperature in the normal and superconducting states in $YBa_2Cu_3O_{7-y}$ and lanthanum cuprates. This comparison with experiment allowed us to associate the resonance peak in $YBa_2Cu_3O_{7-y}$ with the magnon branch modified in the short-range antiferromagnetic order. The lack of the resonance peak in $La_{2-x}Sr_xCuO_4$ was connected with an increased damping of spin excitations at the antiferromagnetic wave vector $\mathbf{Q} = (\pi, \pi)$. One of the possible reasons for the overdamped excitations is an increased hole damping in this crystal. It was shown that for low frequencies the susceptibility is peaked at incommensurate momenta $(\pi \pm 2\pi\delta, \pi)$ and $(\pi, \pi \pm 2\pi\delta)$. The incommensurability is the consequence of the dip in the momentum dependence of the spin excitation damping at **Q**. The dip appears due to the Fermi surface nesting for low hole concentrations x and due to small hole-magnon interaction constants for moderate concentrations. In agreement with experiment for $x \lesssim 0.12$ the incommensurability grows nearly proportional to x and tends to saturation for $x > 0.12$. This is connected with the concentration dependence of the spin excitation frequency at **Q**. Also in agreement with experiment the incommensurability decreases with increasing temperature. The dispersion of the maxima in the susceptibility resembles two converging parabolas. The upper parabola with branches pointed up reflects the dispersion of spin excitations, while the lower parabola with branches pointed down stems from the momentum dependence of the spin excitation damping. In the considered mechanism the magnetic incommensurability is not accompanied by the inhomogeneity of the carrier density.

Acknowledgments

This work was supported by the ESF grant No. 5548.

References

[1] P. Bourges, in *The Gap Symmetry and Fluctuations in High Temperature Superconductors*, edited by J. Bok, G. Deutscher, D. Pavuna, and S. A. Wolf (Plenum Press, 1998), p. 349; H. He, Y. Sidis, P. Bourges, G. D. Gu, A. Ivanov, N. Koshizuka, B. Liang, C. T. Lin, L. P. Regnault, E. Schoenherr, and B. Keimer, Phys. Rev. Lett. **86**, 1610 (2001).

[2] G. Aeppli, T. E. Mason, S. M. Hayden, H. A. Mook, and J. Kulda, Science **279**, 1432 (1997).

[3] M. Arai, T. Nishijima, Y. Endoh, T. Egami, S. Tajima, K. Tamimoto, Y. Shiohara, M. Takahashi, A. Garrett, and S. M. Bennington, Phys. Rev. Lett. **83**, 608 (1999); D. Reznik, P. Bourges, L. Pintschovius, Y. Endoh, Y. Sidis, T. Matsui, and S. Tajima, cond-mat/0307591.

[4] F. C. Zhang and T. M. Rice, Phys. Rev. B **37**, 3759 (1988).

[5] H. Mori, Progr. Theor. Phys. **34**, 399 (1965); A. V. Sherman, J. Phys. A **20**, 569 (1987).

[6] J. H. Jefferson, H. Eskes, L. F. Feiner, Phys. Rev. B **45**, 7959 (1992); A. V. Sherman, Phys. Rev. B **47**, 11521 (1993).

[7] A. Sherman and M. Schreiber, Phys. Rev. B **65**, 134520 (2002); **68**, 094519 (2003); Eur. Phys. J. B **32**, 203 (2003).

[8] J. Kondo and K. Yamaji, Progr. Theor. Phys. **47**, 807 (1972); H. Shimahara and S. Takada, J. Phys. Soc. Jpn. **61**, 989 (1992); S. Winterfeldt and D. Ihle, Phys. Rev. B **58**, 9402 (1998).

[9] A. K. McMahan, J. F. Annett, and R. M. Martin, Phys. Rev. B **42**, 6268 (1990); V. A. Gavrichkov, S. G. Ovchinnikov, A. A. Borisov, and E. G. Goryachev, Zh. Eksp. Teor. Fiz. **118**, 422 (2000) [JETP (Russia) **91**, 369 (2000)].

[10] D. Z. Liu, Y. Zha, and K. Levin, Phys. Rev. Lett. **75**, 4130 (1995); N. Bulut and D. J. Scalapino, Phys. Rev. B **53**, 5149 (1996).

[11] J. Bonča, P. Prelovšek, and I. Sega, Europhys. Lett. **10**, 87 (1989).

[12] D. Forster, Hydrodynamic Fluctuations, Broken Symmetry, and Correlation Functions (W. A. Benjamin, Inc., London, 1975).

[13] J. L. Tallon, C. Bernhard, H. Shaked, R. L. Hitterman, and J. D. Jorgensen, Phys. Rev. B **51**, 12911 (1995).

[14] B. Keimer, N. Belk, R. G. Birgeneau, A. Cassanho, C. Y. Chen, M. Greven, M. A. Kastner, A. Aharony, Y. Endoh, R. W. Erwin and G. Shirane, Phys. Rev. B **46**, 14034 (1992).

[15] N. D. Mermin and H. Wagner, Phys. Rev. Lett. **17**, 1133 (1966); P. C. Hohenberg, Phys. Rev. **158**, 383 (1967).

[16] A. Sherman, cond-mat/0409379.

[17] D. K. Morr and D. Pines, Phys. Rev. Lett. **81**, 1086 (1998).

[18] A. Sherman and M. Schreiber, Phys. Rev. B **69**, 100505(R) (2004); A. Sherman, phys. status solidi (b) **241**, 2097 (2004).

[19] X. J. Zhou, T. Yoshida, D.-H. Lee, W. L. Yang, V. Brouet, F. Zhou, W. X. Ti, J. W. Xiong, Z. X. Zhao, T. Sasagawa, T. Kakeshita, H. Eisaki, S. Uchida, A. Fujimori, Z. Hussain, and Z.-X. Shen, cond-mat/0403181.

[20] K. Yamada, C. H. Lee, K. Kurahashi, J. Wada, S. Wakimoto, S. Ueki, H. Kimura, Y. Endoh, S. Hosoya, G. Shirane, R. J. Birgeneau, M. Greven, M. A. Kastner, and Y. J. Kim, Phys. Rev. B **57**, 6165 (1998).

[21] S. Wakimoto, G. Shirane, Y. Endoh, K. Hirota, S. Ueki, K. Yamada, R. J. Birgeneau, M. A. Kastner, Y. S. Lee, P. M. Gehring, S. H. Lee, Phys. Rev. B **60**, R769 (1999).

[22] H. Yoshizawa, S. Mitsuda, H. Kitazawa, and K. Katsumata, J. Phys. Soc. Jpn. **57**, 3686 (1988); R. J. Birgeneau, Y. Endoh, Y. Hidaka, K. Kakurai, M. A. Kastner, T. Murakami, G. Shirane, T. R. Thurston, and K. Yamada, Phys. Rev. B **39**, 2868 (1989).

[23] J. M. Tranquada, H. Woo, T. G. Perring, H. Goka, G. D. Gu, G. Xu, M. Fujita, and K. Yamada, Nature **429**, 534 (2004).

[24] M. Fujita, H. Goka, K. Yamada, and M. Matsuda, Phys. Rev. Lett. **88**, 167008 (2002).

QUANTUM CRITICAL PHENOMENA AND THE ANOMALOUS LOW-TEMPERATURE PROPERTIES OF THE U(PT,PD)$_3$ SYSTEM

J.J.M. Franse
Van der Waals – Zeeman Instituut Valckenierstraat 65, 1018 XE Amsterdam
jjmfranse@chello.nl

Abstract: Recently, quantum critical phenomena related to a pressure induced loss of magnetic order and at the same time the emergence of superconductivity have been reported in several Ce and U-based compounds. The paper discusses the possibility that similar phenomena are present as well in the heavy-fermion compound UPt_3 for which compound superconductivity and antiferromagnetism both have been observed at low temperatures.

Keywords: quantum critical phenomena, heavy-fermion compound, U(Pt,Pd)$_3$ system

1. Introduction

The hexagonal compound UPt_3 is known for almost twenty years for its anomalous low-temperature properties and for the co-existence of magnetic fluctuations and superconductivity. [1] In the superconducting state two different superconducting phases have been discovered in zero magnetic field. [2] A third phase emerges at applying an external magnetic field to the sample. [3, 4] This multiple superconducting phase diagram is thought to be due to a symmetry-breaking field for which the weak antiferromagnetic order with moments per uranium atom as small as 0.02 μ_B is found to be the most probable candidate. We shall review these basic properties of UPt_3 in the light of the quantum critical phenomena that have been claimed to be present in several Ce- and U-based compounds. Examples are the antiferromagnetically ordered compound $CePd_2Si_2$ [5] with a Néel temperature of 10 K that looses antiferromagnetic order at a pressure of 28 kbar. Above and below 28 kbar, superconductivity develops with a maximal ordering temperature of 0.3 K near 28 kbar. A second example is the ferromagnetic compound UGe_2 [6] with a Curie temperature of 52 K that looses its ferromagnetism at a pressure of 16 kbar whereas in the pressure

J. Franse et al. (eds.), Smart Materials for Ranging Systems, 129–151.

range between 10 kbar and 16 kbar superconductivity develops with a maximal value for T_c of 0.7 K that is reached for a pressure of 12.5 kbar. In both compounds the Cooper-pair bonding is thought to be mediated by strong magnetic fluctuations in the region where long range magnetic order disappears or becomes weaker, respectively. Many other compounds have recently been found to belong to the same category. [8]

Substitution experiments have been found to further elucidate the anomalous low-temperature properties of UPt_3. We shall review the characteristic properties of the Pd-substituted UPt_3 samples as well. It turns out that in the $U(Pt_{1-x}Pd_x)_3$ compounds for x-values between 0.0 and 0.10 two critical compositions can be distinguished. [9] In the first one near $x = 0.01$ where superconductivity is already lost, static magnetic order appears in the compounds with a maximum value for T_N of 5.8 K and a maximum value for the staggered moment per uranium atom of 0.6 μ_B for $x = 0.05$. [10] The second critical composition is found in the neighbourhood of x = 0.10. At that composition, the static magnetic order is lost. Resistivity measurements for these compounds produce $\rho(T)$ curves with the typical low-temperature upturn that is characteristic for Kondo behaviour. [11] It seems that near this composition the Kondo shielding overrules the RKKY interaction and that no magnetic order occurs. Exactly near that transition, heavy-fermion behaviour is expected, with, in the neighborhood of this critical composition, Non Fermi Liquid (NFL) properties. In the following paragraphs we shall discuss these anomalous properties with special interest for the question whether a quantum critical point does exist in this alloy system. Some of these unusual low-temperature properties have recently been reviewed. [12] In the present paper, we shall look to the compound from a different point of view.

2. Basic electronic properties of UPt_3

2.1 Low-temperature properties of heavy-fermion UPt_3

UPt_3 belongs to the class of uranium-based heavy-fermion compounds. That classification is due to the large value for γ^* in the expression for the specific heat at low temperature:

$$c = \gamma^* T + \beta T^3 \tag{1}$$

where γ^* and β represent the electronic and phonon contributions to the specific heat far below the Debye temperature, θ_D, respectively. In eq.(1), γ^* is the experimentally observed value for the electronic term and includes the enhanced mass of the electrons caused by electron-phonon interactions and by the strong interactions with magnetic fluctuations:

$$\gamma^* = \gamma_o\, m^* / m_0 \qquad (2)$$

with γ_o and m_o the electronic coefficient and the mass of the free electron, respectively. Apart from the effective electron mass, m^*, we speak as well of an effective Fermi velocity, v_F^*, and an effective Fermi temperature, T_F^*. The following relations hold:

$$m^* / m_o = v_F / v_F^* = T_F / T_F^* \qquad (3)$$

with in case of the cerium- and uranium-based heavy-fermion compounds, values for this ratio that can be as large as of the order of hundred.

Not only the specific heat coefficient, γ^*, takes such a large value, also the coefficient of the quadratic term in the temperature dependence of the resistivity is anomalously large:

$$\rho(T) = \rho_o + A^* T^2 \qquad (4)$$

Finally, the susceptibility is enhanced as well, but at low temperature its value is a constant. The following scaling holds:

$$\chi^* / \gamma^* \approx 1 \qquad (5)$$

and, as pointed out by Kadowaki and Woods [13]:

$$A^* / (\gamma^*)^2 = \text{constant} \qquad (6)$$

For the compound UPt_3, the value for γ^* amounts to 420 mJ/mol K^2, whereas the values for A^* and χ^* depend on the direction in which the resistivity and the susceptibility have been measured, although these results do not differ more than roughly a factor of two. To illustrate these findings, the specific heat of an artificial heavy-fermion compound is shown in fig.1 in a plot of c/T versus T^2 and compared with the result for, again, an artificial normal metal. For the effective Fermi velocity, v_F^*, one deduces values of order 5×10^3 m/sec and for the effective Fermi temperature values between 10 and 100 K.

The anomalous low-temperature behaviour is reflected in other thermodynamic properties as well. The thermal expansion, for instance, is unusually large at low temperature. At writing:

$$\alpha(T) = a^* T + b\, T^3 \qquad (7)$$

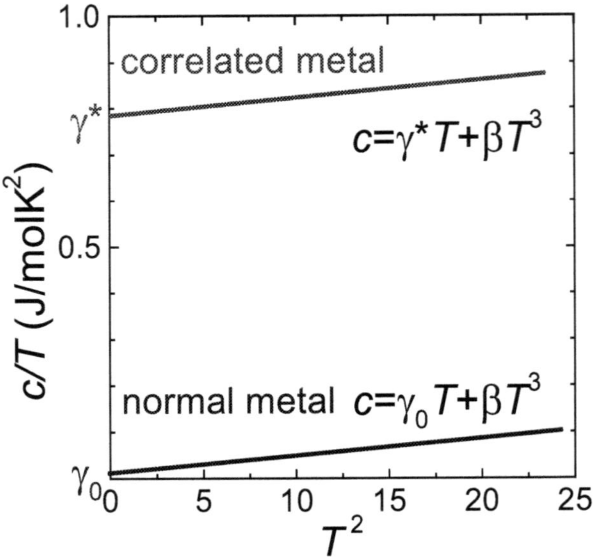

Figure 1. The effect of strong electron correlations on the low-temperature specific heat coefficient γ.

we have a similar expression as for the specific heat in eq.1. The parameters a and b in eq.7 reflect the same contributions as the parameters γ and β, respectively. The following relations exist:

$$a^* = \Gamma_e \gamma^* \quad (8)$$

$$b = \Gamma_{ph} \beta \quad (9)$$

with Γ_e and Γ_{ph} the Grüneisen parameter for the electron and phonon contributions to the thermal expansion, respectively.

Typical for heavy-fermion systems is the large value for Γ_e that can take values which are several orders of magnitude larger than what is found in normal metals. It implies that the effective mass has a large volume dependence as does the effective Fermi temperature.

We conclude this summary by pointing to the anomalies in the magnetic susceptibility as a function of temperature and, at low temperature, as a function of an applied magnetic field. In the low-field susceptibility of the hexagonal plane, we observe a maximum around $T_{max} \approx 15$ K, whereas the high-field susceptibility measured at 1.4 K develops a sharp peak at $B_{max} \approx$ 20 tesla. We note that the two phenomena are related, since:

$$k_B\, T_{max} \approx \mu_B\, B_{max} \tag{10}$$

2.2 Weak antiferromagnetism in UPt_3 below 6 K

Weak antiferromagnetism in UPt_3 has first been demonstrated in neutron diffraction experiments by Aeppli et al. [14-16] Antiferromagnetism shows up in the weak additional reflections below 6 K that extrapolate to very small staggered moments of the order of 0.02 μ_B per uranium atom near zero temperature. The intensities develop below 6 K and have a, for a magnetically ordered material, unusual quasi-linear temperature dependence. The linewidth of the Bragg peaks is not resolution limited and points to a coherence length of antiferromagnetic interactions of the order of 35 nm. It is interesting to note that the weak antiferromagnetism has not been detected

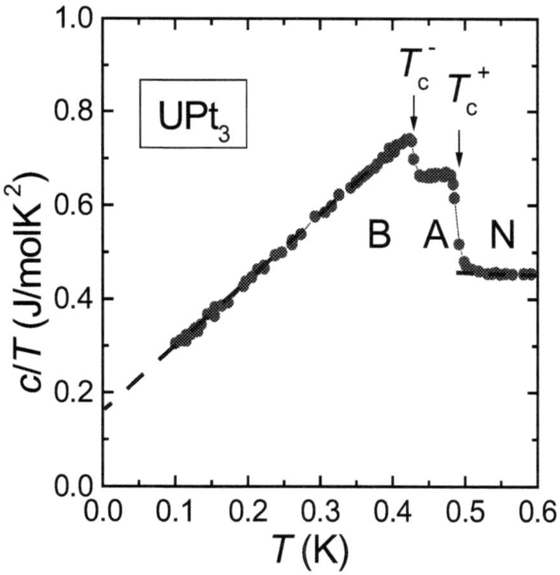

Figure 2. The transition to the superconducting state as observed in specific heat measurements for UPt_3; plotted is c/T vs T; two superconducting states can be distinguished at T_c^+ and T_c^- ; the superconducting phases are denoted by A and B; c/T vs T shows power-law behaviour below T_c^- instead of the usual exponential temperature dependence; moreover, at zero temperature, a finite intersection with the vertical axis is found; figure taken from ref. [22].

in μSR experiments by Dalmas de Réotier et al. [17] and also not by De Visser et al. [18], although Heffner et al. [19] did observe in earlier μSR studies a change in the depolarization rate at 6 K. This lack of signal in the subsequent μSR experiments has been ascribed to the differences in the time window of the μSR and neutron diffraction experiments. [20]

2.3 Superconductivity in UPt_3 below 0.5 K

At temperatures below 0.5 K, superconductivity has first been observed in UPt_3 by Stewart et al. [21] The precise value of T_c depends on sample quality as does the temperature width of the superconducting transition. Most remarkable is the observation that the jump in the specific heat at T_c reveals that heavy-mass electrons are involved in the formation of the superconducting state.

Moreover, the temperature dependence of the specific heat below T_c was found certainly not to be exponential. Rather a power-law behaviour is observed, pointing to line or point nodes in the superconducting gap. Accurate specific-heat measurements by Fisher et al. [2] indicate that in zero field there are two subsequent superconducting transitions which are considered as a sign for a multiple superconducting phase diagram. This phase diagram has been explored in specific-heat as well as thermal-expansion measurements in applied magnetic fields up to a few tesla. An example of the specific heat data in zero field is shown in fig.2 where a plot of c/T *vs* T is presented for $T < 0.6$ K. The double transition, as well as the linear temperature dependence of c/T below T_c^-, give a clear proof of the anomalous superconducting properties of this compound. The multiple superconducting phase diagram has been discussed in terms of a Ginzburg-Landau description. [23-25] Essential for the double transition in zero field is the presence of a field that breaks the hexagonal symmetry. Lattice distortion as well as the presence of weak magnetic ordering could act as the symmetry-breaking field. At entering the superconducting state, the Bragg peak intensity as a function of temperature reveals a change in slope signifying the mutual interaction between superconductivity and magnetism.[26] The relation between symmetry breaking field and double superconducting transition has further been explored in neutron diffraction and specific heat studies under applied pressure up to 5 kbar. [27,28] In these experiments it is found that the weak antiferromagnetic reflections pointing to magnetic moments of the order of 0.02 μ_B per uranium atom, vanish at pressures above 4.5 kbar. At the same pressure the double superconducting transition disappears. Further support for the intimate relation between the double superconducting transition and the weak antiferromagnetism originates from a plot of $m^2(T_c)$ versus ΔT_c that reveals a linear relationship.

The superconducting phase diagram with three different superconducting phases in different field-temperature regimes is, according to Huxley et al. [28], characterized by three different types of symmetry of the superconducting gap function.

Finally, NMR experiments by Tou et al. [29] show that there is no change in the Knight shift at passing the superconducting transition temperature. These observations are in a first approach considered as arguments in favor of odd-parity superconductivity in UPt_3.

2.4 Is there a Quantum Critical Point in UPt_3?

The co-existence and coupling between the weak antiferromagnetic order and the superconductivity in UPt_3 may hint to a Quantum Critical Point (QCP) as observed in several other heavy-fermion compounds like, for instance, $CePd_2Si_2$ [6] and UGe_2 [7]. The generic phase diagram for quantum phase transitions in strongly correlated metals is shown in fig. 3. As a function of the control parameter, δ, the compound can be either in the

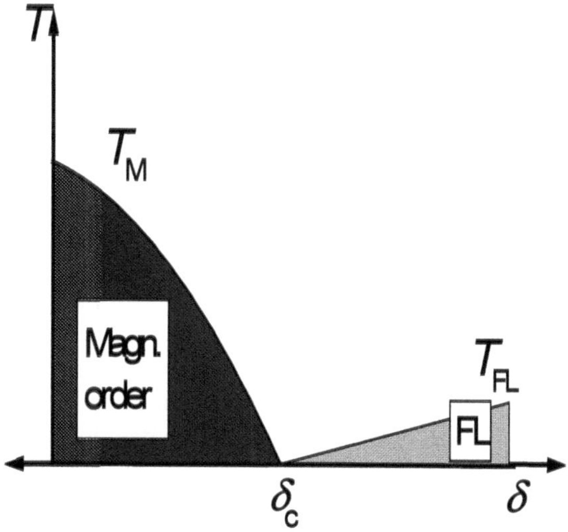

Figure 3. Generic phase diagram in the neighbourhood of a Quantum Critical Point. As a function of increasing values of the control parameter δ, antiferromagnetism is suppressefd bringing the system towards a region near δ_c where superconductivity may occur; at δ-values well above δ_c , a Fermi-liquid description holds and heavy-fermion behaviour is to be expected; at decreasing δ-values, the Fermi temperature takes extremely small values causing the Fermi-liquid description to be no longer applicable; the system is brought in the Non-Fermi-Liquid (NFL) regime.

magnetically ordered state or behave as a Fermi liquid at low enough temperatures. Near the critical value, δ_c, the Fermi liquid temperature approaches zero temperature with the consequence that the effective mass, according to eq.(3), grows beyond limit. In that neighbourhood, the electronic properties do not follow any more a Fermi-liquid description.

The experiments for several compounds reveal the appearance of superconductivity exactly in this region of non-Fermi-liquid (NFL) behavior. The control parameter for the above mentioned compounds $CePd_2Si_2$ and UGe_2 turns out to be pressure. Since we are dealing in the heavy-fermion compound UPt_3 with both antiferromagnetic order and superconductivity and since both phenomena are sensitive to applied pressures, it is interesting to investigate whether the application of pressure is leading to quantum criticality in the region where both phenomena compete. It turns out, however, that, although antiferromagnetism becomes weaker under pressure, there is no sign that the antiferromagnetic ordering temperature decreases to zero with increasing pressure.[30] And the superconducting transition temperature as well reveals a weak pressure dependence.[31] It is the splitting in temperature between the two superconducting transitions that is strongly affected by pressure just as the amplitude of the antiferromagnetic Bragg reflections. The conclusion must be that pressure is not a good choice for the above introduced control parameter that brings the system towards a quantum critical point. In the next section it will be shown that varying the Pd concentration in the $U(Pt,Pd)_3$ system brings us closer to this point.

3. Antiferromagnetism and superconductivity in heavy-fermion $U(Pt,Pd)_3$

One of the first reports on the development of long-range antiferromagnetic order and the disappearance of superconductivity with increasing Pd concentration in the $U(Pt_{1-x}Pd_x)_3$ system already dates from more than fifteen years ago.[32] It has been found that for Pd concentrations with $0.01 < x < 0.1$, long-range magnetic order develops with values for the Néel temperature and for the magnetic moment per uranium atom increasing with increasing Pd content, reaching maximal values for T_N of 5.8 K and for the magnetic moment per uranium atom of 0.6 μ_B for the $x = 0.05$ compound. At further increasing the Pd content, the magnetic ordering temperature decreases and becomes zero around $x = 0.10$. At the same time superconductivity was found to disappear for x-values around 0.005. Other characteristics of the $U(Pt,Pd)_3$ system are the following:

- In the concentration range where superconductivity is found, it turns out that the splitting between the two superconducting transitions,

ΔT_c (= T_c^+ - T_c^-), increases with increasing Pd content as does the weak antiferromagnetic moment;

- All $U(Pt_{1-x}Pd_x)_3$ compounds keep large effective electron masses for Pd concentrations at least up to $x = 0.10$;
- The resistivity *vs* temperature curves of the $U(Pt_{1-x}Pd_x)_3$ system develop from a "normal" quadratic temperature dependence in pure UPt_3 to a Kondo-type of temperature dependence for $x \geq 0.10$;
- At $x = 0.006$, Non-Fermi-Liquid behaviour is found in the temperature dependence of the resistivity.

These observations justify the conjecture that there is a better chance to find a quantum critical point in the $U(Pt,Pd)_3$ system with the Pd concentration as the control parameter. In order to verify these suggestions, accurate neutron diffraction and μSR experiments have been performed in the composition range close to $x = 0.006$ where superconductivity disappears and long-range magnetic order starts to develop. We subdivide the discussion in compounds with $0 \leq x \leq 0.005$ and with $0.006 \leq x \leq 0.01$.

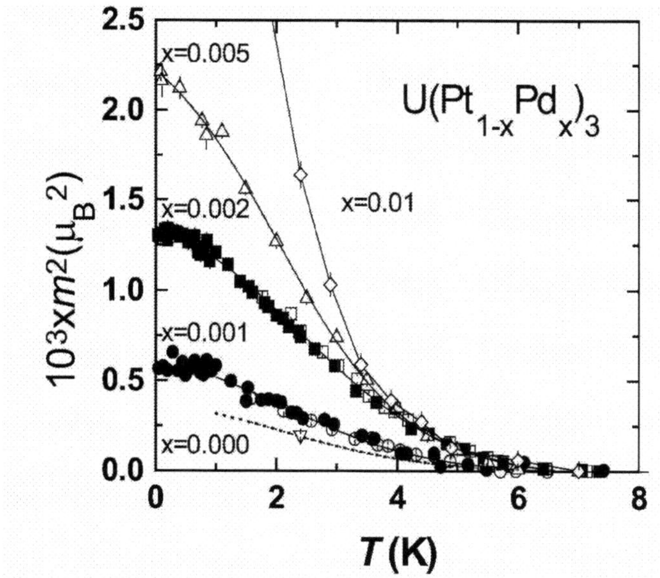

Figure 4. Temperature variation of m^2 derived from the intensity of the magnetic Bragg peak at Q = (1/2, 1. 0) for annealed (open symbols) and unannealed (closed symbols) $U(Pt_{1-x}Pd_x)_3$ samples with x-values up to 0.01; the solid lines are guides to the eye; figure taken from ref.[20].

3.1 Weak antiferromagnetism in $U(Pt,Pd)_3$ for $0.000 \leq x \leq 0.005$

In fig.4, the magnetic intensities are presented for Pd concentrations with $x \leq 0.005$ [32].

We note that in all cases, the intensities start to develop below 6 K. They do not have the characteristic temperature dependence one expects for a compound that shows long-range magnetic order below a magnetic transition temperature. In contrast, the intensities slowly start below 6 K reaching values for $m^2(0)$ between 0.4 and 2.2 x10^{-3} μ_B^2. We already indicated that the increase of $m^2(0)$ goes parallel with an increase of ΔT_c, the splitting in temperature between the two superconducting transitions.

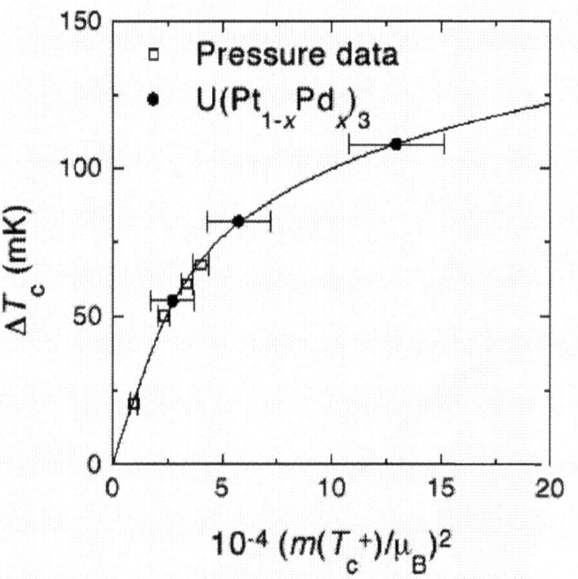

Figure 5. A plot of ΔT_c (= $T_c^+ - T_c^-$) vs $m^2(T_c^+)$ with data obtained from high-pressure studies on pure UPt_3 (specific heat and neutron diffraction, open symbols) and data for a series of $U(Pt,Pd)_3$ compounds with Pd concentrations up to 0.5 at% (closed symbols); figure from ref.[20].

This point is worked out in fig. 5, where the values for ΔT_c and $m^2(0)$ are plotted for different applied pressure to UPt_3 and for different x-values for the $U(Pt_{1-x}Pd_x)_3$ compounds.[34] The results from both approaches nicely coincide on a single curve that starts linearly from zero but that deviates from linear behaviour at higher values for ΔT_c and $m^2(0)$. The linear relationship follows from a Ginzburg-Landau approach in which terms up to the fourth power in the magnetization are included. Deviations at higher values urge to include higher terms in this expansion.

Anyhow, the assumption that the fluctuating antiferromagnetism acts as the symmetry breaking field giving rise to a multiple superconducting phase diagram, is supported by the experimental results for the $U(Pt_{1-x}Pd_x)_3$ compounds with x-values below 0.005. The small-moment weak antiferromagnetism, however, is again not the dominant factor in mediating superconductivity in the $U(Pt,Pd)_3$ system. The weak antiferromagnetism becomes stronger with increasing Pd-content whereas superconductivity falls off. The characteristic temperature for the weak antiferromagnetism does not change with Pd-content whereas that for superconductivity certainly does. In this context, we once again remark that the Knight shift measurements point to odd-parity superconductivity. Odd-parity superconductivity is not likely to be mediated by antiferromagnetic fluctuations. In that sense, the different results and conclusions are consistent.

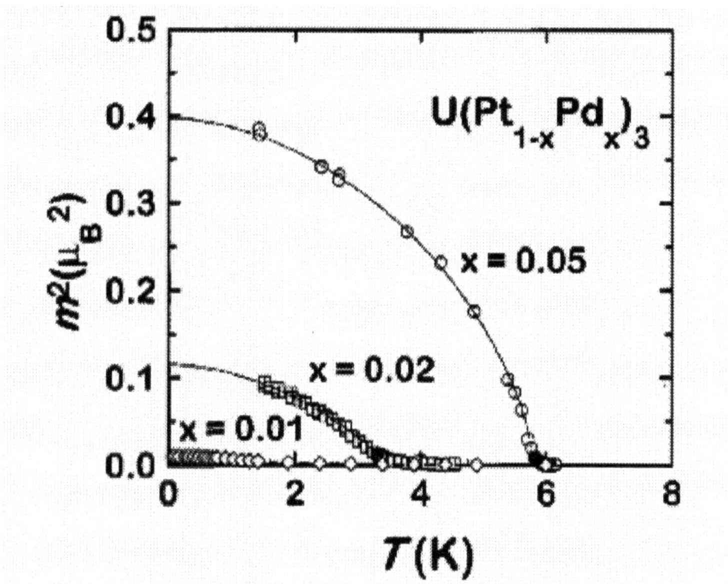

Figure 6. Temperature variation of m^2 for annealed $U(Pt_{1-x}Pd_x)_3$ samples with x-values of 0.01, 0.02 and 0.05; the solid lines represent fits to $(1 - (T/T_N)^{\alpha})^{2\beta}$ with $\alpha =$ and $\beta =$; figure from ref.[20].

3.2 Long-range antiferromagnetic order for x ≥ 0.006

Thermodynamic (specific-heat) and transport (resisitivity) studies for the $U(Pt_{1-x}Pd_x)_3$ compounds have revealed magnetic transitions for x-values of 0.02 and higher, in contrast to the compounds below $x = 0.01$ that did not show anomalies in specific heat and resistivity that point to magnetic order. The sharpest transition in the specific heat studies is found for $x = 0.05$. For this compound, the resistivity *vs* temperature curve exhibits a Cr-type of anomaly just below the magnetic ordering temperature suggesting the presence of a spin-density wave below T_N. A few results from neutron diffraction studies are shown in fig. 6.

In contrast to the results for the compounds with $x \leq 0.005$, the temperature dependence of the square of the deduced magnetic moment follows the expected temperature dependence for a magnetically ordered system below the magnetic ordering temperature rather well. The Néel temperature itself decreases with decreasing x-values, in contrast to the "apparent" ordering temperature of the compounds with x-values smaller than 0.005 that stays at 6 K. For x-values above 0.05, the Néel temperature decreases approaching zero temperature for $x = 0.1$. These results, together with the observations for the compounds with x-values below 0.005, are summarised in fig. 7 where SMAF stands for small moment antiferromagnetism, LMAF for large moment antiferromagnetic order and SC for superconductivity.

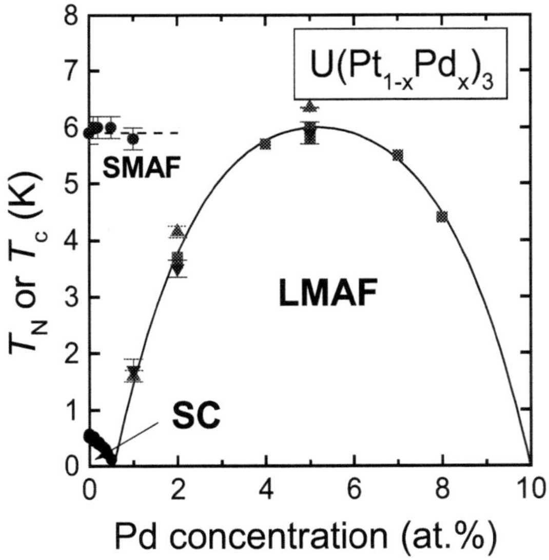

Figure 7. Low-temperature phase diagram of the $U(Pt_{1-x}Pd_x)_3$ system; SC denotes superconductivity, SMAF small-moment antiferromagnetism and LMAF long-range antiferromagnetic order; data from ref.[42].

Summarizing the evolution of superconductivity and antiferromagnetism as function of the Pd-concentration in $U(Pt_{1-x}Pd_x)_3$, we arrive at the following remarkable points:

- Superconductivity is suppressed for $x \geq 0.006$;
- The small-moment antiferromagnetism below 6 K becomes stronger with increasing Pd content but does not result in a shift in the temperature below which the antiferromagnetism appears;
- Large-moment long-range antiferromagnetic order develops for x-values above 0.006 with maximum values for the magnetic moment (0.6 μ_B) and the magnetic ordering temperature (5.8 K) at $x = 0.05$;
- At further increase of x, the long-range antiferromagnetic order is lost and a Kondo-type of behaviour is found in the resistivity curves of compounds with $x \geq 0.10$.

3.3 The equivalence of applied pressures to UPt_3 and chemical pressure in $U(Pt,Pd)_3$

Since Pt and Pd, in first approximation, can be considered to be iso-electronic in the compound $U(Pt,Pd)_3$, a question may arise concerning the

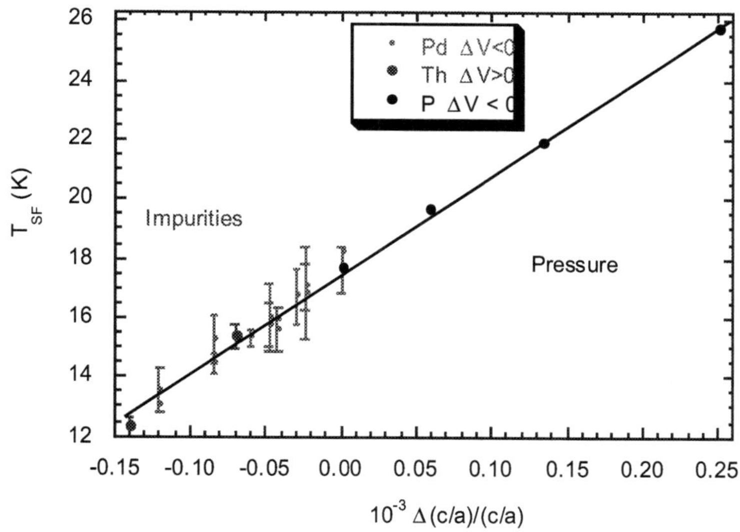

Figure 8. The spin-fluctuation temperature, T_{SF}, of UPt_3 as a function of the pressure-induced relative change in the c/a ratio; a comparison is made with the shift in T_{SF} by the relative change in the c/a ratio induced by substituting Pt by Pd and Th in $U(Pt,Pd)_3$, respectively; data from ref.[35].

mechanism that drives the system from superconductivity to magnetic order upon increasing the Pd concentration. Substitution of Pt by Pd leads to a contraction of the unit cell volume as pressure does. However, the *c/a* ratio increases with increasing Pd content whereas this ratio decreases upon applying hydrostatic pressure. This may be the clue at comparing the effects of pressure and of Pd substitution on the electronic properties of UPt_3. In fig. 8, values for the characteristic (heavy-fermion) temperature, T_{SF}, as determined from either high pressure experiments on pure UPt_3 or from substituted $U(Pt,Pd)_3$ and $(U,Th)Pt_3$ compounds, are plotted as a function of the *c/a* ratio.[34] Whereas the molar volume decreases under pressure and at substituting Pt by Pd, the volume increases with Th substitutions for U.

Nevertheless, the whole set of experimental data falls on one straight line at plotting T_{SF} versus the relative change of *c/a*. It is evident that it is the *c/a* ratio that governs the characteristic energy scales, in accordance with observations of Battlogg et al. for different types of substituting atoms in UPt_3.[36]

3.4 Disappearance of antiferromagnetic order in $U(Pt_{1-x}Pd_x)_3$ for x ≈ 0.1 and the Doniach diagram

The disappearance of antiferromagnetic order in $U(Pt_{1-x}Pd_x)_3$ for $x \approx 0.1$ coincides with the appearance of Kondo behaviour in the resistivity. This is illustrated in fig. 9, where for the concentration range $0.00 \leq x \leq 0.15$, resistivity curves are displayed in the temperature range below 50 K.[11] For x=0, i.e. pure UPt_3, the data in the lowest temperature interval can be represented by:

$$\rho(T) = \rho_o + A\,T^n \qquad (11)$$

with a value for the exponent n of 2. The exponent n in eq.(11) decreases with increasing x-values; for $x = 0.01$ the expression becomes:

$$\rho(T) = \rho_o + A\,T^{1.6} \qquad (12)$$

In the resistivity curve for $x = 0.05$, the Cr-type of anomaly, pointing to the possible presence of a spin density wave below 5.8. K, is clearly visible. At higher x-values, the resistivity increases towards lower temperatures, with, for the $x = 0.15$ compound, a maximum around 5 K which is, with reference to other heavy-fermion compounds like, for instance, UBe_{13}, considered to be caused by coherence effects.

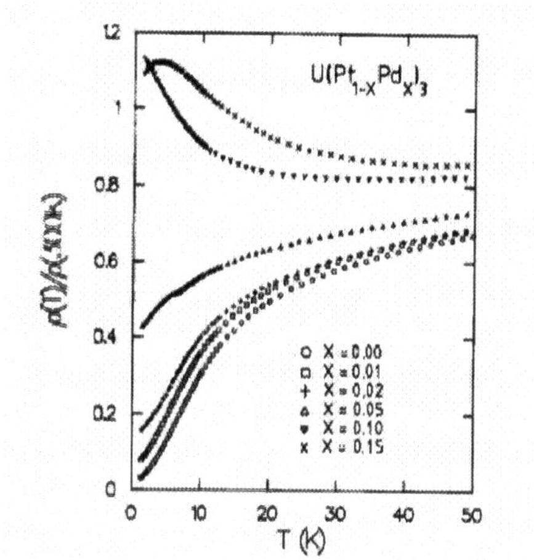

Figure 9. Resistivity of $U(Pt_{1-x}Pd_x)_3$ polycrystalline samples as a function of temperature; the data have been normalised to 1 at 300 K; note the change from Fermi-liquid to Kondo behaviour for x-values between 0.00 and 0.15; figure from ref.[11].

For the $x = 0.10$ compound, the temperature dependence can be approximated in a certain temperature interval by a logarithmic function, a trademark of Kondo behaviour. This concentration dependence of the different phenomena reminds to the Doniach diagram [38] in which, for dense Kondo systems, the competing interactions are characterised as the on-site Kondo effect reducing the magnetic moment by the antiparallel polarisation of conduction electrons in the direct neighbourhood of the magnetic ion and the intersite RKKY interactions resulting in magnetic order. The energies related to these two phenomena depend in an exponential (Kondo) and quadratic (RKKY) way on the exchange-interaction parameter, J,:

$$E_K \sim e^{-1/|J|} \tag{13}$$

and

$$E_{RKKY} \sim J^2 \tag{14}$$

respectively. The two curves for E_K and E_{RKKY} as a function of J intersect at a critical value, J_c. Around the critical value, a heavy-fermion state is expected to exist with for $J < J_c$ magnetic order and for $J > J_c$ Fermi liquid behaviour. At J_c, quantum critical phenomena are expected to be found.

Unfortunately, $U(Pt_{1-x}Pd_x)_3$ compounds with x-values around 0.10 have not been studied so far in great detail. This part of the phase diagram is open for further scrutiny.

3.5 Competition between superconductivity and antiferromagnetic order in $U(Pt_{1-x}Pd_x)_3$ around x = 0.006

Let us return to the other part of the phase diagram of the $U(Pt_{1-x}Pd_x)_3$ system in the concentration range where superconductivity disappears and long-range antiferromagnetic order appears. Such a competition between superconductivity and antiferromagnetic order is not necessarily the consequence of the existence of a quantum critical point. The reasoning behind the occurrence of superconductivity in the Non-Fermi-Liquid regime

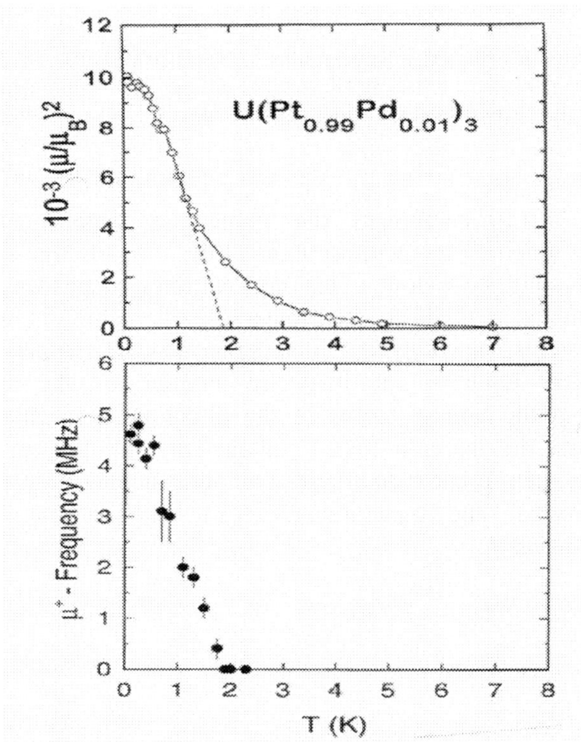

Figure 10. Antiferromagnetism in $U(Pt_{0.99}Pd_{0.01})_3$ below 6 K as studied in neutron diffraction (upper figure) and μSR experiments (lower figure); the long-range antiferromagnetic order below 2 K is visible in both experiments; the weak antiferromagnetism that does not show up in the μSR data gradually weakens with increasing temperature and disappears above 6 K; data from ref.[20].

of fig. 4 is that magnetic order is weakened by the increasing magnetic fluctuations and that at the same time Cooper-pair formation is promoted by the easy polarisation of electron spins in the sea of antiferromagnetic fluctuations. Superconductivity already appears in the generic phase diagram of fig. 3 before the magnetic ordering temperature reaches its extrapolated zero-temperature value. Do the same systematics hold for the occurrence of superconductivity and magnetic order in the $U(Pt_{1-x}Pd_x)_3$ system? To answer this question, detailed neutron diffraction and muon spin rotation experiments have been carried out just above the critical Pd-concentration at $x = 0.006$, whereas superconductivity has been studied in great detail just below this critical composition.

To illustrate the information that can be distracted from the neutron diffraction and muon spin rotation experiments, experimental data from both techniques are combined in fig. 10 for the compound $U(Pt_{0.99}Pd_{0.01})_3$. The lowest temperature data of the neutron diffraction results [20] extrapolate to

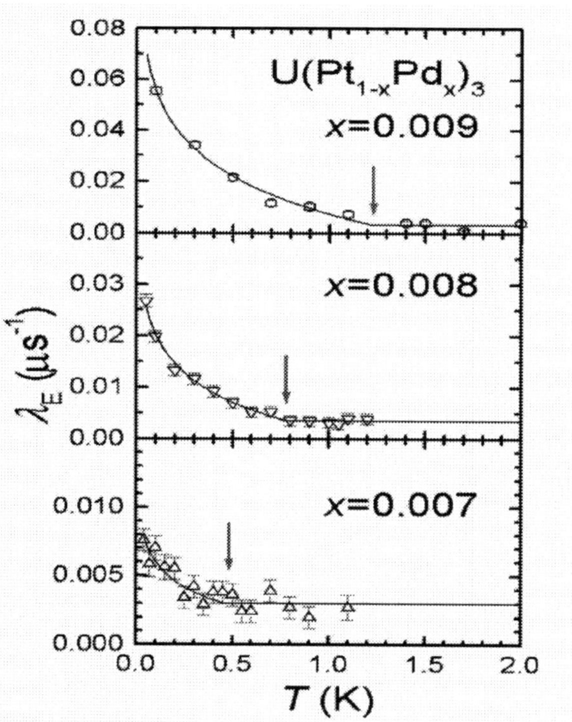

Figure 11. The long-range antiferromagnetic and the superconducting ordering temperatures as a function of Pd concentration in the $U(Pt_{1-x}Pd_x)_3$ system; data from ref.[43].

a magnetic ordering temperature of about 1.8 K, whereas above this temperature the intensity of the Bragg reflections displays a different temperature dependence with a characteristic temperature of 6 K. In the lower part of fig. 11, the muon spin rotation data confirm the presence of a magnetic ordering temperature of 1.8 K.[39] Above this temperature, however, the muon frequency is zero and no static magnetic order is indicated. The conclusion must be that the weak antiferromagnetism above 1.8 K is due to fluctuating magnetic moments which are invisible on timescales in which the muon spin rotation data are collected. Apparently, neutron experiments are fast enough to detect these magnetic correlations. The small-moment fluctuating antiferromagnetic state turns out to be invisible for slow techniques like muon spin rotation (10^{-4} – 10^{-11}) s^{-1} or NMR (10^{-7}) s^{-1} but neutrons (10^{-9} – 10^{-14}) s^{-1} are able to detect it. Similar conclusions have been reached in the study of stripe phase ordering in the perovskite $La_{1.475}Nd_{0.4}Sr_{0.125}CuO_4$ by Nachumi et al.[41] The muon depolarization function has been measured by De Visser et al. for Pd concentrations approaching the critical concentration, see fig. 11. [42]

At x = 0.007, there is still a magnetic signal pointing to a magnetic ordering temperature of 0.45 K. Extrapolation of the magnetic ordering

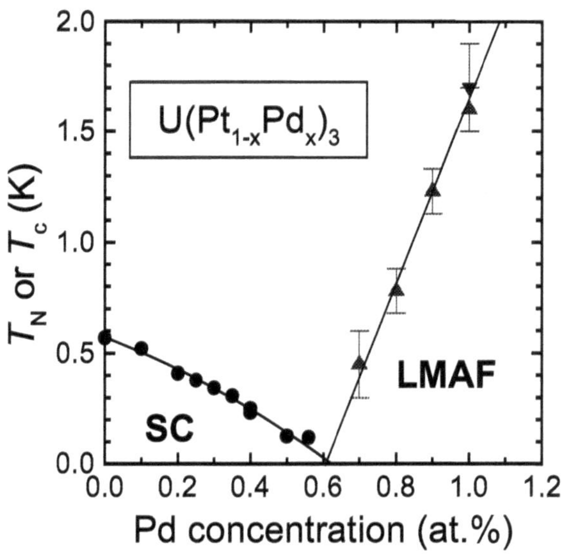

Figure 12. The long-range antiferromagnetic and the superconducting ordering temperatures as a function of Pd concentration in the $U(Pt_{1-x}Pd_x)_3$ system; data from ref.[43].

temperatures as a function of composition leads to a critical composition for antiferromagnetic order of $x_c^{AF} \approx 0.006$, see fig. 12. The same figure shows that at the superconducting side of the phase diagram, an extrapolation of T_c to zero temperature leads to a critical composition for superconductivity of $x_c^{SC} \approx 0.006$, almost the same value as just mentioned for x_c^{AF}. From these results we have to conclude that superconductivity and long-range (static) antiferromagnetic order do not coexist and that the magnetic and superconducting critical concentrations coincide. In other words: where static antiferromagnetic order becomes weak and antiferromagnetic fluctuations are strong, superconductivity is absent.

In the previous section, we stressed the importance of the *c/a* ratio in describing pressure and substitution effects on the characteristic heavy-fermion temperature of UPt_3. As far as superconductivity is concerned, both pressure and Pd substitution induce a decrease in the superconducting transition temperature. Still it is useful to plot also in this case T_c *vs* the relative change in the *c/a* ratio.

In fig. 13 this is realised by expressing the relative change in *c/a* in terms of hydrostatic pressure.[43] The reduction of long-range antiferromagnetic order is also shown in this figure. It illustrates again that near $T_N = 0$ K, superconductivity is suppressed. There is a clear change in slope in the T_c *vs* pressure curve at zero pressure. In fact, on the basis of the data at negative pressures (Pd substitution for Pt), one would expect a further increase of T_c at positive pressures. The other way around, on the basis of the high pressure

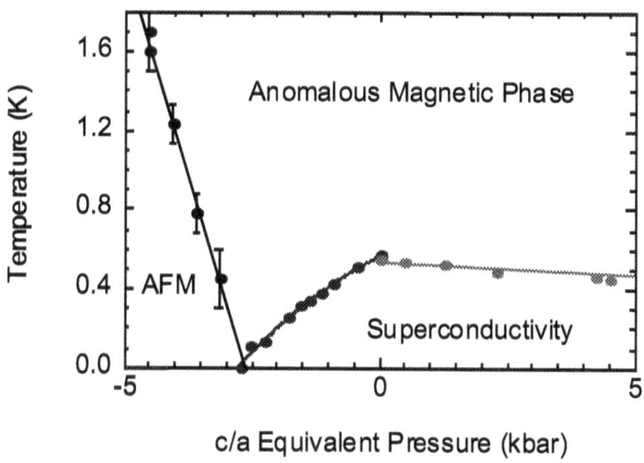

Figure 13. The long-range antiferromagnetic and the superconducting ordering temperatures as a function of the pressure that is equivalent to a certain relative change in the c/a ratio; data have been obtained from direct pressure experiments on UPt_3 and from Pd substitution experiments on the $U(Pt_{1-x}Pd_x)_3$ system.

data for UPt_3, one expects T_c to grow with increasing Pd content. There is, however, starting from zero pressure an abrupt change in slope in the "T_c *vs* equivalent pressure" curve that might be due to impurity scattering on Pd atoms or to an increase of antiferromagnetic fluctuations that become stronger at higher Pd concentrations and that, instead of mediating the pair bonding, weaken the superconductivity. In these pictures, it is not easy to understand that the critical concentrations for long-range antiferromagnetic order and superconductivity coincide. So the question remains what type of critical point do we find at a Pd concentration of 0.6 at.% where both long-range antiferromagnetic order and superconductivity have their critical concentration?

This question has been addressed in careful resistivity measurements from which the exponent of the temperature dependence of the resistivity has been deduced, according to the expression:

$$\rho(T) = \rho_o + A\,T^{\alpha} \tag{15}$$

Values for the exponent α are presented in fig. 14.

We like to mention that at a quantum critical point where antiferromagnetic order is on the verge to disappear, the value of α is expected to be 3/2, whereas a value of 5/3 hints to the presence of ferromagnetic fluctuations. The actual experimental value for α for $x \geq 0.006$ is equal to 1.6, i.e. in between the two above given values. Specific-heat

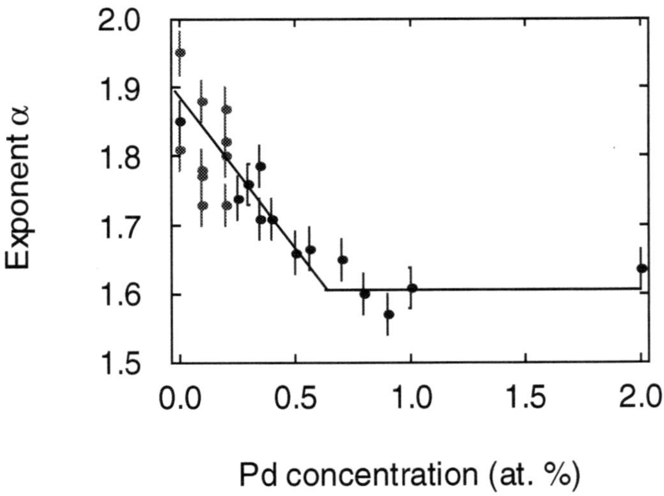

Figure 14. Values for the exponent α in the description of the temperature dependence of the resistivity of $U(Pt_{1-x}Pd_x)_3$ compounds by the relation $\rho(T) = \rho_o + A\,T^{\alpha}$ for x-values between 0.00 and 0.02; data from ref.[40].

studies and susceptibility measurements could help to further unravel the remarkable physics of the $U(Pt_{1-x}Pd_x)_3$ system for $x \approx 0.006$.

4. Concluding remarks

As is the case with the several high-temperature superconductor systems and with some of the heavy-fermion compounds, the superconducting phase in $U(Pt,Pd)_3$ is located close to the region where long-range antiferromagnetic order is observed. Still we meet different scenarios in the different classes of compounds. In $La_{2-x}Sr_xCuO_4$, the antiferromagnetic and superconducting phases are well separated as function of composition: AF order disappears near $x = 0.02$ [44], whereas superconductivity starts near $x = 0.06$. In $YBa_2Cu_3O_{6+\delta}$, it is the oxygen concentration that leads to the different phases. For $\delta \leq 0.4$, AF-order is destroyed with increasing δ [45], whereas for $\delta \geq 0.4$, superconductivity appears up to $x = 1$ with two different regimes: $0.4 \geq \delta \geq 0.75$ ($T_c \approx 60$ K) and $0.75 \geq \delta \geq 1$ ($T_c \approx 90$ K). So, AF-order and superconductivity meet at $\delta = 0.4$. The situation is somewhat different in $CePd_2Si_2$, where pressure drives the system from magnetic order towards superconductivity and where superconductivity and antiferromagnetic order seem to coexist in a narrow pressure-temperature region. Anyhow, the superconducting phase is always neighbouring the antiferromagnetic one in these compounds, suggesting that antiferromagnetic correlations play an essential role in the formation of the superconducting state. The compound UGe_2, however, reveals a different picture. Here, we find ferromagnetic order below 53 K with a large magnetic anisotropy keeping the uranium magnetic moments along the shortest (c-) axis of the orthorhombic unit cell, the direction of the shortest distance between neighbouring uranium atoms. Ferromagnetic order is suppressed at 1.6 GPa, superconductivity is found in the pressure interval between 1 and 1.6 GPa. In this compound, the two low-temperature ordering phenomena coincide. This fact may suggest that in this compound ferromagnetic rather than antiferromagnetic correlations induce superconductivity. Remarkable enough, in UPt_3 both antiferromagnetic and ferromagnetic correlations have been reported. At doping UPt_3 with Pd, the ferromagnetic fluctuations are depressed by the increasing tendency for antiferromagnetic order. For the 0.6 at% Pd compound, the composition where long-range antiferromagnetic order starts, these ferromagnetic correlations are most likely completely suppressed. In such a scenario, it is attractive to suggest, as it was put forwards by De Visser [34], that in the case of $U(Pt,Pd)_3$, the odd-parity superconductivity is induced by ferromagnetic fluctuations. In order to justify such a picture, the spin-fluctuation spectrum has to be investigated in detail as a function of Pd concentration in the interval $0 \leq$ at% Pd ≤ 0.6.

Acknowledgments

The author has profited very much from discussions with A. de Visser and from presentations by M.J. Graf and C.P. Opeil.

References

[1] G.R. Stewart, Rev. Mod. Phys. 56 (1984) 755.
[2] R.A. Fisher, S. Kim, B.F. Woodfield, N.E. Philips, L. Taillefer, K. Hasselbach, J. Flouquet, A.L. Giorgi, J.L. Smith, Phys. Rev. Lett. 62 (1989) 1411.
[3] K. Hasselbach, L. Taillefer and J. Flouquet, Phys. Rev. Lett. 63 (1989) 93.
[4] N.H. van Dijk, A. de Visser, J.J.M. Franse, S. Holtmeier, L. Taillefer and J. Flouquet, Physica B 186-188 (1993) 267.
[5] N.H. van Dijk, A. de Visser and J.J.M. Franse, J. Low Temp. Phys. 93 (1993) 101.
[6] N.D. Mathur, F.M. Grosche, S.R. Julian, I.R. Walker, D.M. Freye, R.K.W. Haselwimmer and G.G. Lonzarich, Nature 394 (1998) 39.
[7] S.S. Saxena, P. Agarwal, K. Ahilan, F.M. Grosche, R.K.W. Haselwimmer, M.J. Steiner, E. Pugh, I. Walker, S.R. Julian, P. Monthoux, G.G. Lonzarich, A. Huxley, I. Sheikin, D. Braithwaite and J. Flouquet, Nature 406 (2000) 587.
[8] S.R. Julian, International Conference on Strongly Correlated Electron Systems, Program and Abstracts (2004) 1.
[9] A. de Visser, A. Menovsky and J.J.M. Franse, Physica B 147 (1987) 81.
[10] P.H. Frings, B. Renker and C. Vettier, J. Magn. Magn. Mat. 63-64 (1987) 202.
[11] R. Verhoef, A. de Visser, A. Menovsky, A.J. Riemersma and J.J.M. Franse, Physica B 142 (1986) 11.
[12] J.J.M. Franse, Acta Physica Polonica A 100 (2001) 153.
[13] K. Kadowaki and S.B. Woods, Solid State Commun. 58 (1986) 507.
[14] G. Aeppli, E. Bucher, G. Shirane, Phys. Rev. B 32 (1985) 7579.
[15] G. Aeppli, A. Goldman, G. Shirane, E. Bucher, M.-Ch. Lux-Steiner, Phys. Rev. Lett. 58 (1987) 808.
[16] G. Aeppli, E. Bucher, C. Broholm, J.K. Kjems, J. Baumann, J. Hufnagl, Phys. Rev. Lett. 60 (1988) 615.
[17] P. Dalmas de Réotier, A. Huxley, A. Yaouanc, J. Flouquet, P. Bonville, P. Imbert, P. Pari, P.C.M. Gubbens and A. M. Mulders, Phys. Lett. A 205 (1995) 239.
[18] A. de Visser, R.J. Keizer, R. van Harrevelt, A.A. Menovsky, J.J.M. Franse, A. Amato, F.N. Gygax, M. Pinkpank and A. Schenck, Physica B 230-232 (1997) 53.
[19] R. Heffner, D.W. Cooke, A.L. Giorgi, R.L. Hutson, M.E. Schillaci, H.D. Rempp, J.L. Smith, J.O. Willis, D.E. McLaughlin, C. Boekema, R.L. Lichti, J. Oostens, A.B. Denison, Phys. Rev. B 39 (1989) 11345.
[20] R.J. Keizer, A. de Visser, A.A. Menovsky, J.J.M. Franse, B. Fåk and J.-M. Mignot, Phys. Rev. B 60 (1999) 6668.
[21] G.R. Stewart, Z. Fisk, J.O. Willis, J.L. Smith, Phys. Rev. Lett. 52 (1984) 679.
[22] T. Vorenkamp et al. Phys. Rev. B (1993)
[23] R. Joynt, Supercond. Sci Technol. 3 (1988) 210.
[24] K. Machida, M. Ozaki, T. Ohmi, J. Phys. Soc. Jpn. 58 (1989) 4116.
[25] D.W. Hess, T. Tokuyasu, J.A. Sauls, J. Phys. : Cond. Matt. 1 (1989) 8135.
[26] G. Aeppli, D. Bishop, C. Broholm, E. Bucher, K. Siemensmeyer, M. Steiner and N. Stüsser, Phys. Rev. Lett. 63 (1989) 676.
[27] T. Trappmann, H. von Löhneysen, L. Taillefer, Phys. Rev. B 43 (1991) 13714.

[28] A. Huxley, P. Rodière, D. McK. Paul, N. van Dijk, R. Cublitt and J. Flouquet, Nature 406 (2000) 160.
[29] H. Tou, Y. Kitaoka, K. Asayama, N. Kimura, Y. Onuki, E. Yamamoto, Y. Haga and K. Maezawa, Phys. Rev. Lett. 77 (1996) 1374. and H. Tou, Y. Kitaoka, K. Ishida, K. Asayama, N. Kimura, Y. Onuki, E. Yamamoto, Y. Haga and K. Maezawa, Phys. Rev. Lett. 80 (1998) 1178
[30] S.M. Hayden, L. Taillefer, C. Vettier and J. Flouquet, Phys. Rev. B 46 (1992) 8675.
[31] J.O. Willis, J.D. Thompson, Z. Fisk, A. de Visser, J.J.M. Franse and A. Menovsky, Phys. Rev. B 31 (1985) 1654.
[32] J.J.M. Franse, K. Kadowaki, A. Menovsky, M. van Sprang and A. de Visser, J. Appl. Phys. 61 (1987) 3380.
[33] R.J. Keizer, A. de Visser, M.J. Graf, A.A. Menovsky and J.J.M. Franse., Phys. Rev. B 60 (1999) 10527.
[34] A. de Visser, Physica B 319 (2002) 233.
[35] C.P. Opeil and M.J. Graf, Physica B 319 (2002) 246.
[36] B. Battlogg, D.J. Bishop, E. Bucher, B. Golding, A.P. Ramirez, Z. Fisk, J.L. Smith and H.R. Ott, J. Magn. Magn. Mat. 63-64 (1987) 441.
[37] R. Verhoef, A. de Visser, A. Menovsky, A.J. Riemersma and J.J.M. Franse, Physica B 142 (1986) 11.
[38] S. Doniach, Physica B 91 (1977) 231.
[39] R.J. Keizer, A. de Visser, A.A. Menovsky, J.J.M. Franse, A. Amato, F.N. Gygax, M. Pinkpank and A. Schenck, J. Phys.: Condens. Matter 11 (1999) 8591.
[40] M.J. Graf, R.J. Keizer, A. de Visser and S.T. Hannahs, Physica B 284-288 (2000) 1281.
[41] Nachumi et al. Phys. Rev. B 58 (1998) 8760.
[42] A. de Visser, M.J. Graf, P. Estrela, A. Amato, C. Baines, D. Andreica, F.N. Gygax and A. Schenck, Phys. Rev. Lett. 85 (2000) 3005.
[43] M.J. Graf, private communication (2002).
[44] F. Borsa et al., Phys. Rev. B 52 (1995) 7334.
[45] J. Rossat-Mignod et al., in : G. Reiter, P. Horsch, G.C. Psaltakis (Eds), Dynamics of Magnetic Fluctuations in High-Temperature Superconductors, Plenum, New York, 1991.

THE PERIODIC ANDERSON MODEL IN THE GENERATING FUNCTIONAL APPROACH

Yu.A. Izyumov
Institute for Metal Physics of the RAS, Ural Division, 620219 Ekaterinburg, Russia
Yuri.Izyumov@imp.uran.ru

D.S. Alexeev
Institute for Metal Physics of the RAS, Ural Division, 620219 Ekaterinburg, Russia

Abstract: The periodic Anderson model at arbitrary values of the on-site Coulomb repulsion is studied by the generating functional approach. For the Green's function of the localized d-electrons an exact equation is derived with variational derivatives with respect to fluctuating fields. It coincides with the one for the Hubbard model if the hopping matrix element t_{ij} is replaced by the effective one $\Delta_{ij}(\omega)$, that depends on frequency and is proportional to the hybridization parameter squared. It is shown that the first-order correction with respect to $\Delta_{ij}(\omega)$ for the terminal part of the Green's function contains a singular term originating from the interaction of d-electrons with spin fluctuations. The dynamical susceptibility of these fluctuations is calculated in the hydrodynamic limit.

Keywords: Anderson model, Kondo lattice, magnetic susceptibility, strong correlation.

1. Introduction

Likewise the Hubbard model the periodic Anderson model (PAM) is a basic model in the theory of strongly correlated electron systems. It is destined for the description of the transition metals, lanthanides, actinides and their compositions including the heavy-fermion compounds. The model consists of two groups of electrons: itinerant and localized ones (s and d electrons), the hybridization between them is admitted. The model is described by the following parameters: the width of the s-electron band W, the energy of the atomic level ε_d, the on-site Coulomb repulsion U of d-electrons with opposite spins, the parameter V of the

J. Franse et al. (eds.), Smart Materials for Ranging Systems, 153–163.

sd-hybridization and the electron concentration n. The common form of the Hamiltonian of PAM is [1]:

$$\mathcal{H} = \sum_{ij\sigma} t_{ij} c^+_{i\sigma} c_{j\sigma} + \varepsilon_d \sum_{i\sigma} d^+_{i\sigma} d_{i\sigma} + U \sum_{i} n^d_{i\uparrow} n^d_{i\downarrow} + \sum_{ij\sigma} (V_{ij} c^+_{i\sigma} d_{j\sigma} + V^*_{ij} d^+_{j\sigma} c_{i\sigma}), \tag{1}$$

where $c(d)$ and $c^+(d^+)$ are the annihilation and creation operators of an electron in a non-degenerate conduction band state (localized d-state). $n^d_{i\sigma} = d^+_{i\sigma} d_{i\sigma}$ is the occupation number operator for the d-states. When the hopping matrix elements t_{ij} are different from zero for the nearest neighbors, then the first term in Eq.(1) forms the band with the bandwidth $W = 2zt$, where z is the coordination number.

Physics described by the model with so many parameters is very rich and the model is able particularly to treat heavy fermion systems. To study the model many approaches were suggested (see reviews [2–5]). They are successful for particular regions of the parameter space but no one is totally universal. In this paper we apply to PAM the generating functional approach (GFA) developed first by Kadanoff and Baym [6] for conventional systems and generalized for strongly correlated electron systems [7–10]. In particular it has been applied to the Hubbard model with arbitrary U in the X-operators formalism [10]. The approach makes it possible to derive equations for the electron Green's function (GF) in terms of variational derivatives with respect to fluctuating fields.

The main equation for the d-electron GF in PAM coincides with the equation for the Hubbard model if the hopping matrix elements t_{ij} in the Hubbard model are replaced by the effective ones $\Delta_{ij}(\omega)$, that are $\sim V^2$ and depend on frequency. By iteration of this equation with respect to $\Delta_{ij}(\omega)$ one can construct a perturbation theory near the atomic limit. A singular term in the expansions, describing the interaction of d-electrons with spin fluctuations, was found. This term leads to a resonance peak near the Fermi-level with a width of the order of the Kondo temperature. The dynamical spin susceptibility in the paramagnetic phase in the hydrodynamic limit was also calculated.

2. Electron Green's functions

Hybridized d- and s-electrons are characterized by the following set of GFs:

$$(g_{dd})^{\sigma_1\sigma_2}_{12} = -\langle T d_{1\sigma_1} d^+_{2\sigma_2} \rangle, \qquad (g_{sd})^{\sigma_1\sigma_2}_{12} = -\langle T c_{1\sigma_1} d^+_{2\sigma_2} \rangle, \tag{2}$$
$$(g_{ds})^{\sigma_1\sigma_2}_{12} = -\langle T d_{1\sigma_1} c^+_{2\sigma_2} \rangle, \qquad (g_{ss})^{\sigma_1\sigma_2}_{12} = -\langle T c_{1\sigma_1} c^+_{2\sigma_2} \rangle.$$

Here all definitions are standard. Symbol $\langle \ldots \rangle$ means averaging over Gibbs ensemble, T is the ordering operator with respect to the thermodynamical time τ. The numerical indexes mean site numbers and time: $1 = (i_1, \tau_1)$, so for example

$$d_{1\sigma_1} = e^{(\mathcal{H}-\mu N)\tau_1} d_{i_1\sigma_1} e^{-(\mathcal{H}-\mu N)\tau_1} \tag{3}$$

is the Heisenberg representation of the operator $d_{i_1\sigma_1}$; μ is the chemical potential and N is the operator of the total electron number. In the most general case GFs are not diagonal with respect to spin indexes σ.

It is easy to show that there are linear relations between all components of GFs (2)

$$\begin{aligned} (g_{ds})_{12}^{\sigma_1\sigma_2} &= (g_{dd})_{11'}^{\sigma_1\sigma_2} V_{1'2'}^{*} (g_{ss}^{0})_{2'2}, \\ (g_{sd})_{12}^{\sigma_1\sigma_2} &= (g_{ss}^{0})_{11'} V_{1'2'} (g_{dd})_{2'2}^{\sigma_1\sigma_2}, \\ (g_{ss})_{12}^{\sigma_1\sigma_2} &= (g_{ss}^{0})_{12} + (g_{ss}^{0})_{11'} V_{1'2'} (g_{dd})_{2'3'}^{\sigma_1\sigma_2} V_{3'4'}^{*} (g_{ss}^{0})_{4'2}, \end{aligned} \tag{4}$$

where g_{ss}^{0} is the zeroth approximation of the GF of s-electrons respective to the hybridization,

$$(g_{ss}^{0})_{12}^{-1} = -\frac{\partial}{\partial \tau_1}\delta_{12} - t_{12}, \tag{5}$$

The GF of d-electrons takes into account Coulomb interaction on a site, therefore it has a self-energy $\Sigma_{12}^{\sigma_1\sigma_2}$ and obeys the Dyson equation. It can be written in momentum representation as:

$$g_{dd}^{\sigma}(k) = \frac{1}{i\omega_n - \varepsilon_d - \Sigma^{\sigma}(k) - \dfrac{\left|V(\boldsymbol{k})\right|^2}{i\omega_n - \varepsilon(\boldsymbol{k})}}. \tag{6}$$

Here $\varepsilon(\boldsymbol{k})$ and $V(\boldsymbol{k})$ are Fourier components of the hopping t_{ij} and the hybridization V_{ij}. All energies in (6) are counted from μ. Due to the relations (4) the problem of calculation of all electron GFs reduces to the calculation of d-electron GF with strong Coulomb interaction.

It is convenient to use the X-operators for the description of strongly correlated d-electrons

$$\begin{aligned} X_i^{\sigma 0} &= d_{i\sigma}^{+}(1 - n_{i\bar\sigma}^{d}), & X_i^{2\sigma} &= \sigma d_{i\bar\sigma}^{+} n_{i\sigma}^{d}, \\ X_i^{\sigma\bar\sigma} &= d_{i\sigma}^{+} d_{i\bar\sigma}, & X_i^{20} &= \sigma d_{i\bar\sigma}^{+} d_{i\sigma}^{+}, \\ X_i^{\sigma\sigma} &= n_{i\sigma}^{d}(1 - n_{i\bar\sigma}^{d}), & X_i^{22} &= n_{i\sigma}^{d} n_{i\bar\sigma}^{d}, \\ X_i^{00} &= (1 - n_{i\sigma}^{d})(1 - n_{i\bar\sigma}^{d}). \end{aligned} \tag{7}$$

Electron fermi-operators are linear combinations of fermi-like X-operators (f-type):

$$c_{i\sigma}^{+} = X_i^{\sigma 0} - \sigma X_i^{2\bar\sigma}, \qquad c_{i\sigma} = X_i^{0\sigma} - \sigma X_i^{\bar\sigma 2}. \tag{8}$$

As for the Hubbard model [10] we introduce two-component spinors

$$\Psi(i\sigma) = \begin{pmatrix} X_i^{0\sigma} \\ \bar{\sigma}X_i^{\bar{\sigma}2} \end{pmatrix}, \qquad \Psi^+(i\sigma) = \begin{pmatrix} X_i^{\sigma 0} & \bar{\sigma}X_i^{2\bar{\sigma}} \end{pmatrix}. \tag{9}$$

Then the Hamiltonian (1) for PAM is read:

$$\mathcal{H} = \sum_{ij\sigma} t_{ij} c_{i\sigma}^+ c_{j\sigma} + \varepsilon_d \sum_{i\sigma} X_i^{\sigma\sigma} + (2\varepsilon_d + U)\sum_i X_i^{22} + \\ + \sum_{ij\sigma}\sum_{\alpha}\left[V_{ij} c_{i\sigma}^+ \Psi_\alpha(j\sigma) + V_{ij}^* \Psi_\alpha^+(j\sigma) c_{i\sigma}\right]. \tag{10}$$

Here indexes $\alpha = 1, 2$ numerate spinor components. Because we ignore direct hopping of d-electrons on the lattice, in the Hamiltonian (10) a quadratic term with the X-operators is absent, and X-operators enter linearly via the hybridization term. However in the second order perturbation with respect to hybridization such a term should appear; it describes the induced hopping on the lattice.

Let us introduce a GF for d-electrons

$$\mathcal{G}^{\sigma_1\sigma_2}(12) = -\langle T\Psi(1\sigma_1)\Psi^+(2\sigma_2)\rangle. \tag{11}$$

It is a 2×2 matrix respective to spinor indexes. For its calculation we use GFA by introducting the generating functional

$$Z[V] = \mathrm{Tr}\left(\mathrm{e}^{-\beta\mathcal{H}} T \mathrm{e}^{-V}\right), \tag{12}$$

where V is the operator taking into account the interaction of the system with fluctuating fields; the trace is taken over all variables of the system. As usual [10] we choose V as a linear form of boson-like and diagonal X-operators

$$V = v_{1'}^{00} X_{1'}^{00} + v_{1'}^{22} X_{1'}^{22} + v_{1'}^{\sigma_2'\sigma_1'} X_{1'}^{\sigma_1'\sigma_2'} + v_{1'}^{02} X_{1'}^{20} + v_{1'}^{20} X_{1'}^{02} + w_{1'}^{\sigma_2'\sigma_1'} c_{1'\sigma_1'}^+ c_{1'\sigma_2'}, \tag{13}$$

(the summation over repeated primed indexes is implied). If the operators X^{22}, X^{20} and X^{02} are neglected, the expression (13) reduces to the expression used for PAM in the $U \to \infty$ limit [9].

We have to introduce GF with the fluctuating fields

$$\mathcal{G}^{\sigma_1\sigma_2}(12) = -\langle T\Psi(1\sigma_1)\Psi^+(2\sigma_2)e^{-V}\rangle \equiv -\langle T\Psi(1\sigma_1)\Psi^+(2\sigma_2)\rangle_V. \tag{14}$$

Boson-type GFs can be written as functional derivatives of the generating functional $Z = e^{\Phi}$:

$$\mathcal{N}^{\sigma_1\sigma_2}(12) = -\langle T\big(X_1^{\sigma_1\sigma_1} - \langle X_1^{\sigma_1\sigma_1}\rangle\big)\big(X_2^{\sigma_2\sigma_2} - \langle X_2^{\sigma_2\sigma_2}\rangle\big)\rangle_V = -\frac{\delta^2\Phi[V]}{\delta v_1^{\sigma_1\sigma_1}\delta v_2^{\sigma_2\sigma_2}},$$

$$\mathcal{D}^{\sigma\bar{\sigma}}(12) = -\langle TX_1^{\sigma\bar{\sigma}}X_2^{\bar{\sigma}\sigma}\rangle_V = -\frac{\delta^2\Phi[V]}{\delta v_1^{\sigma\bar{\sigma}}\delta v_2^{\bar{\sigma}\sigma}}, \tag{15}$$

$$\mathcal{D}^{02}(12) = -\langle TX_1^{02}X_2^{20}\rangle_V = -\frac{\delta^2\Phi[V]}{\delta v_1^{02}\delta v_2^{20}}.$$

For the calculation of a d-electron GF we have to write an equation of motion using definition (14).

3. Equation of motion for d-electron Green's function

First of all we write the equation of motion for electron operators

$$\dot{c}_{1\sigma} = -\, t_{11'}c_{1'\sigma} - V_{11'}\sum_{\alpha}\Psi_\alpha(1'\sigma), \tag{16}$$

$$\begin{aligned}\dot{\Psi}(1\sigma) = &-E_1^\sigma\Psi(1\sigma) - F_1^\sigma\widehat{\Delta}_{11'}\Psi(1'\sigma) - \\ &- X_1^{\bar{\sigma}\sigma}\tau^z\widehat{\Delta}_{11'}\Psi(1'\bar{\sigma}) + \sigma X_1^{02}i\tau^y\widehat{\Delta}_{11'}\widetilde{\Psi}^+(1'\bar{\sigma}).\end{aligned} \tag{17}$$

In the second equation we introduce a matrix quantity

$$\widehat{\Delta}_{12} = \Delta_{12}\Im, \qquad \Im = \begin{pmatrix} 1 & 1 \\ 1 & 1 \end{pmatrix}, \tag{18}$$

where

$$\Delta_{12} = V_{11'}^*(g_{ss}^0)_{1'2'}V_{2'2}. \tag{19}$$

In Eq.(17) τ^α are Pauli matrices and the quantities E_1^σ and F_1^σ are 2×2 matrices with respect to spinor indexes, the same as in the Hubbard model [10].

It is remarkable that Eq.(17) has exactly the same form as in the Hubbard model, only instead of a bare hopping matrix element $\hat{t}_{12}$ now we have the induced hopping element $\widehat{\Delta}_{12}$. Because equations of motion for Ψ-operators for PAM and the Hubbard model have the same form (with the only change $t_{12} \to \Delta_{12}$) the equation of motion for a d-electron GF in PAM has the same form as in the Hubbard model [10].

Determine a matrix GF for d-electrons

$$\mathcal{L}(\underline{1}\underline{2}) = \begin{pmatrix} -\langle T\Psi(1\sigma_1)\Psi^+(1\sigma_2)\rangle_V & -\langle T\Psi(1\sigma_1)\Psi(1\sigma_2)\rangle_V \\ -\langle T\Psi^+(1\sigma_1)\Psi^+(1\sigma_2)\rangle_V & -\langle T\Psi^+(1\sigma_1)\Psi(1\sigma_2)\rangle_V \end{pmatrix}. \tag{20}$$

Underlined numerical indexes in $\mathcal{L}(\underline{1}\,\underline{2})$ are a set of indexes $\underline{1} = (1\sigma_1\alpha_1\nu_1)$, where α_1 is the spinor index and ν_1 is the index of the particle-hole representation, numerating matrix elements in (20). Thus $\mathcal{L}(\underline{1}\,\underline{2})$ is a matrix 8×8 with respect to discrete indexes $\sigma\alpha\nu$.

A GF (20) for PAM obeys the same equation of motion as for the Hubbard model [10], namely

$$\left[L_{0V}^{-1}(\underline{1}\,\underline{1}') - (\hat{A}\Phi Y)(\underline{1}\,\underline{1}') - (\hat{A}Y)(\underline{1}\,\underline{1}')\right]\mathcal{L}(\underline{1}'\underline{2}) = (\hat{A}\Phi)(\underline{1}\,\underline{2}), \tag{21}$$

where $\hat{A}(\underline{1}\,\underline{2})$ is a matrix including functional derivatives over fluctuating fields, and $L_{0V}^{-1}(\underline{1}\,\underline{2})$ is a matrix depending on the fluctuating field (both expressions one can be found in [10]).

Matrix $Y(\underline{1}\,\underline{2})$ involves matrix elements of the induced hopping

$$Y(\underline{1}\,\underline{2}) = \begin{pmatrix} \widehat{\Delta}(12) & 0 \\ 0 & -\widehat{\Delta}(12) \end{pmatrix}, \tag{22}$$

So, for a d-electron GF (14), determined on X-operators, we have a matrix equation with functional derivatives. The electron GF for d-electrons g_{dd}, determined in (2), is expressed through the matrix elements of the GF determined in (11) or (14).

$$(g_{dd})_{12}^{\sigma_1\sigma_2} = -\langle T d_{1\sigma_1} d_{2\sigma_2}^{+}\rangle = \sum_{\alpha\beta}(\mathcal{G}_{12}^{\sigma_1\sigma_2})_{\alpha\beta}. \tag{23}$$

Of cause in all expressions for $(\mathcal{G}_{12}^{\sigma_1\sigma_2})_{\alpha\beta}$ all fluctuating fields should be put zero.

4. Iteration of equations

By iteration of the matrix equation (21) for the GF $\mathcal{L}(\underline{1}\,\underline{2})$ over Δ one can obtain expressions for the self-energy and the terminal part [7–10]. A GF (11), which is a matrix element $\mathcal{L}^{11}$ with respect to index ν, can be represented in a multiplicative form:

$$\mathcal{G} = G\Lambda, \tag{24}$$

$$G^{-1} = G_{0V}^{-1} - \Sigma, \qquad \Sigma = \Sigma' + \Lambda\widehat{\Delta}. \tag{25}$$

Iterations in the limit of two first orders determine the terminal part $\Lambda = \Lambda_0 + \Lambda_1 + \ldots$. In the momentum representation one has

$$\Lambda_0^{\sigma} = \begin{pmatrix} 1 - \langle n_{1\bar{\sigma}}^{d}\rangle & 0 \\ 0 & \langle n_{1\bar{\sigma}}^{d}\rangle \end{pmatrix}, \tag{26}$$

$$\Lambda_1^\sigma(k) = \begin{pmatrix} \lambda_1^\sigma(k) & \lambda_2^\sigma(k) \\ -\lambda_1^\sigma(k) & -\lambda_2^\sigma(k) \end{pmatrix}, \tag{27}$$

where

$$\lambda_1^\sigma(k) = -\sum_q \Delta(k+q)\Big[(G_{11}^\sigma + G_{21}^\sigma)(k+q)\mathcal{N}^{\bar\sigma\bar\sigma}(q) + \tag{28}$$

$$+ (G_{11}^{\bar\sigma} + G_{21}^{\bar\sigma})(k+q)\mathcal{D}^{\sigma\bar\sigma}(q) + (G_{21}^{\bar\sigma} + G_{22}^{\bar\sigma})(-k-q)\mathcal{D}^{02}(q)\Big],$$

and the expression for $\lambda_2^\sigma(k)$ is obtained from $\lambda_1^\sigma(k)$ by changing the sign and indexes $1 \leftrightarrow 2$.

Contributions of the first and second orders in the uncutable self-energy part Σ' is given by the matrices:

$$\Sigma_1'^\sigma = -\eta^\sigma \begin{pmatrix} 1 & -1 \\ -1 & 1 \end{pmatrix}, \tag{29}$$

$$\Sigma_2'^\sigma = \begin{pmatrix} \varphi_1^\sigma(k) & \varphi_2^\sigma(k) \\ -\varphi_1^\sigma(k) & -\varphi_2^\sigma(k) \end{pmatrix}, \tag{30}$$

where

$$\eta^\sigma = \sum_k \Delta(k)\left[G_{11}^{\bar\sigma}(k) - G_{22}^{\bar\sigma}(k)\right], \tag{31}$$

$$\varphi_1^\sigma(k) = \sum_q \sum_{k_1} \Delta(k+q)\Delta(k_1+q)\Big[G_{11}^\sigma(k_1)\gamma^{\bar\sigma}(k+q)G_{11}^{\bar\sigma}(k_1+q) + \tag{32}$$

$$+ G_{11}^{\bar\sigma}(k_1)\gamma^\sigma(k+q)G_{11}^{\bar\sigma}(k_1+q) + \sum_{\sigma'} G_{22}^{\bar\sigma}(-k_1)\gamma^{\sigma'}(k+q)G_{11}^{\bar\sigma'}(-k_1-q)\Big],$$

An expression for $\varphi_2^\sigma(k)$ is obtained from $\varphi_1^\sigma(k)$ by changing indexes $1 \leftrightarrow 2$.

5. Structure of quasiparticle spectrum

We will show how a low-energy scale appears in PAM in the framework of the GFA.

For simplicity we consider now the limit $U \to \infty$. In this case only one matrix element of the $\mathcal{G}_{\alpha\beta}$ survives

$$G_{11}^\sigma(k) = \frac{1}{i\omega_n - \varepsilon_d - \Sigma^\sigma(k)},$$

where

$$\Sigma^{\sigma}(k) = \Sigma'^{\sigma}(k) + \Lambda^{\sigma}(k)\Delta(k),$$

and the terminal part $\Lambda^{\sigma} = \Lambda_0^{\sigma} + \Lambda_1^{\sigma} + \ldots$, and the uncutable self-energy part $\Sigma'^{\sigma} = \Sigma_1'^{\sigma} + \Sigma_2'^{\sigma} + \ldots$ are expanded into a power series of $\Delta(k)$:

$$\Lambda_0^{\sigma}(k) = 1 - \langle n_d^{\bar{\sigma}} \rangle, \tag{33}$$

$$\Lambda_1^{\sigma}(k) = -\sum_q \Delta(k+q)\Big[G_{11}^{\sigma}(k+q)\mathcal{N}^{\bar{\sigma}\bar{\sigma}}(q) + G_{11}^{\bar{\sigma}}(k+q)\mathcal{D}^{\sigma\bar{\sigma}}(q)\Big]. \tag{34}$$

We omit $\Sigma_1'^{\sigma}$ because it does not depend on 4-momentum, and $\Sigma_2'^{\sigma}$ weekly depends on it.

After the analytical continuation $i\omega_n \to E + i\delta$ we can write the imaginary part of the d-electron GF as $\mathcal{G}^{\sigma} = G_{11}^{\sigma}(k)\Lambda^{\sigma}(k)$. In the limit of small imaginary part of $\Lambda_1^{\sigma} \equiv \Lambda_1'^{\sigma} + i\Lambda_1''^{\sigma}$ we can write approximately

$$Im\mathcal{G}^{\sigma}(E) = -\frac{V^2\big[Re\Lambda^{\sigma}(E)\big]^2}{(E-\varepsilon_d)^2}\,\pi\,\delta\left(\varepsilon(\boldsymbol{k}) - E + \frac{V^2 Re\Lambda^{\sigma}(E)}{E-\varepsilon_d}\right).$$

(we neglect $\boldsymbol{k}$-dependence of $Re\Lambda^{\sigma}(\boldsymbol{k}, E)$). Then by summing this expression over $\boldsymbol{k}$, we obtain the density of d-states

$$\rho_d^{\sigma}(E) = \frac{V^2\big[Re\Lambda^{\sigma}(E)\big]^2}{W(E-\varepsilon_d)^2}. \tag{35}$$

This expression is valid when an equation yields

$$\varepsilon(\boldsymbol{k}) - E + \frac{V^2 Re\Lambda^{\sigma}(E)}{E-\varepsilon_d} = 0 \tag{36}$$

for all $\boldsymbol{k}$ from the Brillouin zone.

Now we must calculate expression (34) for $\Lambda_1^{\sigma}(k)$. The first term in it, describing the interaction of d-electrons with spin fluctuations, contains a singular term near the Fermi-energy $E = 0$:

$$Re\Lambda_1(E) \approx -\frac{V^2 b}{W|E-\varepsilon_d|}\,ln\left|\frac{E}{E-E_l}\right|, \tag{37}$$

where E_l is the lower edge of the d-electron band, and b is the instant spin correlator on a site (we consider a paramagnetic phase). We supposed that the spectrum of spin fluctuations is rather narrow.

We see that the first order correction to the terminal part contains a Kondo-like singularity, which produces a resonance peak at the Fermi-level. The width of the peak $\bar{E}$ is determined from Eq.(36) by summing it up over all $\boldsymbol{k}$. We come then to the equation

$$\frac{\bar{E}-\varepsilon_d}{Re\Lambda_1(\bar{E})} = V^2 g(\bar{E}), \tag{38}$$

where $g(E) = \sum_{\boldsymbol{k}} \dfrac{1}{E-\varepsilon(\boldsymbol{k})}$ is a zeroth approximation single-site GF of s-electrons. We write its solution near the Fermi-level in the form

$$\bar{E} \approx |E_l|\, e^{-\dfrac{1}{\rho_d(0) J^*}}, \tag{39}$$

where $\rho_d(0)$ is the density of d-states (with $Re\Lambda = \Lambda_0$) on the Fermi-level (35) and J^* is an effective exchange integral

$$J^* = \frac{V^2\, b}{\Lambda_0^2}\, |g(0)|. \tag{40}$$

Relation (39) determines a low-energy scale for the spectral density of d-electrons, which is of the order of the Kondo temperature kT_K for PAM. Because this scale results from the interaction with spin fluctuations of localized moments we also study the spin susceptibility of d-electrons.

6. Spin fluctuations in the hydrodynamic regime

For the paramagnetic phase it is convenient to calculate the spin susceptibility χ through the memory function M [11]:

$$\chi_d^{\sigma\bar{\sigma}}(\boldsymbol{q},\omega) = -\frac{M_d^{\sigma\bar{\sigma}}(\boldsymbol{q},\omega)}{\omega - \dfrac{M_d^{\sigma\bar{\sigma}}(\boldsymbol{q},\omega)}{\chi(\boldsymbol{q})}}, \tag{41}$$

where $\chi(\boldsymbol{q}) \equiv \chi_d^{\sigma\bar{\sigma}}(\boldsymbol{q},0)$ is the static susceptibility. The memory function is expressed via a retarded GF of the spincurrents [12]

$$M_d^{\sigma\bar{\sigma}}(\boldsymbol{q},\omega) = -\sum_{ij} e^{-i\boldsymbol{q}\boldsymbol{R}_{ij}} \int_{-\infty}^{+\infty} \frac{d\omega'}{\pi}\, \frac{Im\langle\langle i\dot{S}_i^{\sigma}| - i\dot{S}_j^{\bar{\sigma}}\rangle\rangle_{\omega'}^{irr}}{\omega'(\omega-\omega'+i\delta)}. \tag{42}$$

Here $S_i^{\sigma} = X_i^{\sigma\bar{\sigma}}$ and

$$i\dot{S}_i^{\sigma} = [S_i^{\sigma}, \mathcal{H}] = \tag{43}$$
$$= -\widehat{\Delta}_{\alpha'\beta'}(ii')\Psi_{\beta'}(i'\bar{\sigma})\Psi_{\alpha'}^{+}(i\sigma) + \Psi_{\beta'}(i\bar{\sigma})\Psi_{\alpha'}^{+}(i'\sigma)\widehat{\Delta}_{\alpha'\beta'}(i'i).$$

By a simple decoupling of terms in (42) with products of four electron Ψ-operators we find

$$Im\langle\langle i\dot{S}_i^{\sigma}|-i\dot{S}_j^{\bar{\sigma}}\rangle\rangle_{\boldsymbol{q},\omega}^{irr} = -4\pi\sum_{\boldsymbol{k}}\int d\omega'\big[\Delta(k)-\Delta(k-q)\big]^2\times \quad (44)$$
$$\times\big[f(\omega'-\omega)-f(\omega')\big]tr\big\{Im\big(\mathcal{G}^{\sigma}(\boldsymbol{k}-\boldsymbol{q};\omega'-\omega)\big)\,Im\big(\Im\mathcal{G}^{\bar{\sigma}}(\boldsymbol{k};\omega')\Im\big)\big\}.$$

Such decoupling corresponds to the approximation of interacting modes [11].

Now we consider the hydrodynamic regime $\omega \ll \boldsymbol{v}\boldsymbol{q}$, where $\boldsymbol{v}$ is a character momentum of an s-electron on the Fermi-surface. In this case the imaginary part of M is important, and we obtain from (42) and (44)

$$ImM_d^{\sigma\bar{\sigma}}(\boldsymbol{q},\omega)\approx -D_dq^2, \quad (45)$$

where the coefficient of spin diffusion is written

$$D_d = 4\pi\sum_{\boldsymbol{k}}(\boldsymbol{v}(\boldsymbol{k})\boldsymbol{e})\int d\omega\left(\frac{V}{\omega-\varepsilon(\boldsymbol{k})}\right)^4\frac{\partial f(\omega)}{\partial\omega}\times \quad (46)$$
$$\times\, tr\big\{Im\big(\mathcal{G}^{\sigma}(\boldsymbol{k};\omega)\big)\,Im\big(\Im\mathcal{G}^{\bar{\sigma}}(\boldsymbol{k};\omega)\Im\big)\big\},$$

and $\boldsymbol{v}(\boldsymbol{k}) = \boldsymbol{\nabla}_{\boldsymbol{k}}\varepsilon(\boldsymbol{k})$, $\boldsymbol{e} = \boldsymbol{q}/q$. From relations (41) and (45) the diffusion form of the spin susceptibility for d-electrons follows:

$$\frac{1}{\omega}Im\chi^{\sigma\bar{\sigma}}(\boldsymbol{q},\omega) = \chi(\boldsymbol{q})\frac{\widetilde{D}_dq^2}{\omega^2+\big(\widetilde{D}_dq^2\big)^2}, \quad (47)$$

where $\widetilde{D}_d = D_d/\chi(\boldsymbol{q})$. Because $D_d \sim V^4$, the spectrum of spin fluctuations is really narrow as we supposed.

7. Summary

We applied the generating functional approach to the periodic Anderson model. Calculation of the electron GFs g_{dd}, g_{ds}, g_{sd} and g_{ss} reduces to calculation of only the d-electron GF. For this, an exact matrix equation was derived with the variational derivatives. Iterations with respect to the effective matrix element $\Delta_{ij}(\omega)$ allow to construct a perturbation theory near the atomic limit. Along with the self-energy, the terminal part of the GF $\mathcal{G}$ is very important. The first order correction for it describes the interaction of d-electrons with spin fluctuations. In the paramagnetic phase this term contains a logarithmic singularity near the Fermi-level and thus produces a Kondo-like resonance peak in the d-electron density of states. The spin susceptibility of d-electrons in the paramagnetic phase in the hydrodynamic regime has a diffusion form with the diffusion coefficient $\sim V^4$, and therefore with a narrow spectrum of fluctuations.

Acknowledgments

Yu.A. Izuymov and D.S. Alexeev thank the Russian Foundation of Support of Scientific School, grant NS–747.2003.2.

References

[1] P.W. Anderson, Phys.Rev. **124**, 41 (1961).

[2] L. Degiorgi, Rev.Mod.Rhys. **71**, 687 (1999).

[3] H. Tsunetsugu, M. Sigrist, K. Ueda, Rev.Mod.Phys. **69**, 809 (1997).

[4] A.C. Hewson, *The Kondo Problem to Heavy Fermions*, Cambridge University Press, (Cambridge University Press, Cambidge, England, 1993).

[5] A. Georges, G. Kotliar, W. Krauth, M.J. Rozenberg, Rev.Mod.Phys. **68**, 13 (1996).

[6] L. Kadanoff, G. Baym, *Quantum Statistical Mechanics*, (W.A. Benjamin, inc., New York,1962).

[7] Yu. A. Izyumov, in Seventh Training Course in the Physics of Correlated Electron Systems and High-Tc Superconductors, ed. A. Avella, F. Mancini, AIP Conference Proceedings 678 (2003).

[8] Yu. A. Izyumov, N.I. Chaschin, The Physics of Metals and Metallography, **97**, N3, 5 (2004).

[9] Yu. A. Izyumov, D.S. Alexeev, The Physics of Metals and Metallography, **97**, N1, 18 (2004)

[10] Yu.A. Izyumov, N.I. Chaschin, D.S. Alexeev, F. Mancini, EPJB, **45**, 69 (2005).

[11] D. Forster, *Hydrodynamical Fluctuations, Broken Symmetry and Correlation Functions*, (Benjamin, inc., New York, 1975).

[12] G. Jakely, N.M. Plakida, Theor. and Mat. Phys. **114**, N3, 426 (1998).

[13] F. Mancini, A. Avella. Adv.Phys. **53**, 5–6, 537 (2004).

ORBITAL ORDERING AND PHASE SEPARATION PHENOMENA IN LIGHTLY DOPED MANGANITES

I.O. Troyanchuk and V.A. Khomchenko
Institute of Solid State and Semiconductor Physics, NAS, P. Brovka str. 17, 220072 Minsk, Belarus
troyan@ifttp.bas-net.by

H. Szymczak
Institute of Physics, PAS, Lotnikow srt. 32/46, 02-668 Warsaw, Poland

Abstract: We review magnetic and structural phase diagrams of the $La_{0.88}MnO_x$, $La_{1-x}Sr_x(Mn_{1-x/2}Nb_{x/2})O_3$, $Nd_{1-x}Ca_xMnO_3$, and $Bi_{1-x}Ca_xMnO_3$ series constructed on the basis of x-ray, neutron powder diffraction, Young's modulus, magnetization and resistivity measurements. It is shown that the main factor controlling the magnetic properties of the manganites is a type of an orbital state. The results are discussed in the framework of structurally driven magnetic phase separation model.

Keywords: manganites, exchange interactions, magnetic phase transitions, phase separation, orbital ordering.

1. Introduction

Mixed-valence manganites with a perovskite structure are the model objects for the physics of strongly correlated electronic systems. The interest in the study of manganites is due to a variety of phase states and transitions and intrinsic correlation of the crystal structure, magnetic, and transport properties. The nature of the interplay between the crystal structure, magnetic, and transport properties of manganites is still a matter of discussion in spite of numerous investigations. Several models were proposed to explain a magnetic state evolution under hole doping as well as a metal-insulator transition at the Curie point. In the double-exchange model of Zener, simultaneous ferromagnetic and metallic transitions have been qualitatively explained by the fact that electrons tend to move between Mn^{3+}

J. Franse et al. (eds.), Smart Materials for Ranging Systems, 165–178.

and Mn^{4+} ions having the same spin orientation, therefore electron delocalization favors the ferromagnetic order [1]. More recently Millis et al. pointed out that double exchange alone cannot account for many of the experimental results [2]. They showed that a Jahn-Teller-type electron-phonon coupling should play an important role in explanation of the colossal magnetoresistance effect. Another mechanism of antiferromagnet-ferromagnet phase transitions in manganites was proposed by Nagaev [3]. He assumed that the intermediate phase can be described as a inhomogeneous magnetic state driven by an electronic phase segregation. In this scenario the ferromagnetic regions contain an excess of holes and are metallic. Goodenough et al. argued that the magnetic properties of manganites were determined by the type of orbital state [4]. According to the rules for 180° superexchange, if the electronic configuration correlates with vibrational modes, Mn^{3+}-O^{2-}-Mn^{3+} interactions are antiferromagnetic in case of the static Jahn-Teller effect and ferromagnetic when the Jahn-Teller effect is dynamic. Thus, antiferromagnet-ferromagnet phase transitions can occur going through a mixed state of phases with different orbital dynamics.

The recent magnetic phase diagrams of the $La_{1-x}Sr_xMnO_3$ and $La_{1-x}Ca_xMnO_3$ systems were constructed assuming a homogeneous canted magnetic state in a low doping range [5,6]. On the other hand, there are numerous experimental data which indicate the existence of phase separation in manganites. The results of nuclear magnetic resonance [7-11], neutron diffraction [11,12], muon spin relaxation [13], X-ray absorption [14], scanning tunneling spectroscopy [15], and electron microscopy [16] experiments give evidence of magnetic and structural inhomogeneities, but the driving force of magnetic phase separation in manganites is still not fully clear. In order to contribute to the solution of this problem we have investigated the features of the antiferromagnet-ferromagnet phase transition in low-doped $La_{0.88}MnO_x$, $La_{1-x}Sr_x(Mn_{1-x/2}Nb_{x/2})O_3$, $Nd_{1-x}Ca_xMnO_3$, and $Bi_{1-x}Ca_xMnO_3$ manganites.

2. Results and discussion

2.1 *$La_{0.88}MnO_x$ system*

Tentative magnetic phase diagram of the $La_{0.88}MnO_x$ ($2.82 \leq x \leq 2.96$) manganites is shown in Figure 1. The most strongly reduced sample $La_{0.88}MnO_{2.82}$ is antiferromagnet with a Néel temperature of 140 K. Its properties are found to be similar to the properties of stoichiometric $LaMnO_3$. Both compounds have very close unit cell parameters, the same magnetization value, and close temperatures of both magnetic (T_{OO}~140K) and orbital (T_{CO}~750K) orderings. The existence of orbital ordering in the A-type antiferromagnetic structure of $La_{0.88}MnO_{2.82}$ is corroborated by neutron diffraction measurements [17]. With increasing oxygen content up to the

$x = 2.85$ sample, the magnetic and orbital ordering temperatures lower while the magnetization increases slightly. Results of the neutron diffraction measurements carried out for the $x = 2.84$ sample confirm the appearance of a ferromagnetic component. A further increase of the oxygen concentration leads to a significant enhancement of the ferromagnetic contribution. The transition temperature to the paramagnetic state begins to increase and the transition becomes broader. Neutron diffraction data obtained for the $x = 2.87$ sample indicate that ferromagnetic coupling becomes predominant. No long-range antiferromagnetic order has been observed for this compound. At the same time, the refined magnetic moment is lower than that expected for the full spin arrangement. Besides, the relatively large magnetic anisotropy at low temperature assumes the presence of an anisotropic magnetic coupling which differs from the isotropic ferromagnetic one. This can be attributed to existence of either short-range antiferromagnetic clusters or a spin-glass phase. No pronounced thermomagnetic irreversibility indicating the anisotropic magnetic interactions is observed starting from the $x = 2.92$ sample. The values of magnetization estimated for the monoclinic compounds are close to those expected for full spin alignment. The ground state of all the orthorhombic compounds $2.82 \leq x \leq 2.90$ is insulating. It should be noted that the appearance of metallic conductivity does not coincide with the transition to monoclinic phase. Simultaneous first-order magnetic transition and metal-insulator transition at T_C are observed for $x \geq 2.92$ compounds.

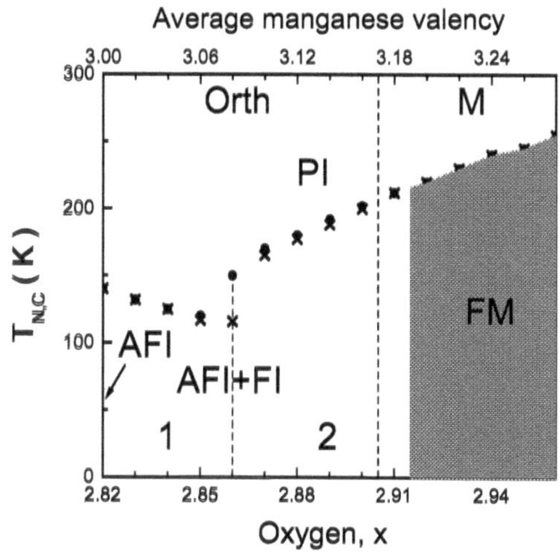

Figure 1. Magnetic phase diagram of the $La_{0.88}MnO_x$ ($2.82 \leq x \leq 2.96$) system. Orth- orthorthorhombic crystal structure, M- monoclinic crystal structure; PI, AFI, FI, and FM- paramagnetic insulating, antiferromagnetic insulating, ferromagnetic insulating, and ferromagnetic metallic states, respectively. Areas 1 and 2 correspond to the concentration regions where an antiferromagnetic or ferromagnetic phase predominates, respectively.

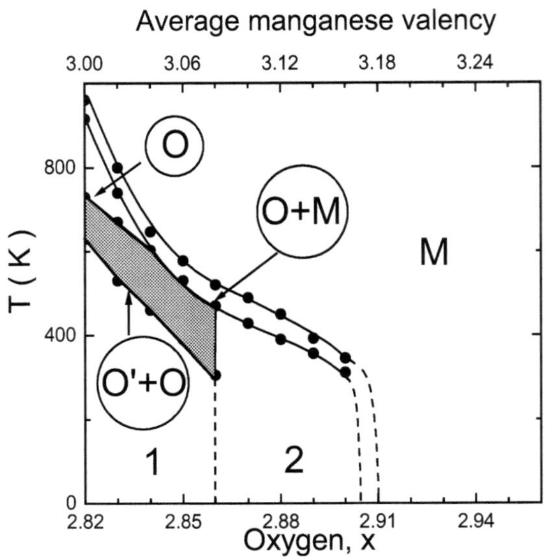

Figure 2. Crystal structure phase diagram of the $La_{0.88}MnO_x$ ($2.82 \leq x \leq 2.96$) system. O^I, O, and M- orbitally ordered orthorhombic, orbitally disordered orthorhombic, and monoclinic phases, respectively. Areas 1 and 2 correspond to the concentration regions where the static Jahn-Teller distortions or dynamic orbital correlations predominate, respectively.

A strong correlation between the magnetic and structural properties of $La_{0.88}MnO_x$ ($2.82 \leq x \leq 2.96$) manganites is observed. The hypothetical structural phase diagram of $La_{0.88}MnO_x$ ($2.82 \leq x \leq 2.96$) constructed using X-ray, neuron diffraction, Young's modulus, resistivity, and DTA data is shown in Figure 2. For $La_{0.88}MnO_{2.82}$, the sharp anomalies of the Young's modulus and resistivity are associated with the removal of cooperative orbital ordering; it is observed at approximately 650 K. The DTA measurements revealed the release of latent heat in the range 650–730 K. Neutron diffraction data indicate the coexistence of orbitally ordered O^I and orbitally disordered O phases at $T = 700$ K. Another thermal anomaly connected with the transition to the monoclinic phase is observed in the temperature range $915 \text{ K} \leq T \leq 960 \text{ K}$. With the increase of the oxygen content to $x = 2.83$, the temperatures of both orbital order-disorder and orthorhombic-monoclinic phase transitions significantly decrease. The range of coexistence of O^I and O phases becomes broader, while the width of the anomaly associated with the temperature-induced orthorhombic-monoclinic transition remains practically constant. Starting from the $x = 2.84$ sample, the differential thermal analysis does not show any significant heat effect, which could be interpreted as a transition to a pure orbitally disordered state; however, the anomaly related to the transition from an orthorhombic to a monoclinic phase remains well pronounced. Neutron diffraction data coupled with Young's modulus measurements indicate the existence of predominantly static Jahn-Teller distortions at room temperature and two-phase character of the crystal structure above $T \sim 470$ K. Inhomogeneous structural states are observed up to 650 K. Above this temperature the monoclinic phase is stabilized. A further increase of the oxygen concentration leads to the broadening and gradual disappearance of the anomaly which relates to the transition to the orbitally ordered state. The neutron diffraction study performed for the $La_{0.88}MnO_{2.87}$ compound indicates that the value of the MnO_6 octahedron distortion increases with decreasing temperature to 200 K. However, even in the case of $T = 200$ K, where the worst agreement factors for one-phase structural model have been observed, the introduction of the second orthorhombic phase was unsuccessful. Apparently, even at 200 K, the orbitally ordered clusters are still too small and separate to distinguish the O^I phase in the diffraction experiment. The temperature of orthorhombic-monoclinic phase transition gradually decreases as the oxygen content increases and starting from the $x = 2.91$ sample, the monoclinic phase is stabilized (Fig. 2). It is necessary to mention that the X-ray and neutron diffraction experiments can reveal a two-phase structural state rather in the case of macroscopic structural phase separation. In the cases of local structural inhomogeneities or nanometer scale structural clusters, these experiments give only an average picture of a structural state [18]. Thus, the correlation between the orbital state and magnetic properties of the $La_{0.88}MnO_x$ manganites is prominent. The static Jahn-Teller distortions are responsible for the A-type antiferromagnetic structure, while dynamic orbital correlations lead to ferromagnetism.

It is worth noting that there are two alternative models of orbital state corresponding to ferromagnetic ordering in manganites: 3D dynamic $d_{3z^2-r^2}$ orbital correlations and staggered ordering of $d_{3z^2-r^2}$ and $d_{x^2-y^2}$ orbitals. Neutron diffraction studies have shown that $LaMnO_3$ undergoes a structural transition from O^I -orthorhombic to O-orthorhombic phase at $T_{JT} = 750$ K [19]. The MnO_6 octahedron in the O-orthorhombic phase becomes nearly regular, i.e. the orbital ordering disappears [19]. However, x-ray absorption near the edge structure and the extended x-ray absorption fine structure at the MnK-edge measurements have revealed that the MnO_6 octahedrons in $LaMnO_3$ remain tetragonally distorted at $T > T_{JT}$ [20]. The empty Mn^{3+} electronic d states were shown to be unaltered through the Jahn–Teller transition. The lowest energy for the e_g electron corresponds to the three possible distortions giving rise to three degenerate vibronic states, $d_{x^2-r^2}$, $d_{y^2-r^2}$, and $d_{z^2-r^2}$, being the electronic orbitals of the vibronic state. The thermally excited electron jumps between these states above T_{JT} and is localized in an ordered state below T_{JT}. The orbital ordering proposed for $LaMnO_3$ arises then from the ordering of the local Jahn–Teller distortions.

The high temperature (O-orthorhombic) phase can be described as a dynamical locally distorted phase with the strong antiferrodistortive first neighbour coupling [20].

The similar situation seems to be observed for Mn^{4+}-doped manganites. The atomic pair-density function of $La_{1-x}Sr_xMnO_3$ manganites ($0 \le x \le 0.4$), obtained by pulsed neutron diffraction, indicates the existence of tetragonally distorted MnO_6 octahedrons even in the rhombohedral metallic phase, when the crystallographic structure shows no JT distortions [21]. This is possible only in the case of the dynamic orbital correlations described above. One can assume that when one puts non-Jahn–Teller Mn^{4+} ions in the background of the Mn^{3+} ions, the e_g orbitals of all the Mn^{3+} ions surrounding the localized hole (Mn^{4+}) tend to be directed towards it, forming an orbital polaron [22]. Due to the strong antiferrodistortive Mn^{3+} first neighbour coupling [20], dynamic correlations of the $d_{3z^2-r^2}$ orbitals should arise.

According to the rules for 180° superexchange the dynamic orbital correlations lead to ferromagnetic interaction between the Mn^{3+} ions [4]. Hence, one can expect that ferromagnetism in manganites can arise even in the absence of Mn^{4+} ions, if only the JT effect is dynamic. For instance, Mn substitution with non-Jahn–Teller diamagnetic Nb^{5+}, Al^{3+}, Sc^{3+}, etc, ions should result in the appearance of ferromagnetic order. Below we show that this assumption is correct.

Second possibility lies in description of orbital state as a hybridization of the $d_{3z^2-r^2}$ and $d_{x^2-y^2}$ orbitals as $\cos(\theta/2)|3z^2-r^2\rangle \pm \sin(\theta/2)|x^2-y^2\rangle$. Such an orbital ordering is recently proposed experimentally and theoretically in the ferromagnetic insulating phase of $La_{0.88}Sr_{0.12}MnO_3$ and $Pr_{0.75}Ca_{0.25}MnO_3$ [23-25]. The difficulties in determination of priority of the present models are conditioned by the fact that staggered ordering of $d_{3z^2-r^2}$ and $d_{x^2-y^2}$ orbitals can exhibit itself in experiments in the same way as 3D dynamic $d_{3z^2-r^2}$ orbital correlations.

2.2 $La_{1-x}Sr_x(Mn_{1-x/2}Nb_{x/2})O_3$ system

Hypothetical magnetic phase diagram of $La_{1-x}Sr_x(Mn_{1-x/2}Nb_{x/2})O_3$ is shown in Figure 3. The parent $LaMnO_3$ compound shows the spontaneous magnetization value at 5 K corresponding to magnetic moment of 0.07 μ_B per Mn^{3+} ion. The Néel point where spontaneous magnetization develops is 143 K. According to [26] the spontaneous magnetization has a relativistic nature. Substitution of Mn with Nb leads to an enhancement of the spontaneous magnetization whereas the temperature of transition into paramagnetic state slightly decreases. In accordance with the magnetization data the $La_{0.8}Sr_{0.2}(Mn_{0.9}Nb_{0.1})O_3$ and $La_{0.7}Sr_{0.3}(Mn_{0.85}Nb_{0.15})O_3$ samples are ferromagnets with the magnetic moment per chemical formula around 2.3 μ_B and 2.6 μ_B respectively. Neutron diffraction study has revealed the magnetic moment of Mn^{3+} in the parent $LaMnO_3$ antiferromagnetic compound to be close to 3.5 μ_B [27], whereas Nb^{5+} is a diamagnetic ion. Hence the expected moment should be close to 3 μ_B per formula unit what is in rather good agreement with the observed one. The Nb doped sample ($x = 0.3$) has a well defined Curie point- 123 K. Both Curie point and spontaneous magnetization start gradually to decrease when Nb content exceeds 15 % from total sites number in the manganese sublattice. The magnetic state cardinally changes as the concentration of niobium reaches 25 %. We have observed that the magnetic susceptibility of the $x = 0.5$ sample dramatically decreases. The ZFC-magnetization shows a peak at 30 K. Below this temperature the FC magnetization practically does not change. Taking into account the character of the *M*(*H*) dependence, we have concluded that the sample $x = 0.5$ can be considered as a spin glass with $T_f = 30$ K. We can explain the collapse of the long-range ferromagnetic ordering by a diamagnetic dilution of the Mn-sublattice. According to the resistivity vs. temperature measurements, $La_{1-x}Sr_x(Mn_{1-x/2}Nb_{x/2})O_3$ samples are semiconductors. Below the Curie point a large magnetoresistance is observed.

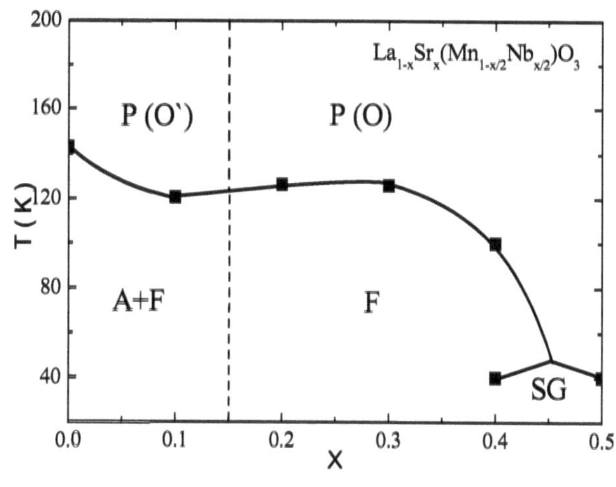

Figure 3. Magnetic phase diagram for $La_{1-x}Sr_x(Mn_{1-x/2}Nb_{x/2})O_3$ series (A- antiferromagnet, F- ferromagnet, P- paramagnet, SG- spin glass; O^I and O- orbitally ordered and orbitally disordered phases, respectively).

The results presented here deal with the facts that the Nb-doped $La_{1-x}Sr_x(Mn^{3+}_{1-x/2}Nb^{5+}_{x/2})O_3$ samples enriched with Mn^{3+} ions are ferromagnetic and show a large magnetoresistance. It is worth noting that the possibility of the existence of ferromagnetic ordering in the manganites, despite the absence of Mn^{3+} ions, refutes the double exchange and the electronic phase separation concepts. The result obtained indicates an important role of the ferromagnetic-superexchange via oxygen scenario for the magnetic interactions in manganites. According to the superexchange mechanism the Mn^{3+}-O-Mn^{3+} and Mn^{3+}-O-Mn^{4+} 180° magnetic interactions are strongly ferromagnetic for the orbitally disordered state whereas the Mn^{4+}-O-Mn^{4+} ones are strongly antiferromagnetic [4]. The Curie point associated with the Mn^{3+}-O-Mn^{3+} positive superexchange may be close to room temperature for manganites with perovskite structure, because our samples contain diamagnetic Nb^{5+} ions which should strongly decrease the Curie point. The stoichiometric $LaMn^{3+}O_3$ compound also shows ferromagnetic interactions between Mn^{3+} ions when cooperative Jahn-Teller distortions have vanished at $T = 750$ K. The orbital ordering changes the character of the superexchange magnetic interactions which in the orbitally ordered state become anisotropic [4, 28].

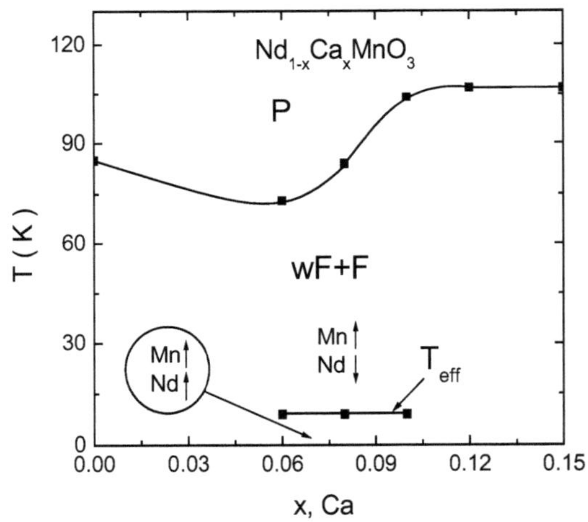

Figure 4. Magnetic phase diagram of the $Nd_{1-x}Ca_xMnO_3$ low-doped manganites. wF- weak ferromagnet, F- ferromagnet, P- paramagnet, T_{eff}- effective temperature of the reorientational phase transition.

2.3 *$Nd_{1-x}Ca_xMnO_3$ system*

The hypothetical magnetic phase diagram of the $Nd_{1-x}Ca_xMnO_3$ system at low Ca doping level is presented in Figure 4. Neutron diffraction shows that the samples with $x < 0.08$ consist mainly of antiferromagnetic phase, whereas at $x \geq 0.08$ the ferromagnetic component dominates. Under hole doping the temperature of the transition to the paramagnetic state at first decreases and then, around $x = 0.1$, increases. We have observed two magnetic phase transitions in the range $0.06 \leq x \leq 0.1$ as the temperature decreases.

The $Nd_{1-x}Ca_xMnO_3$ solid solutions contain two types of magnetically active sublattices: neodymium and manganese ones. At first we discuss the Nd contribution to the magnetic properties. The f-f exchange interaction in the rare-earth sublattice is as a rule rather weak in comparison with the d-d interaction between manganese ions. One can expect that neodymium magnetic moments should order as a result of the f-d exchange interactions between the neodymium and manganese sublattices. Studies of the magnetic properties of $Nd_{1-x}Ca_xMnO_3$ samples confirm this viewpoint. According to neutron diffraction data the magnetic moments of the neodymium ions start to be ordered slightly below T_N. The magnetic moment of a Nd ion is about 1.2 μ_B at 2 K, and is directed opposite to the weakly ferromagnetic vector in $NdMnO_3$, while in the sample $x = 0.12$ the orientation of the Nd and Mn magnetic moments is the same. In the range $0.06 \leq x \leq 0.10$ a metamagnetic behavior was observed in large magnetic fields (triple hysteresis loops with a negative remanent magnetization).

According to our hypothesis, samples in the range $0.06 \leq x \leq 0.10$ consist of antiferromagnetic (weakly ferromagnetic) and ferromagnetic phases, which are exchange coupled at the boundary. The neodymium sublattice in both weakly ferromagnetic and ferromagnetic phases orders nearby the Néel point (Curie point). However, the orientation of the neodymium magnetic moments in both these phases is different: the f-d exchange is positive for the ferromagnetic phase, whereas it is negative in the weakly ferromagnetic phase. The ferromagnetic phase strongly affects th magnetic properties of the weakly ferromagnetic phase due to exchange coupling at the boundary. This interaction may induce a reorientational transition from an antiparallel orientation of the neodymium moments and the weakly ferromagnetic vector to a parallel one. We believe that nearby a certain temperature the ground state of the Nd^{3+} ions becomes degenerate, because opposite contributions from the exchange coupled ferromagnetic and weakly ferromagnetic phases at the Nd site become equal. According to theoretical consideration this state should be unstable, thus leading to a magnetic structure transformation [29]. A neutron diffraction study carried out for the $x = 0.08$ sample is in agreement with this interpretation of the low-temperature phase transition.

On the basis of magnetization data we propose the *H-T* magnetic-phase diagram of $Nd_{0.92}Ca_{0.08}MnO_{2.98}$ shown in Fig. 5. Depending on the prehistory in the wide range of magnetic fields, the phases with a parallel or antiparallel orientation of neodymium and manganese sublattices can be realized in weakly ferromagnetic phase. One can see that the magnetic field required for changing the relative orientation of the Nd and Mn magnetic moments in the weakly ferromagnetic phase increases as temperature rises. The width of the field range in which the hysteresis is observed practically does not depend on the temperature. This type of magnetic phase diagram is in agreement with a crossover of energy sub-levels of the Nd ions.

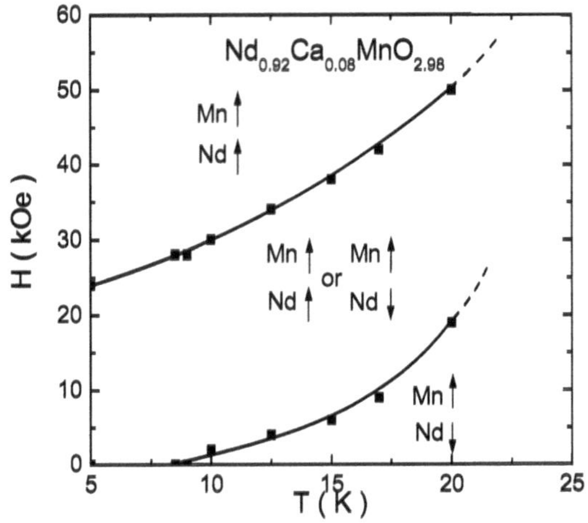

Figure 5. The *H-T* magnetic phase diagram for $Nd_{0.92}Ca_{0.08}MnO_{2.98}$.

2.4 *$Bi_{1-x}Ca_xMnO_3$ system*

Figure 6 presents a magnetic phase diagram of the $Bi_{1-x}Ca_xMnO_3$ manganites. As the calcium content in the $Bi_{1-x}Ca_xMnO_3$ system increases, the latter passes through three different magnetic states, namely, ferromagnetic ($x \leq 0.1$), spin-glass ($0.15 \leq x \leq 0.25$), and antiferromagnetic ($x > 0.25$). In the case of antiferromagnetic compositions, the magnetic-ordering and structural-transformation temperatures vary only weakly within the concentration interval from $x = 0.25$ to 0.6. The ferromagnetic ordering in $BiMnO_3$ is most likely due to cooperative ordering of the $d_{x^2-y^2}$ orbitals [30, 31]. With orbital ordering of this type, according to the Goodenough-Kanamori rules, ferromagnetic ordering becomes energetically more favorable than antiferromagnetic ordering. We may recall that rare-earth manganites exhibit orbital ordering of the d_{z^2} type, which stabilizes the A-type antiferromagnetic structure [4]. Orbital disorder in $BiMnO_3$ sets in, apparently, at a fairly high temperature, near 760 K. Replacement of bismuth ions by calcium results in the formation of quadrivalent manganese ions, which should be accompanied by destruction of orbital ordering due to the appearance of non-Jahn-Teller Mn^{4+} ions in the lattice. However, the orbitally disordered phase in manganites should be ferromagnetic [4, 28], whereas we observed a state of the spin-glass type. A direct transition from the antiferromagnetic to the spin-glass phase without passing through the ferromagnetic state was observed to occur in the rare-earth manganites $Sm_{1-x}Ba_xMnO_3$ and $Y_{1-x}Ca_xMnO_3$ ($x \sim 0.12$) [32, 33]. It should be pointed out that at approximately this concentration of rare-earth ions, the ferromagnetic–spin glass transition takes place in $Bi_{1-x}Ca_xMnO_3$.

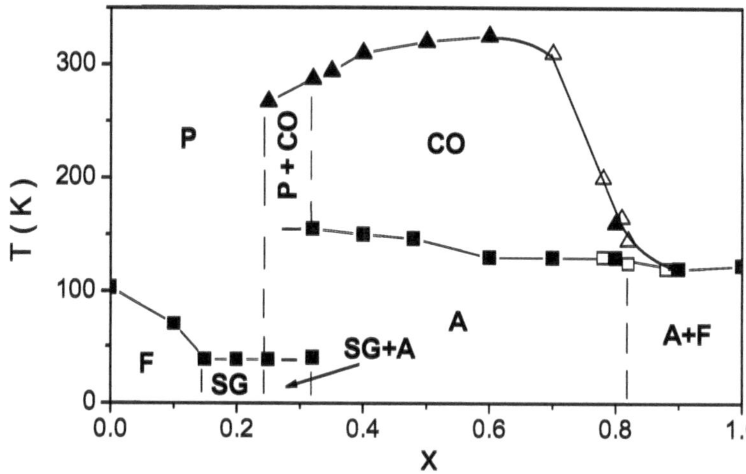

Figure 6. Magnetic phase diagram of the $Bi_{1-x}Ca_xMnO_3$ manganites. A- antiferromagnet, F- ferromagnet, P- paramagnet, SG- spin glass, CO- charge-ordered state.

There is more than one opinion on the nature of exchange interactions in manganites. The antiferromagnetic state certainly forms through oxygen-mediated superexchange interactions of the type Mn-O-Mn. Most researchers believe that the ferromagnetic state in manganites is created through double exchange, i.e., via direct carrier transfer between various lattice sites. In order for such an exchange mechanism to operate, manganese ions in different valence states must be present and the electrical conductivity must be high. The presence of manganese ions of different valencies is not a sufficient condition for high electrical conductivity; indeed, the 3d orbitals of manganese and the 2p orbitals of oxygen should also overlap strongly. It is believed that this parameter is controlled by the Mn-O-Mn bond angle [4, 30]. The larger the lanthanide ion, the larger should be the Mn-O-Mn angle, the wider the 3d band, and, accordingly, the higher the magnetic ordering temperature and the electrical conductivity. It was observed that the magnetic state of the manganites also depends on the difference between the ionic radii of the rare-earth and the lanthanide ions. A large difference between the radii lowers, as a rule, the magnetic ordering temperature as a result of competition between various exchange interactions characterized by a large difference in the Mn-O-Mn angles. This is why the spin-glass state sets in in the $Sm_{1-x}Ba_xMnO_3$ system, wherein the average radius of the Sm and Ba ions is far larger than that between the Y and Ca cations in the $Y_{1-x}Ca_xMnO_3$ system [32, 33]. However, in all rare-earth manganites, the Mn^{3+}-O-Mn^{4+} exchange coupling in the orbitally disordered phase is apparently ferromagnetic. The Mn-O-Mn angles in bismuth-based manganites are fairly large, which is supported by structural studies [31] and the quite high Curie temperature of $BiMnO_3$. Hence, in the case of an orbitally disordered phase, one can expect the ferromagnetic part of the exchange interactions to be dominant, which is at odds with experiment. Therefore, we believe that, in contrast to the rare-earth manganites, no orbitally disordered phase forms in the $BiMnO_3$ system in the concentration interval $0.1 \leq x \leq 0.3$. The spin-glass state forms in the $Bi_{1-x}Ca_xMnO_3$ system most likely as a result of competition between ferromagnetic interactions in $BiMnO_3$-type clusters and antiferromagnetic coupling in clusters in which the Mn^{3+} orbitals are frozen in random orientations. As the Ca^{2+} concentration increases, a new type of antiferromagnetic clusters, apparently due to charge ordering, appears. The existence. in $Bi_{0.75}Ca_{0.25}MnO_3$, of large clusters, charge-ordered in a similar way to those in $Bi_{0.5}Ca_{0.5}MnO_3$, is suggested in studies of its elastic properties. Despite the presence of the spin-glass-type ground state, there is a certain fraction of states characterized by short-range order of the type of a charge-ordered phase, which is indicated by the fact that the Young modulus minima for the $x = 0.25$ and 0.35 compositions are close in temperature. We believe that the extremely high stability of the orbitally and charge-ordered states in bismuth-based manganites derives from the strongly anisotropic character of the Bi-O covalent bonding.

3. Conclusions

The magnetic and structural phase diagrams $La_{0.88}MnO_x$, $La_{1-x}Sr_x(Mn_{1-x/2}Nb_{x/2})O_3$, $Nd_{1-x}Ca_xMnO_3$, and $Bi_{1-x}Ca_xMnO_3$ manganites have been proposed. It has been shown that the magnetic properties of the samples under study are determined by the type of their orbital state. The dynamic correlations of $d_{3z^2-r^2}$ orbitals favour ferromagnetic ordering in the manganites, while an A-type antiferromagnetic structure is typical for the static Jahn-Teller distortions. It has been argued that concentrational transition from an antiferromagnetic to a ferromagnetic state occurs via the formation of an inhomogeneous state due to a structural phase separation mechanism.

Acknowledgments

This work was partly supported by the Belarus Fund for Basic Research (grant F04MS-004).

References

[1] C. Zener, Phys. Rev. **82**, 403 (1951).
[2] A.J. Millis, B.I. Shraiman, and R. Müller, Phys. Rev. Lett. **77**, 175 (1996).
[3] E.L. Nagaev, Phys. Rep. **346**, 387 (2001).
[4] J.B. Goodenough, A. Wold, R.J. Arnott, and N. Menyuk, Phys. Rev. **124**, 373 (1961).
[5] G.-L. Liu, J.-S. Zhou, and J.B. Goodenough, Phys. Rev. B **64**, 144414 (2001).
[6] G. Biotteau, M. Hennion, F. Moussa, J. Rodriguez-Carvajal, L. Pinsard, A. Revcolevschi, Y.M. Mukovskii, and D. Shulyatev, Phys. Rev. B **64**, 104421 (2001).
[7] G. Allodi, R. De Renzi, and G. Guidi, Phys. Rev. B **57**, 1024 (1998).
[8] G. Papavassiliou, M. Fardis, M. Belesi, T.G. Maris, G. Kallias, M. Pissas, D. Niarchos, C. Dimitropoulos, and J. Dolinsek, Phys. Rev. Lett. **84**, 761 (2000).
[9] G. Papavassiliou, M. Belesi, M. Fardis, and C. Dimitropoulos, Phys. Rev. Lett. **87**, 177204 (2001).
[10] M.M. Savosta, V.I. Kamenev, V.A. Borodin, P. Novak, M. Marysco, J. Hejtmanek, K. Dorr, and M. Sahata, Phys. Rev. B **67**, 094403 (2003).
[11] P.A. Algarabel, J.M. De Teresa, J. Blasco, M.R. Ibarra, Cz. Kapusta, M. Sikora, D. Zajac, P.C. Riedi, and C. Ritter, Phys. Rev. B **67**, 134402 (2003).
[12] M. Hennion, F. Moussa, G. Biotteau, J. Rodriguez-Carvajal, L. Pinsard, and A. Revcolevschi, Phys. Rev. Lett. **81**, 1957 (1998).
[13] R.H. Heffner, L.P. Le, M.F. Hundley, J.J. Neumeier, G.M. Luke, K. Kojima, B. Nachumi, Y.J. Uemura, D.E. MacLaughlin, and S.-W. Cheong, Phys. Rev. Lett. **77**, 1869 (1996).
[14] C.H. Booth, F. Bridges, G.H. Kwei, J.M. Lawrence, A.L. Cornelius, and J.J. Neumeier, Phys. Rev. Lett. **80**, 853 (1998).
[15] M. Fath, S. Freisem, A.A. Menovsky, Y. Tomioka, J. Aarts, and J.A. Mydosh, Science **285**, 1540 (1999).
[16] M. Uehara, S. Mori, C.H. Chen, and S.-W. Cheong, Nature (London) **399**, 560 (1999).
[17] B.C. Hauback, H. Fjellvag, and N. Sakai, J. Solid State Chem. **124**, 43 (1996).
[18] T. Shibata, B. Bunker, J.F. Mitchel, and P. Schiffer, Phys. Rev. Lett. **88**, 207205 (2002).
[19] J. Rodriguez-Carvajal, M. Hennion, F. Moussa, A.H. Moudden, L. Pinsard, and A. Revcolevschi, Phys. Rev. B **57**, R3189 (1998).
[20] M.C. Sanchez, G. Subias, J. Garcia , and J. Blasco, Phys. Rev. Lett. **90**, 045503 (2003).
[21] D. Louca, T. Egami, E.L. Brosha, H. Roder, and A.R. Bishop, Phys. Rev. B **56**, R8475 (1997).
[22] T. Mizokawa, D.I. Khomskii, and G.A. Sawatzky, Phys. Rev. B **63**, 024403 (2001).
[23] Y. Endoh, K. Hirota, S. Ishihara, S. Okamoto, Y. Murakami, A. Nishizawa, T. Fukuda, H. Kimura, H. Nojiri, K. Kaneko, and S. Maekawa, Phys. Rev. Lett. **82**, 4328 (1999).
[24] S. Okamoto, S. Ishihara, and S. Maekawa, Phys. Rev. B **61**, 14647 (2000).
[25] R. Kajimoto, H. Mochizuki, and H. Yoshizawa, Physica B **329-333**, 738 (2003).
[26] R.L. Rasera and G.L. Catchen, Phys. Rev. B **58**, 3218 (1998).
[27] E.O. Wollan and W.C. Koehler, Phys. Rev. **100**, 545 (1955).

[28] J.B. Goodenough, J.-S. Zhou, F. Rivadulla, and E. Winkler, J. Solid State Chem. **175**, 116 (2003).

[29] A.K. Zvezdin, V.M. Matveev, A.A. Mukhin, A.A. Popov, Rare-Earth Ions in Magnetically Ordered Crystals, Moscow, Science (1985).

[30] I.O. Troyanchuk, N.V. Kasper, O.S. Mantytskaya, and S.P. Pastushonok, Zh. Eksp. Teor. Fiz. **105**, 239 (1994) [JETP **78**, 212 (1994)].

[31] T. Atou, H. Chiba, K. Ohoyama, Y. Yamaguchi, and Y. Syono, J. Solid State Chem. **145**, 639 (1999).

[32] I.O. Troyanchuk, D.D. Khalyavin, S.V. Truhanov, and H. Szymczak, J. Phys.: Condens. Matter **11**, 8707 (1999).

[33] R. Mathieu, P. Nordblad, D.N.H. Nam, N.X. Phuc, and N.V. Khiem, Phys. Rev. B **63**, 174405 (2001).

INTERFACE AND MAGNETIC CHARACTERIZATION OF FM/AF/FM MULTILAYERS

V. R. Shah, Christian Schanzer and Peter Böni
Physik Department E21, Technische Universität München, James-Franck Strasse, D-85747 Garching, Germany
shah@frm2.tum.de

Hans-Benjamin Braun*
Institut Laue-Langevin, 6 rue Jules Horowitz, BP 156-38042 Grenoble Cedex 9, France
beni.braun@ucd.ie

Abstract: We use polarized neutron reflectivity to study the evolution of the magnetization profile during the hysteresis of FeCoV (20 nm)/NiO (t_{NiO}) /FeCoV (20 nm) trilayers. For small t_{NiO}, we find that the magnetization in the two ferromagnetic layers remains parallel, indicating a strong exchange coupling across the anti-ferromagnetic NiO layer. Beyond a film thickness t_{NiO} = 40 nm, the observed step in the hysteresis is seen to correspond to an antiparallel orientation of the two ferromagnetic layers. Since this length scale agrees with the domain wall width of NiO, our results suggest that the interlayer exchange coupling involves a partial domain wall formation in the antiferromagnetic layer.

Keywords: Magnetic multilayers, interlayer exchange coupling, exchange bias, magnetization reversal, X-ray reflectivity, polarized neutron reflectivity, domain walls, magnetic roughness.

* Permanent address:
UCD School of Physics, University College Dublin, Belfield, Dublin 4, Ireland

J. Franse et al. (eds.), Smart Materials for Ranging Systems, 179–194.

1. Introduction

Interlayer exchange coupling between two ferromagnetic (FM) layers mediated by materials like metals, semiconductors, insulators etc. has been a fascinating topic in the recent past. Such nanostructures provide interesting physics and innovative ideas for a new generation of magnetic storage devices, magnetic field sensors and state of the art 'spintronic' devices [1]. In general, the nature of interlayer exchange coupling depends on the type of the spacer material and its thickness. In the case of nonmagnetic metals, the Ruderman-Kittel-Kasuya-Yosida (RKKY) type of exchange interaction and the critical spanning vectors of the Fermi surface of the spacer layer explain quite satisfactorily the observed oscillation period of the FM and AFM exchange between the two adjacent ferromagnetic layers [2, 3]. The oscillatory nature of the exchange interaction has been exploited to obtain giant magneto-resistance (GMR) in metallic multilayers [4] and found application in GMR read-heads that had a significant impact on magnetic data storage. Later, it was discovered that the use of insulating spacer layers (like MgO, Al_2O_3 etc.) in the tunneling regime (layer thickness ~ few Å), leads to a further enhancement of the MR ratio [5, 6]. Devices based on the tunneling magneto-resistance (TMR) are being implemented in a magnetic random access memory (MRAM) which is non-volatile and thus expected to bring significant improvements over the conventional RAM. The mechanism of interlayer exchange in insulators can be understood as spin-polarized tunneling [7]. It is clear that the tunneling mechanism operates only for a very small thickness of the spacer layer, and the tunneling current decays exponentially as the spacer thickness increases. The strength of the exchange coupling coefficient (J) correspondingly decays exponentially with increasing spacer thickness, and the nature of coupling changes from an AF type for very thin spacer layers to a FM type for larger spacer layers [7]. However, no oscillatory type of exchange interaction was observed in insulating spacer layers. An intermediate scenario is realized in a semi-conducting spacer layer. Using $Fe_{1-x}Si_x$ as a spacer layer, Grünberg et al. [8] found a mixed type of exchange interaction between two Fe layers. With increasing Si concentration the exchange coupling changes from an oscillatory behaviour to an exponentially decaying one. The study of ferromagnet-semiconductor interfaces and diluted magnetic semiconductors (DMS) is currently a hot topic as such systems are envisaged to play a key role in prospective 'spintronic' devices that make use of the electron's charge and spin degrees of freedom [9].

The phenomenon of exchange bias was discovered in 1956 by Meiklejohn and Bean [10]. For systems exhibiting a FM/AF interface, the cooling in an external field typically results in a shifted or 'biased' hysteresis loop. Upon field cooling from above the Néel temperature of the AF, interface ordering sets in at the so called blocking temperature (T_B), and the

AF moments at the interface are pinned. Below T_B, due to the direct FM coupling between the AF and FM interface moments, the AF moments introduce a torque and hinder the rotation of the FM moments during the magnetization reversal when the applied field is opposite to the cooling field direction. When the FM is rotated into the direction of the cooling field, this torque assists the external field resulting in an early reversal of the FM magnetization. As a result, the M-H loop of the FM is shifted in the field axis opposite to the cooling field direction, the so called negative exchange bias. This feature of the FM/AF interface has been exploited to pin one FM layer in the artificially engineered GMR heads (spin-valves), while the other layer is free to align with the external magnetic field. Even after four decades since its discovery, the microscopic origin of the exchange bias phenomenon remains a theoretical challenge since a quantitative prediction of the bias field from simple interface direct exchange models yields values that are two orders of magnitude larger than the experimental values [11]. In order to account for this discrepancy, the formation of domain walls in the FM or in the AF was postulated. According to this view, not only interface moments of the AF and FM are involved but also the internal spin structure of the FM and AF is affected during magnetization reversal. Malozemoff [12] suggested that, due to the random fields (arising from the random interface roughness) acting at the FM/AF interfaces, the AF breaks up into lateral domains forming transverse domain walls. Contrary to Malozemoff's model, Mauri *et al.* [13] and Kiwi *et al.* [14] predicted the formation of planar domain walls parallel to the interface in both AF and FM. In both cases, the formation of the domain wall reduces the interface energy and yields exchange bias fields of the same order as obtained in experiments. However, until recently there was no experimental evidence for the formation of such AF domain walls at the FM/AF interfaces. Scholl et al. using x-ray magnetic linear dichroism technique (XMLD) in Co/NiO [15] demonstrated that indeed the AF spins are twisted internally and that the twist angle could be varied by rotating the magnetization of the Co layer by applying an external magnetic field. These experiments hint at the fact that the internal spin structure of the AF is affected by the interface exchange coupling between the FM and AF.

In this article, we report investigations on the related system of a trilayer where the two ferromagnetic layers are separated by an antiferromagnet. We explore the AF thickness dependence on the magnetization reversal of trilayers FeCoV/NiO/FeCoV by bulk magnetization measurements and polarized neutron reflectivity (PNR). Previous investigations using various FM (Fe, Co, permalloy etc.) and AF layers (CoO, NiO, FeMn) revealed a rich variety of phenomena [16-20]. Both metallic and insulating antiferromagnets showed a nonvanishing exchange interaction between the two FM layers. Yang and Chien [20] investigated the dependence of this exchange coupling on the AF thickness within a series of field cooling experiments. The trilayers were field cooled from the AF ordering ("Néel-'') temperature down to room temperature, resulting in a pinning of the magnetization in the FM layers by the field cooling process. The angles between the two FM layers as a function of the AF thickness were determined at room temperature from the anisotropy of the M-H loops. In fact, the authors derived an empirical relation between the twist angle (θ) and the AF thickness (t_{AF}) as $\theta = 1.76\ t_{AF}$, for $53 < t_{AF} < 90$ Å. This result suggests that the AF plays an important role in mediating the interlayer exchange coupling, presumably by accommodating a spin twist within the AF layer. In addition, other researchers have also noticed different types of magnetization reversal in FM/AF/FM systems possibly hinting towards an interlayer exchange interaction [16, 17, 19, 20]. Since PNR is proved to be a powerful and a direct technique to unravel vectorial and layer resolved magnetization profiles of thin films [21], we have performed PNR experiments to observe the spin configurations during the magnetization reversal.

2. Experimental details

Multilayers of [Ti (5 nm) / FeCoV (20 nm) / NiO (t_{NiO}) / FeCoV (20 nm) / Ti (5 nm)], with $0.9 \leq t_{NiO} \leq 100$ nm are prepared on glass substrates using the DC magnetron sputtering facility 'TIPSI' at Paul Scherrer Institute, Villigen, Switzerland. The FeCoV layers are sputtered from an alloy target with a composition of $Fe_{50}Co_{48}V_2$. The sputtering of FeCoV and Ti is performed in an Ar atmosphere whereas the NiO layers are prepared from a Ni target by reactive sputtering in an appropriate Ar:O_2 atmosphere. It is well known that the (111) surface of NiO is a spin uncompensated surface (all the spins are parallel in the [111] planes) and it is expected that the termination of such an AF surface at the FM/AF interface promotes the highest interfacial exchange coupling with the FM and results in a large exchange bias. Appropriate sputter conditions for the formation of stoichiometric and highly (111) textured NiO were obtained by preparing NiO single films. Depending on the Ar:O_2 ratio and total pressure, the Ni-O_x calibration films possess different chemical composition and surface roughness. From the analysis of the critical angle for total reflection in specular x-ray reflectivity (XRR), we estimated the stoichiometry of the films. X-ray diffraction from the single films revealed the texture. The optimum Ar:O_2 ratio and the total pressure for sputtering to obtain stoichiometric, highly (111) textured and low surface roughness was found to be 86:14 and 6 μbar, respectively. However, when NiO was deposited on the top of FeCoV, additional (200) texture was also observed and the (200) surface is a spin compensated surface that can reduce the exchange bias. Nevertheless, due to interface roughness, a 'net' uncompensated spin density is present at the interface and these net uncompensated spins establish the interfacial exchange coupling and result in an average exchange bias [11]. The layer structure as well as interface roughness of our trilayers are obtained from the refinement of the XRR results. DC magnetization loops are measured at room temperature by the extraction technique using a quantum design physical property measurement system (PPMS). Polarized neutron reflectivities are measured at the time-of-flight reflectometer AMOR at the Swiss spallation neutron source SINQ. The polarization of the reflected beam is analyzed in order to distinguish non spin flip (NSF) and spin flip (SF) contributions. In the NSF case, the spin of the neutron remains parallel to the quantization axis (here, the applied magnetic field) after the scattering process and is represented by (+,+ or -,-). In the SF case, the incident neutron spin is rotated by 180° with respect to the quantization axis after the scattering process and is represented by (+,- or -,+). The NSF scattering originates from the component of the sample magnetization parallel to the neutron polarization and the SF scattering originates from the component perpendicular to the neutron polarization [21]. Therefore, a simultaneous modeling of the NSF and SF reflectivities reveals the magnitude and orientation of the sample magnetization. Polarizing supermirrors serve as polarizing and analyzing devices. Their efficiency is determined from additional experiments where the direct polarized beam from the polarizer mirror is analyzed by the analyzer mirror, assuming the mirrors to be identical. The wavelength dependent flipping ratio obtained by these experiments is used to correct the experimental data for the finite efficiency of the polarizing and analyzing supermirrors. The q_z range 0.1-1 nm^{-1} was covered by measuring the reflected TOF spectra at two angles of incidence (0.5° and 1.4°) of the neutron beam. A magnetic field is applied in the plane of the samples in order to magnetize the samples to the desired magnetic state. The x-ray and polarized neutron reflectivity analysis was performed using a code *'Simulreflec'* [22] based on a recursive matrix formalism.

3. Experimental results

Fig. 1 depicts the experimental specular x-ray reflectivity (XRR) data and the simulated reflectivites resulting from the best fit by modeling the chemical structure and layer profile of the trilayers. Prior to the fitting of the XRR of the trilayers, monolayers (Ti, NiO and FeCoV) and bilayers [(Ti/NiO(t_{NiO})/FeCoV) and Ti/FeCoV/NiO(t_{NiO})] were analyzed for the thickness calibration, roughness and scattering length densities (SLD) etc. It was possible to model all the XRR data in the series with a unique set of parameters, except the variation of interface roughness that increased with increasing NiO thickness. In all samples, the formation of an oxide layer is observed at the two FM/AF interfaces. The oxide layer at the first interface (FeCoV-NiO) is slightly thicker compared to that at the second interface (NiO-FeCoV). This difference may originate from the sputtering process. After the deposition of FeCoV and prior to the sputtering of Ni, the chamber is exposed to the Ar:O_2 mixture and consequently the FeCoV surface becomes oxidized. For the second interface, this situation does not occur, rather a minor oxidation that happens due to the interface diffusion. The interface oxide may play a significant role in the interface exchange. Very recent investigations on the magnetic properties of $Fe_{1-x}Co_x$/native oxide multilayers prove that a very thin (~13 Å) $Fe_{1-x}Co_x$-oxide layer acts as a medium for exchange coupling [23]. The fit results also show that up to t_{NiO} = 20 nm, the interfaces are rather smooth and the r.m.s roughness is < 0.7 nm. When t_{NiO} is thicker, the r.m.s roughness becomes larger (~1.4 nm for NiO = 40 nm and ~3 nm for NiO = 100 nm).

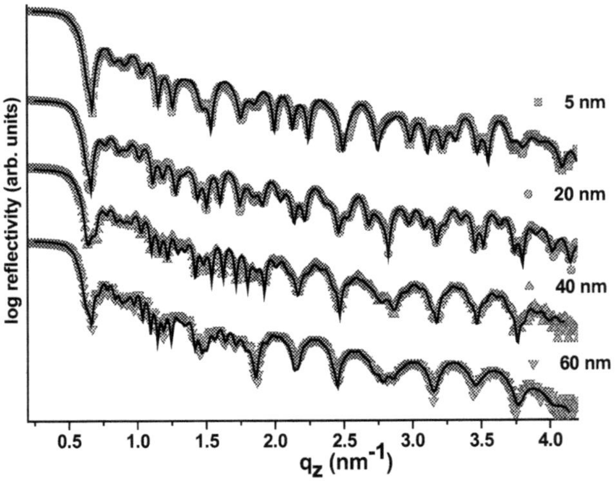

Figure 1. X-ray reflectivity analysis of FeCoV(20 nm)/NiO (t_{NiO})/FeCoV (20 nm) trilayers for t_{NiO} =5, 20, 40, 60 nm. The symbols represent the experimental data and the solid lines the best fit from the modeling.

Hysteresis of the trilayers as shown in Fig. 2 reveals a strong dependence on the AF thickness. We may distinguish three regimes according to the nature of the magnetization reversal. For $t_{NiO} \leq 10$ nm, a single and gradual reversal process of both layers occurs with an average coercivity of ~155 Oe. For $t_{NiO} \geq 20 - 40$ nm, a plateau region is established in the MH loop that widens with increasing t_{NiO}. For $t_{NiO} \geq 40$ nm, magnetization reversal occurs in two steps, one with a reduced coercivity (~ 60 Oe). In order to distinguish between the reversals of the top and bottom FeCoV layers, we investigated the magnetization reversal of a 20 nm thick

FeCoV layer placed either on top or bottom of a NiO layer of varying thickness t_{NiO}. For NiO on top of FeCoV, the coercivity remains unaltered. However, when FeCoV is on top of NiO, the coercivity of FeCoV is dramatically reduced by a factor of ~ 8 (Fig. 3).

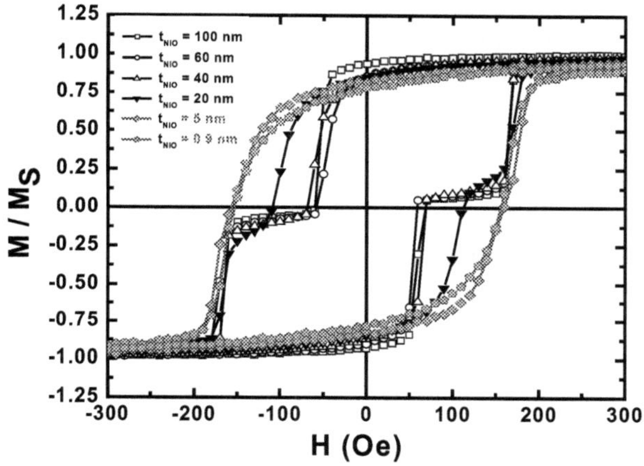

Figure 2. M-H loops of the FM/AF/FM trilayers at 300 K for various AF thicknesses showing the three types of reversals. For $t_{NiO} \leq 10$ nm both FM layers reverse in a single process. The formation of an intermediate plateau between two distinct reversal fields starts at $t_{NiO} \sim 20$ nm. For $t_{NiO} \geq 40$ nm, the plateau remains independent of t_{NiO}.

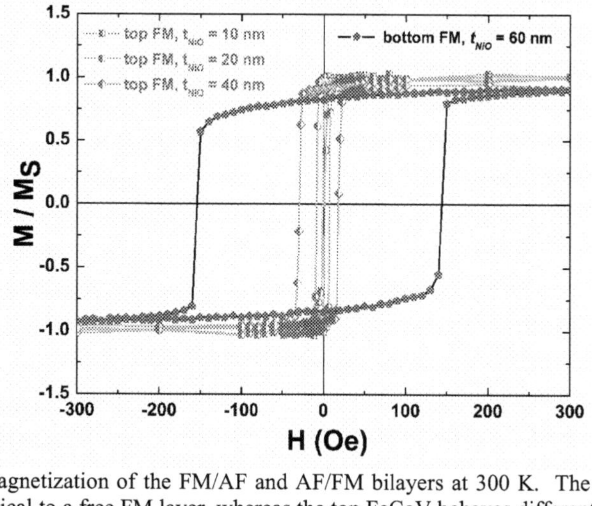

Figure 3. Magnetization of the FM/AF and AF/FM bilayers at 300 K. The bottom FeCoV behaves identical to a free FM layer, whereas the top FeCoV behaves differently as a function of the AF thickness.

Comparing the *M-H* loops of trilayers and their bilayer counterparts one would, at first instance, conclude the existence of an exchange interaction between the FM, mediated by the AF. We can't infer from the bulk magnetization only measurements whether this behaviour originates (i) due to the coupling through the AF or (ii) from Néel's 'orange-peel effect' [24] caused by the dipolar coupling due to magnetic roughness at the interfaces or (iii) whether the two FM layers are entirely decoupled and the step in the M-H loop of the trilayers simply represents the individual magnetization reversals of the two FeCoV layers with different coercivities. In order to check the second possibility, we estimated the 'orange-peel field (H_N)' at the NiO-FeCoV interface as follows. For a sinusoidal interface roughness profile (vertically coherent), with amplitude h and wavelength λ, the coupling field is given by [24]:

$$H_N = \frac{\pi^2 h^2}{\sqrt{2}\lambda t_f} M_S \exp\left(\frac{-2\pi\sqrt{2}t_b}{\lambda}\right) \tag{1}$$

This corresponds to the field that acts on the free layer in a FM (free)/non-magnetic/FM (pinned) system. In Eqn (1), t_f is the thickness of the free layer, t_b that of the non-magnetic spacer layer and M_S is the saturation magnetization of the pinned layer. The r.m.s. roughness at NiO-FeCoV interface obtained from the XRR analysis varies from 0.4 nm to 3nm for t_{NiO} = 1.5 nm – 100 nm, respectively. The wavelength of the roughness

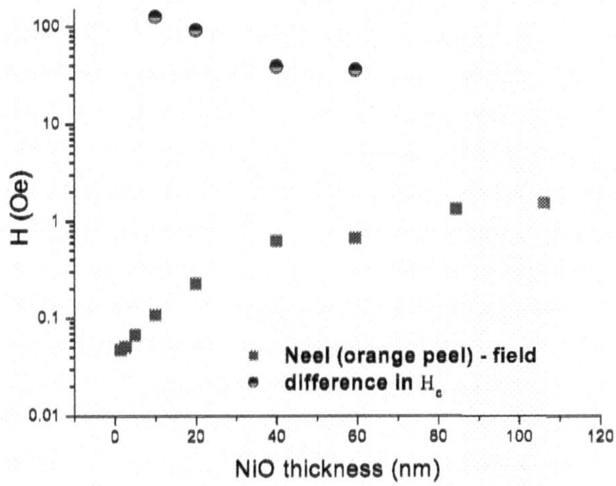

Figure 4. Contribution from Néel's 'orange peel effect' (H_N), originating from the magnetic roughness, compared to the difference in reversal fields of the top FeCoV layer in bilayer and its trilayer counterpart.

oscillation λ obtained from the AFM analysis is ~ 4 μm. From Eqn (1) it then follows that H_N accounts only for a few Oe (< 3 Oe) within the thickness range investigated [Fig. 4]. In addition, it has been shown that a roughness induced biquadratic coupling results in a 90° orientation between the FM layer magnetizations [25]. However, such a configuration is not consistent with our PNR data.

In order to distinguish between (i) and (iii) i.e., whether the FeCoV is exchange coupled via NiO or not, we performed field-cooling experiments and measured the M-H loops at 2 K. The results for t_{NiO} = 20, 40 and 60 nm are shown in Fig. 5. Initially, a sufficiently large magnetic field was applied to orient both magnetizations in one direction, and then the field was set at an appropriate value (as depicted in Fig.5) in the plateau region, before the field cooling procedure. Experimental results shown in Fig. 5 can be understood as the hysteresis loops of a soft and hard FM layer each of which experiencing a shift as indicated by thin and thick arrows. This shift is the well-known exchange bias phenomenon. This shift has opposite sign for the

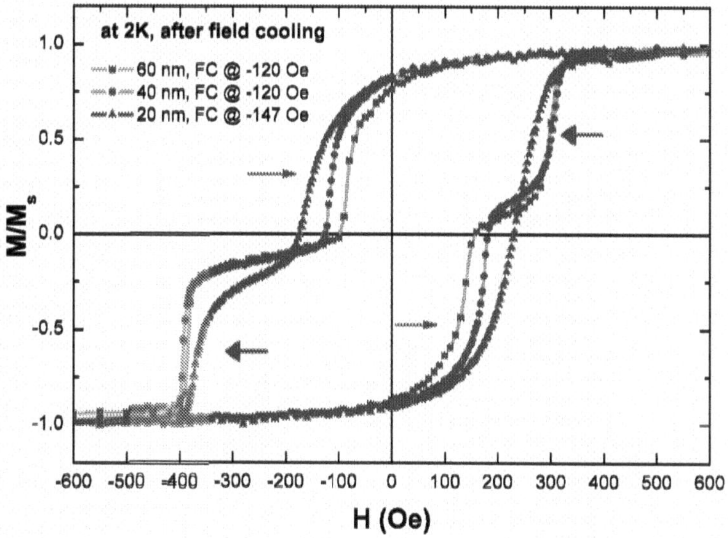

Figure 5. Magnetization loops of FeCoV (20nm) / NiO(t_{NiO} nm) / FeCoV(20 nm) samples at 2 K, after field cooling from room temperature under various cooling fields depicted in the figure after saturating the sample at 5 kOe. The individual ***M-H*** loops from each FeCoV layers are shifted in opposite directions (as shown by thin and thick arrows), indicating that the exchange coupling has opposite nature at the two interfaces.

two FM layers of our trilayer system. Moreover, the direction of the shift could be reversed by reversing the cooling field direction. This implies that one of the FM layers experiences a positive exchange bias and the other of opposite (negative). This is possible only when the exchange coupling at the interface responsible for the biasing has mutually opposite signs at the plateau region. The opposite nature of the exchange coupling may arise from either a spin twist of more than 90° in the AF or a FM driven interface spin configuration of the AF that gets pinned during the field cooling.

In this context, we would like to recall that previous field cooling experiments on permalloy/FeMn/Co trilayers by Yang and Chien [20] showed similar results. The spin configurations at the FM/AF interface in their case were pinned by setting a field at the plateau region of the M-H loop and then cooling from the Néel temperature of the AF (T_N = 400 K for FeMn) to 300 K. Thereby the angles (depending on the AF thicknesses) between the two ferromagnets, arising from an exchange coupling mediated by the AF, are revealed from the M-H loop measurements along various angles within the film plane. Results of our field cooling experiments from room temperature to 2K at reverse fields (Fig. 5) indeed suggest that the interface exchange is also being established at low temperatures for all AF thicknesses. However, it should be noted that the actual magnetic interface order is established at a temperature called 'blocking temperature' (T_B) and not at the Néel temperature. In a very recent theoretical study, Xi [26] predicts that the blocking temperature of a polycrystalline FM/AF system depends on numerous factors like the thickness, Néel temperature, grain size and the uniaxial anisotropy of the AF layer, T_C of the FM layer, the interface exchange coupling, hysteresis loop sweeping time etc. One possible explanation is the distribution of blocking temperatures caused by a distribution of interface exchanges. However, the similarities between our results and those from Yang and Chien and the nature of the shift of M-H loops after field cooling indicates that in our case the AF mediates an interlayer exchange by forming a twist within itself.

Magnetization measurements provide only a rather indirect way to explore magnetization configurations. We performed PNR in order to resolve unambiguously the relative spin orientations of the trilayer at various states of the hysteresis. Fig. 6 shows the spin dependent reflectivities and the magnetization profiles of the FeCoV layers obtained from the analysis, during the reversal process for t_{NiO} = 5, 20 and 60 nm samples. The PNR was measured after a magnetic saturation to –5 kOe and a subsequent increase of the magnetic field to an appropriate positive value. For t_{NiO} = 5 and 20 nm, over the coercive field regime, we observe that the two FM layers are rotating in unison with the applied field. However, for t_{NiO} = 60 nm, our PNR results unambiguously show that the FM layers are magnetized opposite to each other in the plateau region. We also note that the value of

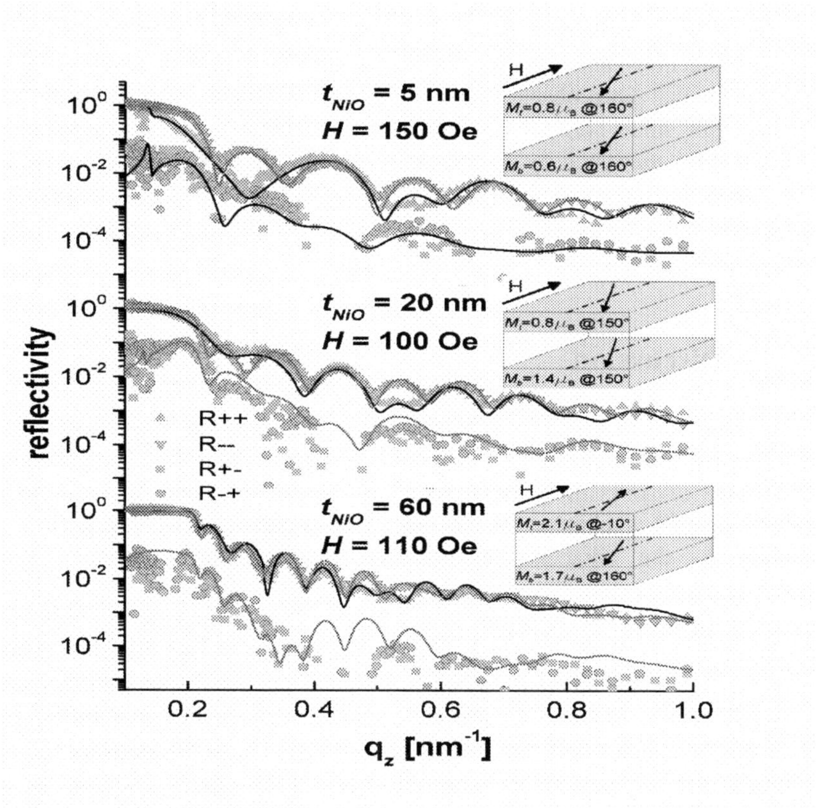

Figure 6. Polarized neutron reflectivities of t_{NiO} = 5 nm, 20 nm and 60 nm samples during the reversal process. The spin flip (R^{+-}, R^{-+}) and non spin flip (R^{++}, R^{--}) reflectivities are simultaneously modeled to obtain the magnetization configuration as shown in the inset. The lines are the computed reflectivities for different scattering cross-sections based on this model.

the effective magnetization is much reduced in the small t_{NiO} compared to large t_{NiO}, suggesting either domain formation or fanning of magnetization in the case of smaller t_{NiO}. Comparing the variation of the net magnetic moment of the top and bottom FeCoV layers as a function of t_{NiO} we infer that the net moment along the field direction approaches the saturation value of FeCoV (~ 2.1 μ_B) as t_{NiO} increases. In particular, this demonstrates that the plateau (~ zero net magnetization) in the case of a thick AF layer with t_{NiO} = 60 nm does not correspond to a net zero magnetization due to formation of domains within each FeCoV layer. The slight reduction of the magnetic moment from the saturation value indicates the existence of small, ripple type variations around the antiparallel state. For small t_{NiO}, the observed reduction of the magnetizations is consistent with either the formation of lateral domains or the existence of a vertical twist across the

FM similar to an exchange spring. The vertical twist is always present as the result of the non-vanishing torque at the interface that is responsible for exchange bias. Our observations are then consistent with the following scenario: For small t_{NiO}, a twist formed in one FeCoV layer may extend to the other layer mediated by a small spin twist in NiO. As $t_{NiO,}$ increases, this twist may be entirely confined within NiO. A nearly antiparallel configuration between the two FeCoV layers might be accompanied by a complete 180° spin twist within the AF.

4. Discussion

In FM/AF/FM trilayers it has been argued that the proximity effects predicted by Slonczewski [27] may play a significant role. He introduced the concept of 'proximity magnetism' as the consequence of combining a modest internal exchange in the AF with a strong coupling across the interface to the adjacent ferromagnet, resulting in a twisted quasi-antiferromagnetic state in the AF.

Semi-quantitative arguments by Néel [28] on partial domain wall formation in the AF of a FM/AF bilayer system predict that for films thinner than $(A_{AF}/K_u)^{1/2}$, the AF remains uniform when the interfacial magnetization is rotated. Here, A_{AF} and K_u are the exchange coefficient and the uniaxial anisotropy constants of the AF, respectively. Upon increasing the AF thickness, the AF is able to accommodate a twist up to 180° as the AF thickness (t_{AF}) approaches $\delta_0 = \pi(A_{AF}/K_u)^{1/2}$, the intrinsic DW width. A schematic illustration of the evolution of the AF spin configuration in a FM/AF/FM trilayer is depicted in Fig.7. The interfaces are not ideally flat and terminated with a fully compensated or uncompensated AF spin distribution as assumed by Néel. According to recent theoretical investigations, this mechanism even persists after taking into account of the interface roughness, polycrystalline grains etc. [29, 30]. The extent and the nature of the partial domain wall formation in the AF depend on the ratio of the AF thickness to its intrinsic domain wall width δ_0 and the strength of the exchange coupling at both interfaces. Furthermore, the AF spins may display a reversible twist for small fields, followed by irreversible switching of the sublattice magnetization at larger fields (i.e, after a certain twist, the AF moments springs forward to complete a 180° rotation) [29]. When t_{AF} is very small compared with δ_0, then the AF moments remain rigid and follow the magnetization reversal of the ferromagnets because they are exchange coupled to both FM.

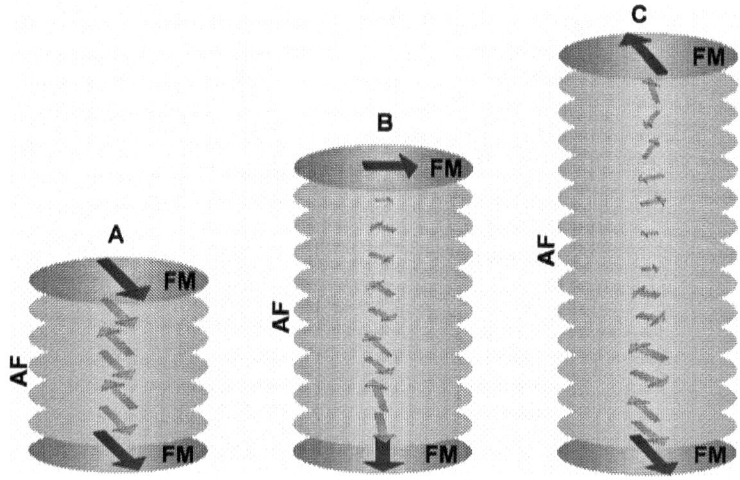

Figure 7. Illustration of partial domain wall formation in an AF layer when the AF is sandwiched between two FM layers. (A) When the AF thickness (t_{AF}) is very small compared to δ_0, no partial DW is formed and the two FM are direct coupled and the AF follows the rotation of the ferromagnets. (B) When t_{AF} is sufficiently large, a partial DW forms with a critical angle starting from 90° up to which the AF moments can be twisted and (C) when $t_{AF} \geq \delta_0$, a 180° DW is established.

Our experimental investigations suggest the existence of an exchange coupling between the two ferromagnets that depends on t_{NiO}. Here, we assume a single grain NiO with a vertical size that scales with the thickness of NiO spacer layer. FM/AF/FM trilayers with exchange coupling at both interfaces can be considered as an AF exchange spring loaded at both ends. The torque applied at either end due to the rotation of the FM magnetization by the application of an external magnetic field results in a winding and unwinding the AF exchange spring. For thin NiO spacers ($t_{NiO} < 20$ nm), we observe a single, gradual reversal process which is consistent with the theoretical prediction of direct coupling between the two FM with the AF spins following the FM spins during the reversal [29]. Our PNR results support this picture. It is worth mentioning that for $t_{NiO} \geq 40$ nm we start observing the plateau region, corresponding to a nearly perfect antiparallel ordering of FM layers. Interestingly, 40 nm is also reported to be a realistic estimate of the width of a 180° DW in NiO [30, 31]. It is interesting to note that other experimental studies using Brillouin Light Scattering [32] and MOKE magnetometry [33] to probe the role of AF in mediating the exchange coupling across the AF spacer concluded that indeed the AF spins play the significant role. In both cases Slonczewski's proximity magnetism only partially explains the observations. In the case of both FeCo/Mn/FeCo

[32], and Fe_3O_4/NiO/Fe_3O_4 [33] trilayers, discrepancies from Slonczewski's formalism were attributed to additional parameters like spin twist in the FM and the AF anisotropy. A 90° coupling was revealed between the FM layers at certain thickness range in these trilayers. Our PNR investigations did not provide evidence of such a coupling. Even though a direct verification of a 180° wall is not possible by PNR due the cancellation of magnetization between the AF sublattices, our PNR investigations suggests that the two FM layers are exchange coupled via NiO and the magnetization behaviour can be explained as a consequence of a spin twist in the AF, as we have also discussed elsewhere [34, 35]. Further investigations using element selective and orientation sensitive magnetization processes in these multilayers are in progress and will be very helpful to verify many recent theoretical predictions of the spin structures in FM/AF/FM trilayers.

5. Applications

There exist various applications in data storage that exploit the particular magnetic properties that arise from FM/AF interfaces and the interlayer exchange coupling between FM layers [36]. For example, in spin valves the unidirectional anisotropy of the exchange bias phenomenon enables the pinning of one of the two FM layers. A uniform external field can then be used to switch between the low magneto-resistance (MR) state with the magnetization in the two FM layers aligned parallel, and a high resistance state where the magnetization in the two layers is antiparallel. MR based devices find applications in magnetic field sensors and read-write heads in computer harddisks. Even for harddisks, there has been a recent tendency to use antiferromagnetically coupled media where the individual bits consist of AF coupled grains. The resulting gain in anisotropy energy allows to push the superparamagnetic limit towards smaller bit sizes, while at the same time the dipolar interaction between the bits is reduced [37].

An intriguing potential future application is based on the idea of using the DW resistance in a multi-turn sensor for automotive industry [38]. It involves the creation, transport and storage of a 360° DW. The creation of a 360° DW is achieved by the 360° rotation of a magnetic field at the junction of a geometrically constrained soft magnetic film. Being topologically stable objects, these 360° DWs can be transported and stored in an appropriate wedge-shaped nanostructure [38]. The existence of each 360° DW wall is then detectable through GMR by applying a clockwise and counter clockwise rotating magnetic field. This sensor is able to count and distinguish between the clockwise or counterclockwise twists of a whole array of 360° DWs.

The present study has illustrated the subtle interplay between geometric properties such as film thickness and the spin structure in magnetic trilayers. We hope that our results will lead to a deeper understanding of magnetic heterostructures and will help to design and improve future spintronic devices.

Acknowledgments

The authors are grateful to M. Horisberger for the kind help in the multilayer preparation, Dr. T. Gutberlet and Dr. M. Gupta for their assistance in PNR data collection. This work is based on experiments performed at the Swiss Spallation Neutron Source, Paul Scherrer Institute, Villigen, Switzerland.

References

[1] G. Prinz, Science **282** (1998) 1660.
[2] P. Bruno, Phys. Rev. **B 52** (1995) 411.
[3] S. S. P. Parkin, Phys. Rev. Lett. **67** (1991) 3598.
[4] M.N. Baibich, J. M. Broto, A. Fert, F. Nguyen Van Dau, F. Petroff, P. Etienne, G. Creuzet, A. Friederich, J. Chazelas, Phys. Rev. Lett. **61** (1988) 2472.
[5] M. Julliere, Phys. Lett. A **54** (1975) 225.
[6] J.M. Moodera, J. Magn. Magn. Mater. **200** (1999) 248.
[7] J. Faure-Vincent, C. Tiusan, C. Bellouard, E. Popova, M. Hehn, F. Montaigne and A. Schuhl, Phys. Rev. Lett. **89** (2002) 107206.
[8] P. Grünberg, D.E. Bürgler, R. Gareev, D. Olligs, M. Buchmeier, M. Breidbach, B. Kuanr and R. Schreiber, J. Phys. D: Appl. Phys. **35** (2002) 2403.
[9] Y. Ohno, D. K. Young, B. Beschoten, F. Matsukura, H. Ohno, D.D. Awschalom, Nature **402** (1999) 790.
[10] W. H. Meiklejohn and C. P. Bean, Phys. Rev. **102** (1956) 1413.
[11] A.E. Berkowitz and K. Takano, J. Magn. Magn. Mater. **200** (1999) 552.
[12] A. P. Malozemoff, Phys. Rev. B **35** (1987) 3679.
[13] D. Mauri, H.C. Siegmann, P.S. Bagus and E. Kay, J. Appl. Phys. **62** (1987) 3047.
[14] M. Kiwi, J. Mejia-Lopez, R. D. Portugal and R. Ramirez, Europhys. Lett. **48** (1999) 573.
[15] A. Scholl, M. Liberati, E. Arenholz, H. Ohldag, and J. Stöhr, Phys. Rev. Lett. **92** (2004) 247201.
[16] T. Ambrose and C. L. Chien, Appl. Phys. Lett. **65** (1994) 1967.
[17] F. Ernult, B. Dieny and J.R. Regnard, J. Magn. Magn. Mater. **242-245** (2002) 515.
[18] O. Zaharko, P. M. Oppeneer, H. Grimmer, M. Horisberger, H.-Ch. Mertins, D. Abramsohn, F. Schäfers, A. Bill and H.-B. Braun, Phys. Rev. B **66** (2002) 134406.
[19] Z.Y. Liu and S. Adenwalla, Phys. Rev. Lett. **91** (2003) 037207.
[20] F.Y. Yang and C. L. Chien, Phys. Rev. Lett. **85** (2000) 2597.
[21] H. Zabel and K. Theis Bröhl, J. Phys:. Condens. Matter **15** (2003) S505; G.P. Felcher, S.G.E. te Velthuis, A. Rühm and W. Donner, Physica B **297** (2001) 87.

[22] SimulReflec by F. Ott, http://www.llb.cea.fr/prism/programs/simulreflec/simulreflec.html.
[23] G. S. D. Beach, F. T. Parker, D. J. Smith and A. E. Berkowitz, Proc. of the 5th International Symposium on Metallic Multilayers, Boulder, CO, USA, June 7-11, 2004.
[24] B. D. Schrag, A. Anguelouch, S. Ingvarsson, Gang Xiao, Yu Lu, P. L. Trouilloud, A. Gupta, R. A. Wanner, W. J. Gallagher, P. M. Rice and S. S. P. Parkin, Appl.Phys.Lett. **77** (2000) 2373.
[25] S. Demokritov, E. Tsymbal, P. Grünberg, W. Zinn and Ivan K. Schuller, Phys. Rev. B **49** (1994) 720.
[26] Haiwen Xi, J. Magn. Magn. Matter. (2004) (in press).
[27] J. C. Slonczewski, J. Magn. Magn. Mater. **150** (1995) 13.
[28] Louis Néel, Ann. Phys. (Paris) **2** (1967) 61.
[29] Haiwen Xi and Robert M. White, Phys. Rev. B **62** (2000) 3933.
[30] M. D. Stiles and R. D. McMichael, Phys. Rev. B **59** (1999) 3722.
[31] C. H. Lai, H. Matsuyama, R. L. White, T. C. Anthony and G. G. Bush, J. Appl. Phys. **79** (1996) 6389.
[32] M. Chirita, G. Robins, R. L. Stamps, R. Sooryakumar, M. E. Philipkowski, C. J. Gutierrez and G. A. Prinz, Phys. Rev. B **59** (1998) 869.
[33] P. A. A. van der Heijden, C. H. W. Swüste, W. J. M. de Jonge, J. M. Gaines, J. T. W. M. van Eemeren and K. M. Schep, Phys. Rev. Lett. **82** (1999) 1020.
[34] C. Schanzer, V. R. Shah, P. Böni, H. B. Braun, T. Gutberlet and M. Gupta, Physica B (accepted for publication).
[35] V. R. Shah, C. Schanzer, P. Böni and H. B. Braun, J. Magn. Magn. Mater. (in press).
[36] J. Nogués and I. K. Schuller, J. Magn. Magn. Mater. **192** (1999) 203.
[37] E. E. Fullerton, D. T. Margulies, M.E. Schabes, M. Carey, B.Gurney, A. Moser, M. Best, G. Zeltzer, K. Rubin, H. Rosen, M. Doerner, Appl. Phys. Lett. **77** (2000) 3806.
[38] Marco Diegel, Roland Mattheis and Ernst Halder, IEEE Trans. Magnetics **40** (2004) 2655.

SQUID –BASED MAGNETIC MICROSCOPE

S. Bondarenko
Institute for Low Temperature Physics and Engineering of NASU, Kharkov, 61103, Ukraine
bondarenko@ilt.kharkov.ua

N. Nakagawa
Center of Non Destructive Evaluation, Iowa State University, Ames, USA
nakagawa@cnde.iastate.edu

Abstract: Two types of scanning magnetic microscopes (MM) are known nowadays: force magnetic microscopes (FMM) and non-force MM (NFMM) ones. The second type of MM can be used to measure a distribution of the vertical component of magnetic fields over surfaces, including cases such as in a cavity of solid test objects (TO) and inside a liquid TO, without external forces or magnetic influences on the TO. A NFMM with a SQUID detector (SQUID-MM) has the highest magnetic sensitivity among all types of MM. Because of these SQUID-MM features, it is suitable for use in developments of smart materials. For example, it may be used for: counter-terrorism investigations by surveying and decoding of the slight slots left after deletion of the magnetic records on a computer disc by traditional magnetic "scrubber" procedures; surveying and decoding of minute magnetic indications left after a mechanical scrubber procedure of mechanically stamped records of serial numbers on cars and other objects; surveying and measuring magnetic properties of small magnetic particles of the object under investigation for identification of the object; high-sensitivity reading of magnetic records on perspective superconducting carriers. The design of our three-channel SQUID-MM is described in this article. Magnetic 3D scan-pictures of several TO's are presented also for demonstration.

Keywords: SQUID, magnetic microscope, magnetic record, superconducting carrier.

J. Franse et al. (eds.), Smart Materials for Ranging Systems, 195–201.

1. Introduction

A Superconducting Quantum Interference Device (SQUID) is a typical quantum measuring device. Measurement of the magnetic field of different objects is a way to determine the parameters of the objects, including displacements. One of the first spatial applications of a SQUID as ranging system was the development of an airborne sensitive magnetometer for searching submarines. In the last years, there is the development of different micro SQUID systems for searching micro objects. Two types of scanning magnetic microscopes (MM) are known nowadays: a force magnetic microscope (FMM) [1] and a non-force magnetic microscope (NFMM) [2]. The second type of MM can be used to measure the distribution of the vertical component of magnetic fields over surfaces, including such cases as inside a cavity of solid test objects (TO) and inside a liquid TO without external forces or magnetic influences on the TO. The NFMM with a SQUID detector (SQUID-MM) has the highest magnetic sensitivity among all types of MM.

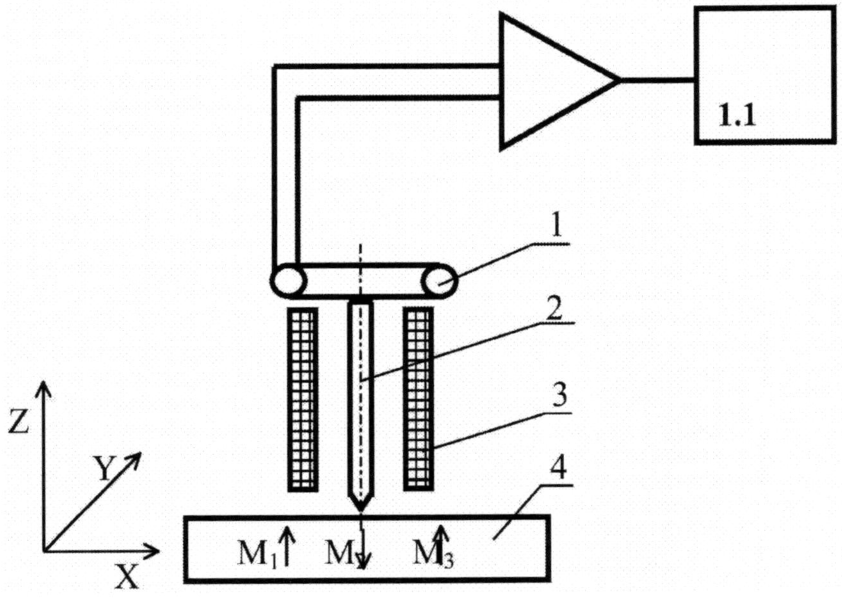

Figure 1. Scheme of a non-force scanning magnetic microscope .1 – magnetic detector; 2 – ferromagnetic concentrator of local magnetic field of the test object (TO); 3 – generator of an alternating magnetic field; 4 – TO with some magnetic dipole sources (M1,M2,M3); the TO can be driven by spatial motors along the X-Y axes.

2. Operation principles of a SQUID magnetic microscope

The principal scheme of a scanning non-force magnetic microscope (NFMM) is shown in Fig.1. The main units of the NFMM are the magnetic detector (1), the ferromagnetic concentrator (FC) of the object's magnetic field (2), the generator of an alternating magnetic field (3), the test object (4), which can be located by step motors near the detector in the X-Y plane, the electronic transformer of the detector signal (6) and the registration device(5).The spatial resolution of a MM depends on the distance between the detector and the object and on the step distance of the mechanical scanner. Usually the step distance can be as small as 0.1-10 microns.

3. A SQUID-MM as a novel magnetic ranging system

The SQUID – MM can have different applications:

- counter-terrorism investigations by surveying and decoding of the slight slots left after deletion of the magnetic records on a computer disc by a traditional magnetic "scrubber" procedure;
- surveying and decoding of minute magnetic indications left after a mechanical scrubber procedure of mechanically stamped records of serial numbers on cars and other objects;
- surveying and measuring magnetic properties of small magnetic particles of an object under investigation for identification of the object;
- a high-sensitivity reading of magnetic records on perspective superconducting carriers.

The general view of MM channels and its basis units is given in Fig. 2.

Figure 2. External view of a magnetic microscope (MM). From left to right are shown: 1-measuring channel with SQUID for "cold" objects study; 2- measuring channel with fluxgate for "warm" objects study; 3-measuring channel with SQUID for "warm" objects study; 4-scanner driving device; 5-SQUID and fluxgate electronic amplifier; 6- personal computer.

Channels 2 and 3 are used for a TO study at T≈300K; channel 1 situated left in fig.2 is used for a TO operating at T≈77K. The first two channels are combined with a two-axis table of the electromechanical scanner. The table with the test object TO can be set under a central fluxgate channel or under a SQUID channel.

During TO research at $T = 300$ K, a fluxgate channel is used first, in order to examine an area of 10x10 mm^2 with a spatial resolution up to 10 microns and with a sensitivity up to 10^{-9} T. After finding the most interesting area, the TO can be moved under the SQUID channel 3, which allows to examine a square having the size of 100x100 $micron^2$ with a spatial resolution of 0,1-1 micron at a sensitivity up to 10^{-12} T. The HTc-SQUID is manufactured from the high-temperature superconductor $YBa_2Cu_2O_{7-\delta}$ and has an operation temperature close to 77 K. The left-hand channel 1 has as detector also a HTc-SQUID with a ferromagnetic concentrator. The channel parameters for sensitivity and spatial resolution coincide with parameters of the right-hand SQUID channel 3. The TO holder is entered into the nitrogen cryostat of this channel 1 from above. The 2D-scanner is situated in the top part of the cryostat. The channels are magnetically shielded by a three-layer permalloy shield, which depresses low-frequency magnetic fields more than 100 times and high frequency more than 1000 times.

The high spatial resolution of the designed MM is provided first of all by application of the ferromagnetic needle between the magnetic field detector and the TO; it acts as a magnetic concentrator.

For providing a high level of channel parameters of the fluxgate channel, a special design of the fluxgate sensor has been applied. The fluxgate sensor looks like a closed ferromagnetic circuit with two pointed parts (the radius of curvature for them is no more than 10 microns). One of these parts is the detecting part of the fluxgate. It approaches the TO surface within 10 microns.

Some scan pictures resulting by the MM are presented in Fig.3 and 4.

4. Scan pictures of some TO's and the possibility of magnetic recording on superconducting carriers

Features of the magnetic recording on a superconducting carrier with reading using a MM are: the possibility of an absolute and total deletion of the information at the heating of the superconductor above its characteristic temperature; the possibility of driving the recording position by the current applied through the superconductor carrier; the possibility of the magnetically forming and using different active electronic structures (Josephson junctions, SQUID structures), while deleting and forming new electronic structures on the same superconducting carrier without heating it.

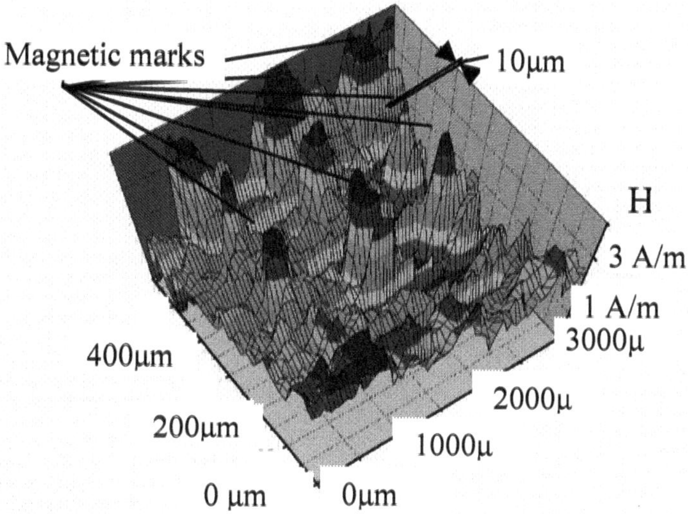

Figure 3. The distribution of the normal component of the magnetic field induction above a surface of a magnetic tape with recorded high density magnetic information.

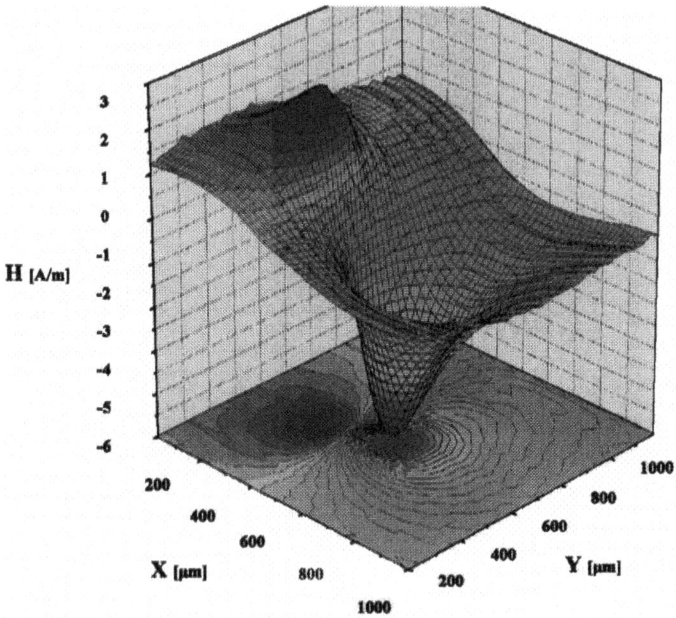

Figure 4. MM- image of the vertical component of a magnetic field over a small magnetic particle (the size is about 400 μm).

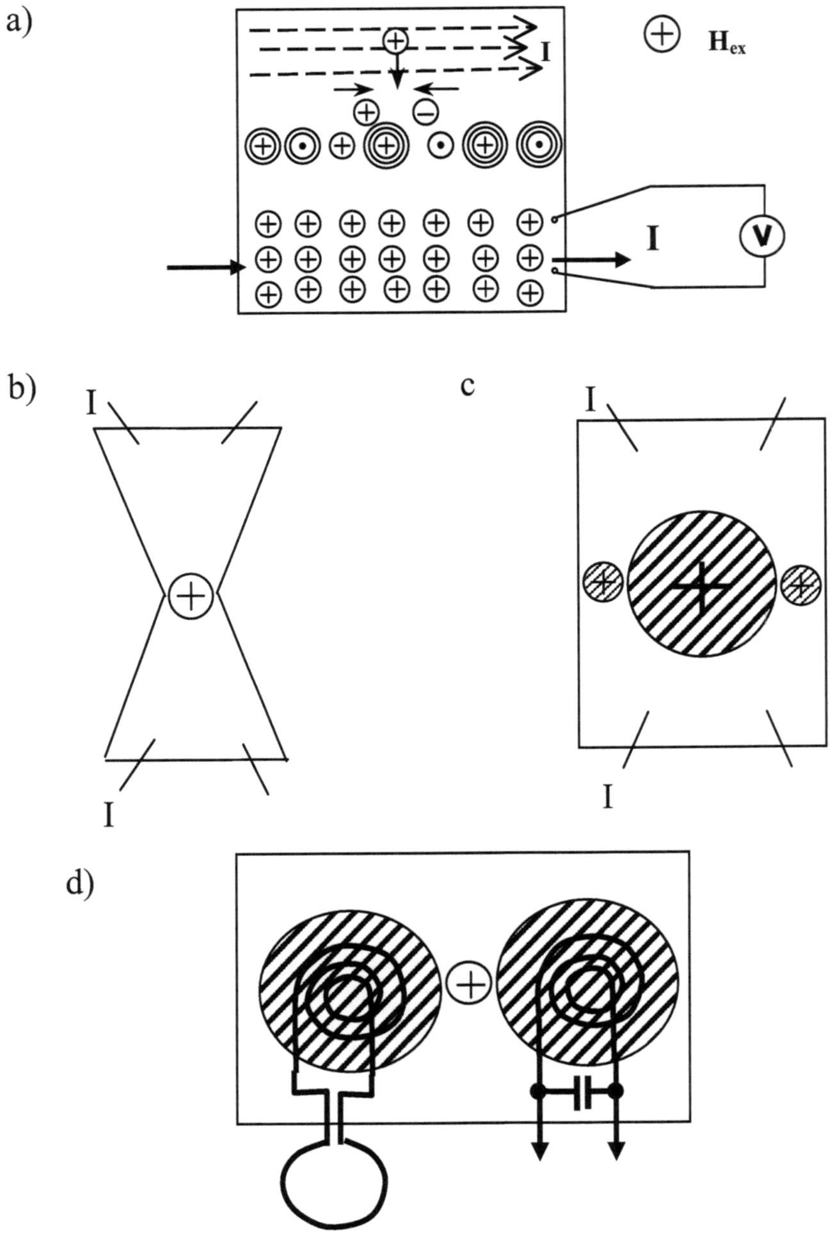

Figure 5. Examples of magnetic recordings: a- the schemes of different vortex systems (from top to bottom: -single vortex in a flow of electrons I; two vortices with opposite magnetic flux; line of vortices with different number of flux quanta; collective flow of a vortex system under influence of a transport current I with generation of the voltage V); b- Josephson junction; c,d- DC- and HF-SQUIDs .

Frozen current vortices (FCV) in superconductors can serve as essential elements of magnetic recording and as a convenient subject for investigations: magnetic flux pinning at a transport current *I* across the superconductor, annihilation of two FCV with opposite magnetic flux, quantum behavior of frozen magnetic flux in the FCV, collective flow of FCV under the influence of the transport current I (Fig.5a).

Formation of Josephson structures by FCV is another interesting field of application of magnetic recording on superconducting carriers.

For example, a single FCV on a narrow place of a superconducting micro bridge can play the role of a Josephson junction (Fig5b). The scheme of a thin-film dc-SQUID is shown in Fig.5c. The "hole" of the SQUID is formed by a FCV with a diameter of 100 microns and two Josephson junctions are formed by two FCV's with smaller diameter.

The scheme of the HF-SQUID with two "holes" is shown in Fig.5d. "Holes" of the SQUID are formed by two big FCV's and the Josephson junctions are formed by a FCV with a smaller diameter.

References

[1] J.R. Barnes, S.J. Oshea and M.E. Welland, J.Appl. Phys., 76, 418 (1994).

[2] T.S. Lee, E. Dantsker, J. Clarke, Rev. Sci. Instr., 76, 4208 (1996).

[3] S.I. Bondarenko, N. Nakagawa, A.A. Shablo, P.P. Pavlov, Physica B, 329-333, 1512 (2003).

FEMTOSECOND ELLIPSOMETRY: A NEW TOOL FOR THE MEASUREMENT OF HOT ELECTRON MOMENTUM RELAXATION TIMES IN METALS

V.V. Kruglyak and R.J. Hicken
School of Physics, University of Exeter, Stocker road, Exeter, EX4 4QL, UK
R.J. Hicken@exeter.ac.uk

Abstract: Optical pump-probe measurements were performed with an elliptically polarised pump to observe the transient nonlinear polarisation due to the Specular Inverse Faraday Effect and the Specular Optical Kerr Effect in thin films of Au, Cu, Ag, Ni, Pd, Ti, Zr, and Hf under identical experimental conditions. The temporal shapes of the signals were used to extract characteristic time scales for the decay of the transient polarisation, which were interpreted in terms of scattering of the linear and angular momentum of hot electrons. A systematic variation of the signal magnitudes and temporal shapes was observed between different metals that reflects their underlying band structure.

Keywords: time-resolved, third-order, nonlinearity, ultrafast, spin, dynamics, dephasing, SIFE, SOKE, noble, transition, hcp.

1. Introduction

Aided by developments in modern laser technology [1], optical pump probe techniques [2] have emerged as a means by which to study a variety of ultrafast electronic processes in metals which were previously inaccessible to the experimentalist. The sample is excited or "pumped" either directly by an intense optical pulse, or else by an optically triggered stimulus such as a pulsed magnetic field. The resulting dynamics are then sensed by measuring changes in the intensity and polarisation state of a time-delayed optical probe pulse that is reflected from the sample surface. In other variations upon the technique, electrons emitted to vacuum [3], or frequency-doubled light [4] may instead be collected.

Optical pump-probe techniques have been particularly useful in the study of magnetic metals. The recent discovery of the ultrafast demagnetisation

J. Franse et al. (eds.), Smart Materials for Ranging Systems, 203–221.

effect [5] may lead to a dramatic increase in the speed of data storage devices. The effect relies upon the collapse of the spontaneous magnetisation following the absorption of an ultrashort laser pulse [6]. The magnetisation vector may then recover in a new direction under the guidance of a small magnetic field [7]. The guiding field might be generated by a circularly polarized light pulse [8], as a realization of all-optical magnetic switching. Alternatively, the laser pulse may be used to locally induce a less drastic modification of a magnetic parameter of the recording medium, such as the anisotropy [9,10] or the exchange bias [11]. In this case the magnetisation reversal of the modified volume is completed by application of a magnetic field pulse with amplitude too small to affect the unpumped volume of the medium. This technique has been called heat assisted magnetic recording [12]. A key issue for both of these approaches is the development of efficient sinks for the excess heat.

Knowledge of the timescales for the linear and angular momentum dynamics of electrons and holes is important in understanding many different fundamental aspects of solid state physics [13]. It underlies topics in applied physics such as electronic transport in metallic nanostructures, where the carrier energy, direction of motion and spin polarization play a significant role [14]. The corresponding spin, momentum and energy scattering times are determined by the interactions between electrons, holes and phonons, and by scattering from defects, and are expected to be of the order of a few femtoseconds [13]. Hence, there is a need for time resolved optical experiments that have sufficient time resolution to resolve these processes [2].

In this paper, a first comparative study of the femtosecond non-linear optical response of a series of noble and transition metal thin films performed under identical experimental conditions is described. The response originates from optical orientation of the linear and angular momentum of electrons. An elliptically polarized pump beam was used so that contributions from the linear and angular non-linear polarisation were present simultaneously in the time resolved signal. This allowed a small yet significant time delay between these two contributions to be observed within signals of bipolar shape. The zero time delay position of the polarisation signal was defined with respect to the onset of the transient reflectivity signal which was measured simultaneously. Since the latter is expected to be independent of the pump polarisation, this also allowed the relative timing of the transient rotation and ellipticity signals to be compared for a particular contribution (due to either the linear or angular polarisation). In some cases, the individual linear and angular polarisation signals had bipolar shapes. The two contributions were separated using the technique developed in Ref. 15 and fitted to a simple theory developed in the relaxation time approximation. This allowed relaxation times to be extracted for the nonlinear polarisation, which were then interpreted in terms of scattering of the linear and angular momentum of hot electrons.

2. Phenomenological description of the transient polarisation

The interaction between light and matter can be viewed as the creation of a coherent quantum superposition of initial and final electron states that has an associated polarization [3], as shown in Figure 1. The coherence between states with different wave vector requires an intermediate virtual state and the presence of a coherent phonon. A transition between the initial and final states may occur when the coherence of the system is broken either due to the finite width of an optical wave packet or by scattering from the environment. The transition results in the absorption of a photon and the creation of a hot electron-hole pair. Otherwise, the photon is re-radiated with a different phase and, perhaps, polarisation.

The decoherence time T_2^*, which has also been called the dephasing time, is equal to $(1/2T_1+1/T_2)^{-1}$, where T_1 and T_2 are the inelastic (energy) and pure dephasing (elastic linear or angular momentum) scattering times [3]. Values for T_2^*, T_1 and T_2 have been deduced from time resolved two-photon photoemission (TR-2PPE), although their physical interpretation is not straightforward [3,16,17]. The induced polarisation may persist for a time T that is greater than T_2^* if the hot electron-hole pair retains some memory of the injected linear and (or) angular momentum after a scattering event.

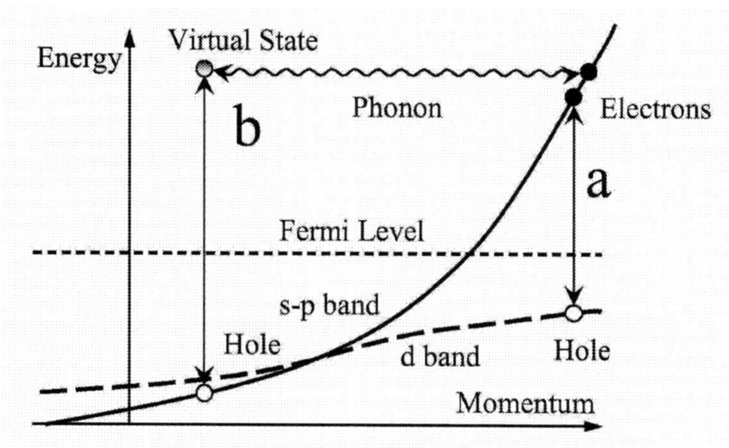

Figure 1. Electron-hole coherent state is schematically illustrated for interband (a) and intraband (b) optical transitions.

The pump induced transient polarisation of the medium modifies the polarisation state of a time delayed probe pulse. Phenomenologically, this process can be regarded as a transient pump induced linear or circular birefringence, also called the Specular Optical Kerr Effect (SOKE) and the Specular Inverse Faraday Effect (SIFE) [18]. These are cubic non-linear effects and are predicted to exist from symmetry arguments. Both effects consist of coherent and incoherent parts. For the coherent part, the pump drives the coherent electron-hole pair that affects the probe polarisation. The effect depends upon the probe phase relative to that of the electron-hole pair, and hence, that of the pump. For the incoherent part of the SIFE and the SOKE, the relative pump-probe phase is not important, since the probe pulse polarisation is modified by the pump induced sample polarisation that survives after the decoherence of the electron-hole pair.

The magnitudes of the incoherent transient SIFE and SOKE, which are the only effects relevant to the present study, may be related to the cubic components χ_{xxyy} and χ_{xyyx} of the nonlinear susceptibility tensor and the polarisation state of the pump pulse. Let us assume that the pump is initially p-polarised. The pump polarisation can be made elliptical by passing the beam through a quarter wave plate. The pump polarisation depends upon the angle φ between the plane of incidence and the fast axis of the quarter wave plate used to convert the pump pulse to an elliptically polarised state. The pump polarisation can be made linear at some angle to the plane of incidence by passing the beam through a polarizer. Let us denote the angle between the plane of incidence and the transmission axis of the pump polarizer as ψ. Then the magnitude of the signal due to the SIFE and SOKE is given by [15]

$$\begin{pmatrix} \Delta\theta \\ \Delta\eta \end{pmatrix} = \frac{1}{2}\sin 4\varphi \begin{pmatrix} \mathrm{Re} \\ \mathrm{Im} \end{pmatrix} F_{+} + \sin 2\varphi \begin{pmatrix} -\mathrm{Im} \\ \mathrm{Re} \end{pmatrix} F_{-}\,, \tag{1}$$

for an experiment with an elliptically polarised pump, and

$$\begin{pmatrix} \Delta\theta \\ \Delta\eta \end{pmatrix} = \sin 2\psi \begin{pmatrix} \mathrm{Re} \\ \mathrm{Im} \end{pmatrix} F_{+}\,, \tag{2}$$

for an experiment with a linearly polarised pump. In both formulae, the amplitudes of SIFE (F_-) and SOKE (F_+) are given by

$$F_{\pm} = -\frac{32\pi^2 I_{pump}}{c\left|1+\hat{n}\right|^2}\left\{\frac{\chi_{xxyy} \pm \chi_{xyyx}}{\hat{n}(1-\hat{n}^2)}\right\}, \tag{3}$$

where I_{pump} is the pump beam intensity, $\hat{n}$ is the complex refraction index, and c is the speed of light. These formulae were derived in the continuous wave approximation for normal incidence of the pump and probe beams.

In order to apply these equations to a femtosecond pump-probe experiment, an additional assumption has to be made regarding the shape of the time resolved signal. We wish to account for the finite relaxation time of the transient polarisation and so the signal must be described by a double convolution of an exponential decay function with the pump and probe intensity envelope functions. We will assume a Gaussian peak shape so that the convolution may be calculated analytically. As we will see, the experimental results require two such contributions, and hence, the following function will be used to fit the experimental data

$$\theta^{SIFE(SOKE)}(\eta^{SIFE(SOKE)}) \propto$$
$$\frac{1}{2}\sum_{j=1}^{2} A_{\theta(\eta),j}^{SIFE(SOKE)}\left(1+erf\left(\frac{t_d\sqrt{2\ln 2}}{\sigma}-\frac{\sigma}{2\tau_j^{SIFE(SOKE)}\sqrt{2\ln 2}}\right)\right)\exp\left\{-\frac{t_d}{\tau_j^{SIFE(SOKE)}}\right\}, \tag{4}$$

where $A_{\theta(\eta),j}^{SIFE(SOKE)}$ are the amplitudes of the SIFE and SOKE signal components in rotation and ellipticity, σ is the full width at half maximum (FWHM) of the Gaussian peak, and t_{d} is the pump-probe time delay. In fitting the experimental data, the amplitudes should be treated as adjustable parameters. Also, for a particular metal, the same four relaxation times $\tau_1^{SIFE(SOKE)}$ and $\tau_2^{SIFE(SOKE)}$ should be used in fitting the SIFE and SOKE rotation and ellipticity signals, while the value of the zero delay time should be the same for all four traces. As the model does not allow us to identify the physical origin of the two contributions, the results of the fitting procedure will require additional interpretation.

This assumption is distinctly different from a commonly used approximation that the signal shape must be described by an intensity cross-correlation of the pump and probe pulses. Let us consider a signal that is described by a single term of the type given in equation (4). The dependence of the signal shape upon the relaxation time is illustrated in Figure 2. Increasing the relaxation time gives rise to four effects: the time resolved signal becomes larger and more asymmetric, it broadens, and the position of the maximum shifts towards positive time delay.

The electron dynamics in metals have been investigated with femtosecond time resolution in two photon photoemission experiments [3,16,17], where the idea that the signal can be described by a double convolution of an exponential decay with the laser pulse shape is commonly

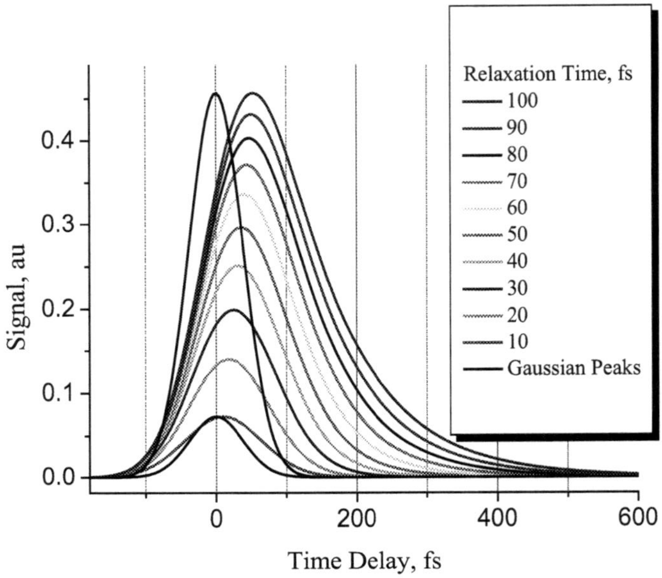

Figure 2. The dependence of the signal shape defined by one of the terms in equation (4) upon the relaxation time is illustrated. The FWHM was set to 90 fs.

used. Similar expressions were also used to fit time resolved polarisation signals obtained from semiconductors [19] and lately metallic Al [15]. However, only the contributions with longer (picosecond) relaxation times were fitted in this way, while the pump-probe crosscorrelation was still used to fit the initial peak that was observed on femtosecond time scales.

3. Experiment

The apparatus used for the experiments is shown in Figure 3. The pulsed light source was a 80 MHz mode-locked Ti: sapphire laser (Spectra-Physics Tsunami) (1) pumped by a continuous wave (cw) diode laser (Spectra-Physics Millennia) (2), that produced 90 fs pulses at a centre photon energy of $\hbar\omega$=1.575±0.005 eV (787 nm). The apertures (3) were used to reset the beam direction following any drift in the laser alignment.

The beam was passed through a pulse picker (4). In order to compensate pulse broadening ("chirp") introduced by the pulse picker and to pre-compensate that caused by other "bulky" optics (lenses, polarizers etc.), the beam was passed through a group velocity dispersion compensation line (5), consisting of a pair of Brewster prisms and a mirror [20]. The temporal width of the pump and probe pulses was checked with a fringe-resolved

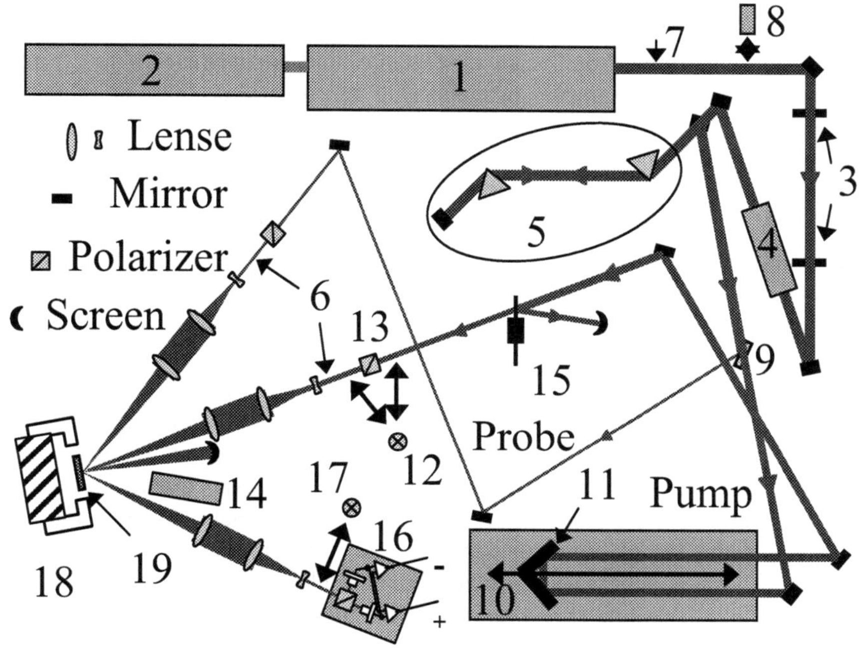

Figure 3. The apparatus used in the experiments is shown in schematic form.

autocorrelator placed at various places in the beam path. The pulse width measured just before the expansion optics (6) was found to be in agreement with that measured just after the laser (7) and with that deduced from bandwidth measurements using a laser spectrum analyzer (8). Measurements were not made at the sample position, since the beam would first need to be re-collimated. This would require the insertion of additional optics that could introduce additional chirp.

Each pulse was split with a beam splitter (9) into pump and probe parts with energies of 2 nJ and 0.04 nJ, respectively, that were directed onto the sample at angles of incidence of 7° and 27°, respectively. A variable optical delay line (10) in the pump path allowed the time delay between the two pulses to be varied with a minimum step size of 1.67 fs. The position of a hollow retro-reflecting prism (11) in the delay line was varied through a few wavelengths at high frequency in order to remove any coherence oscillations around the zero delay position. The maximum time delay was 2 ns.

The probe was always p-polarised. The polarisation of the pump could be varied continuously from a p-polarised state either to a circularly polarised state by rotating a quarter wave plate (12) placed after a polarizer (13) (geometry 1), or to a state in which the beam was linearly polarized at an angle to the plane of incidence (geometry 2). In the latter case a quarter wave plate (12) was placed in front of the polarizer (13) in order to make the beam circularly polarised and, hence, to maintain constant intensity at the sample surface. Both pump and probe beams were expanded by a factor of ten in order to reduce the beam divergence, and focused down to 15 μm diameter spots at the sample surface, using achromatic doublets with focal length of 16 cm. The spots were carefully overlapped while being viewed with a CCD camera with x600 on-screen magnification (14).

The pump was modulated at a frequency of 2 kHz by a mechanical chopper (15). Any transient changes in the intensity and polarisation state of the reflected probe could be simultaneously sensed by an optical bridge detector (16) with its sum (intensity) and difference (polarisation) outputs connected to two lock-in amplifiers. The phase sensitive detection allowed changes in rotation and ellipticity angles as small as 1 μdeg to be resolved. Except for the Cu sample, the spot positions were not adjusted during measurements in either geometry, and measurements were completed within 1 or 2 days for each sample. The transient changes in the probe intensity and polarisation state were recorded simultaneously. In experiments in which the transient ellipticity was measured, a quarter wave plate (17) was placed in front of the bridge detector with its fast axis parallel to the plane of incidence. When necessary, an electromagnet (18) capable of producing a magnetic field of up to 3 kOe was used to magnetise the sample (19).

Measurements were performed upon noble and transition metal thin films, deposited by magnetron sputtering onto Si substrates. Since the $\chi^{(3)}$ tensor was expected to depend upon the detailed structure of the sample, X-ray reflectivity measurements were performed to determine the thickness,

Table 1. A summary of the structural parameters of the investigated samples is given.

	Film Thickness, nm	Film Roughness, nm	Oxide Layer Thickness, nm	Oxide Layer Roughness, nm
Cu	32.60±0.05	0.87±0.05	3.50±0.06	0.94±0.04
Ag	24.2±0.3	2.5±0.3	8.3±0.6	3.6±0.3
Au	28.10±0.03	1.30±0.01	n/a	n/a
Ni	53.6±0.4	0.90±0.15	1.2±0.4	0.46±0.03
Pd	61.8±0.7	2.50±0.26	1.0±0.7	0.500±0.010
Ti	25.6±0.4	0.26±0.05	1.14±0.20	0.320±0.010
Zr	42.50±0.07	0.47±0.06	2.50±0.07	0.446±0.010
Hf	76.60±0.14	1.04±0.08	4.80±0.08	0.872±0.020

roughness and composition of the samples. The results of the X-ray measurements are summarised in Table 1.

For each sample two sets of measurements were performed, using geometry 1 and 2, respectively. Only measurements made in geometry 1 will be described here [21]. The time resolved rotation (or ellipticity) and reflectivity signals were simultaneously recorded for different orientations of the pump quarter wave plate. In this paper, only the polarisation response on time scales of a few hundreds of femtoseconds is discussed. The rise of a transient reflectivity response with much longer relaxation time was used here only to verify the zero time delay position.

The time resolved rotation and ellipticity signals acquired for different orientations of the pump quarter wave plate are shown in Figures 4 and 5, respectively, for different metals. As expected from equation (1), the signals consist of two components varying with periods of 90° and 180° with respect to the angle φ between the fast axis of the pump quarter wave plate and the plane of incidence. The SOKE contribution has period of 90° and

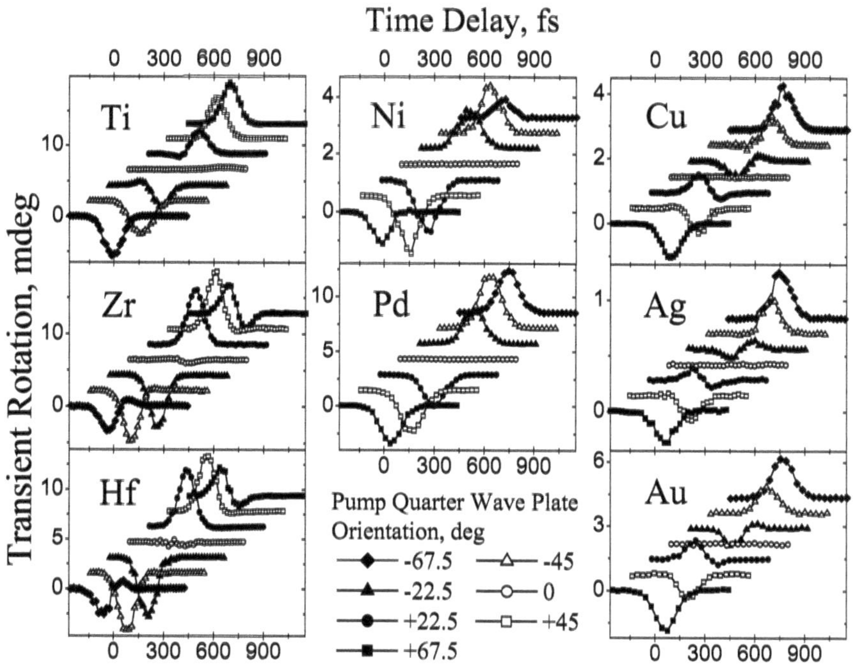

Figure 4. The transient rotation signals obtained in geometry 1 are presented. The signals obtained from different samples are arranged according to their position in the periodic table of elements.

becomes more pronounced as the pump polarisation ellipse axis becomes more asymmetric and is canted at a larger angle to the plane of the probe polarisation (e.g. for $\varphi=\pm22.5°,\pm67.5°$). The SOKE vanishes when the pump is either circularly polarised (e.g. for $\varphi=\pm45°$) or linearly polarised parallel to the probe polarisation (e.g. for $\varphi=0°$). The SIFE contribution has period of 180° and is most pronounced for circular pump polarisation (e.g. for $\varphi=\pm45°$), while vanishing when the pump is linearly polarised (e.g. for $\varphi=0°$).

Let us make the common assumption that both SIFE and SOKE give rise to peaks with centre at the zero time delay position and with shape identical to the cross-correlation of the pump and probe intensity envelopes. Then if both effects have finite amplitude, one would expect to see a signal with shape identical to that of the cross correlation curve for all values of φ, but with varying amplitude as the SIFE and SOKE either add or cancel. Instead, we observe that for some values of φ the signal has a bipolar temporal shape. This suggests that a small time delay exists between the centres of the SIFE and SOKE peaks. In addition, the signals acquired with circular pump are

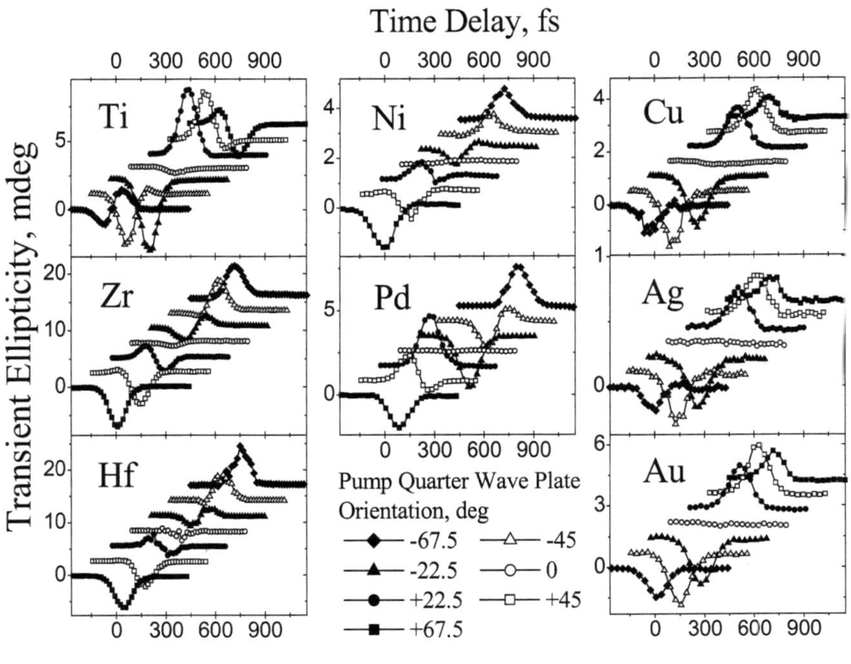

Figure 5. The transient ellipticity signals obtained in geometry 1 are presented. The signals obtained from different samples are arranged according to their position in the periodic table of elements.

themselves slightly bipolar. This indicates the presence of two components shifted in time within the pure SIFE signals. The transient rotation and ellipticity signals are similar in shape inside the group of noble metals. The Ni and Pd samples have transient rotation signals of similar shape, while their transient ellipticity signals are different. In the group of hcp metals, the Zr and Hf signal shapes are similar both in rotation and ellipticity, while the Ti time resolved signal is somewhat different.

4. Analysis and discussion

For the signals measured in geometry 1, both the SIFE and SOKE contributions are present simultaneously. These can be separated since, according to equations (1)-(3), for pairs of signals with φ equal to +22.5° and +67.5° (or alternatively -22.5° and -67.5°) the SIFE component has the same magnitude and sign, while the SOKE component has the same magnitude but opposite sign. This means that the difference divided by 2 of the signals at +22.5° and +67.5° (or -22.5° and -67.5°) contains only the SOKE component, while their sum contains only the SIFE component, but multiplied by $\sqrt{2}$ in comparison with that measured at $\varphi=\pm45°$. In order to improve the signal-to-noise ratio and to remove any remaining background due to the reflectivity breakthrough, the SOKE signals extracted from pairs of signals at +22.5° and +67.5°, and -22.5° and -67.5° were averaged. Similarly, the SIFE components extracted from these pairs of signals and divided by $\sqrt{2}$ were averaged with the SIFE signals measured at $\varphi=\pm45°$. The extracted time resolved SIFE and SOKE contributions to the signal measured in geometry 1 are shown in Figure 6. We note that, strictly speaking, formulae (1)-(2) (3) are valid only at normal incidence. However, they provide a reasonable approximation to the present experimental configuration since the pump and probe beams are strongly refracted towards the film normal within the sample.

There is a striking similarity between the relative signs and relative magnitudes of the SIFE and SOKE signals shown within the group of noble metals. Zr and Hf define a second group to which Ti can also be assigned if one assumes that the bipolar shapes of the Ti SIFE ellipticity and SOKE rotation traces result from the presence of two peak-like contributions (two each for SIFE and SOKE) that have different centre positions and opposite signs. One of these peaks tends to make the SIFE and SOKE signals similar to those of Zr and Hf, while the other peak breaks the similarity. Making similar allowance for the bipolar shape of the Pd SIFE ellipticity and SOKE rotation signals, Pd and Ni form another group. The relative signs and magnitudes of the signals differ substantially between the three different groups. To make this observation even clearer, the complex amplitudes of the SIFE (F_-) and SOKE ($F_+/2$), as defined by equation (3), determined

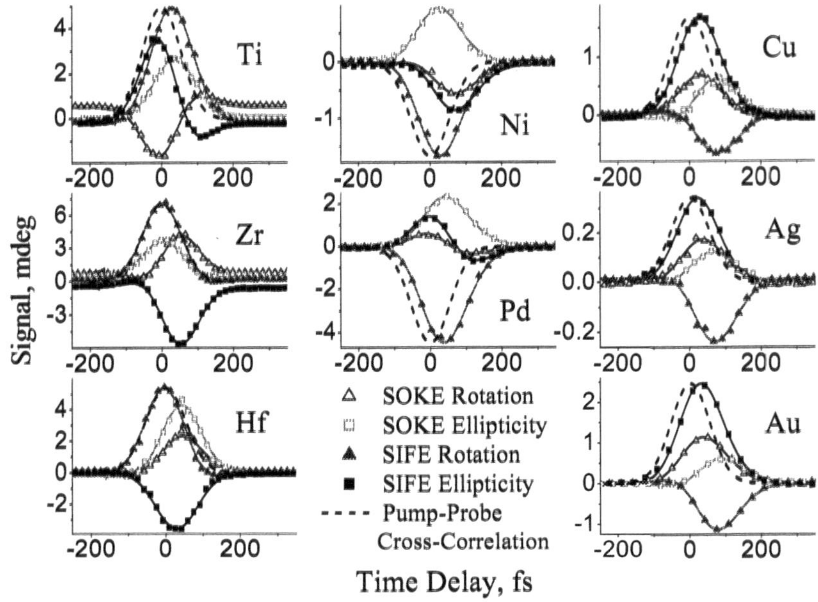

Figure 6. The SIFE and SOKE contributions to the transient polarisation response (symbols are data points, and solid curves are fits to equation (4), as explained below) are shown with the pump-probe intensity cross-correlation for the different samples measured.

from the maximum values of signals shown in Figure 6 are plotted in Figure 7 for the different metals. The "alternative" values of F_- and $F_+/2$ for Ti and Pd are shown by dashed lines. The three groups correspond to three different columns of the periodic table of elements, and so a different colour has been used for each group.

The electronic structure is similar within each group as far as optical properties are concerned. For example, at the photon energy used in our experiments the optical response of the noble metals is entirely due to intraband optical transitions [22, while that of Ti, Zr, and Hf is dominated by interband transitions [23,24]. The optical structures of Ni and Pd are more complicated, and both interband and intraband transitions can contribute [24,25. This interpretation is supported by the observation that the signals of largest and smallest magnitude occur in Zr and Ag, respectively.

The SIFE and SOKE contributions (both in rotation and ellipticity) generally occur at different time delays. However, the zero delay condition must be uniquely defined. Therefore the non-linear response cannot be

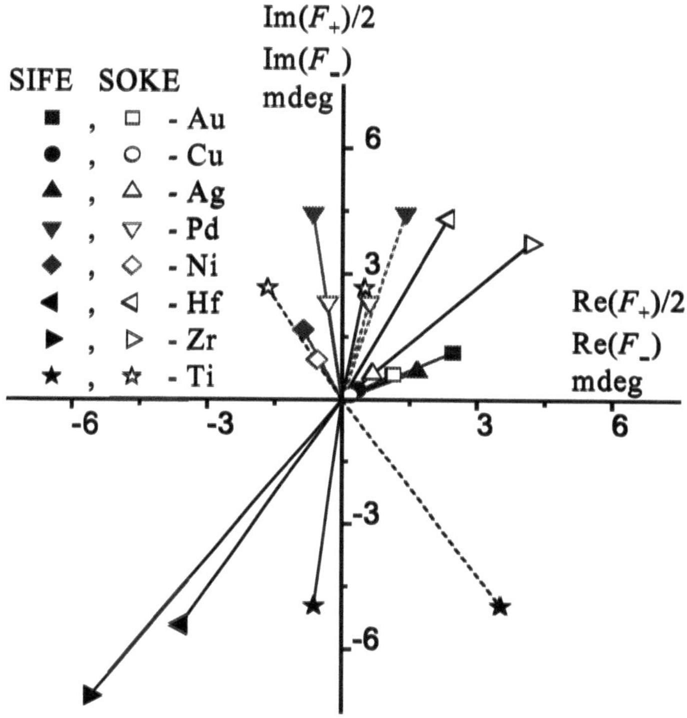

Figure 7. The complex SIFE (F_-) and SOKE ($F_+/2$) magnitudes are shown with filled and open symbols, respectively, on the complex plane for different samples. Different colours were used to denote metals from different groups of the periodical table. Dashed lines correspond to the "alternative" values of F_- and F_+ for Ti and Pd that correspond to the peaks breaking the similarity with the other metals in their groups.

described as a cross-correlation of the pump and probe intensity envelopes, as has been previously assumed (for example see Ref. 26), and hence, cannot be used for the purpose of pulse width measurements. The temporal positions of the SIFE and SOKE peaks are summarised in Table 2. The same allowance for the bipolar shaped signals of Pd and Ti is made. One can see that the relative order of appearance of the SIFE and SOKE peaks is the same within each group of metals, the only exception being the transient ellipticity signal of Hf in which the SIFE and SOKE have opposite order to that in Ti and Zr.

The temporal shapes of the SIFE and SOKE signals may be described by a convolution of the probe intensity envelope with evolution functions for the sample polarisation, as discussed in the previous section. This transient polarisation results from optical orientation of electronic linear and angular momentum and decays on timescales of tens of femtoseconds. Differences

Table 2. The temporal positions of the SIFE and SOKE peaks are summarised. The numbers in parenthesis for Ti and Pd correspond to the "alternative" peaks that break the similarity with the other metals in their groups.

	SIFE Rotation, fs	SIFE Ellipticity, fs	SOKE Rotation, fs	SOKE Ellipticity, fs
Cu	77	27	29	75
Ag	71	21	35	63
Au	77	31	37	85
Ni	27	70	72	25
Pd	39	131 (-1)	117 (-9)	43
Ti	30	110 (-10)	110 (-8)	38
Zr	0	53	57	2
Hf	-3	29	45	43

in the relaxation times of the SIFE and SOKE cause their peak positions to be shifted by different amounts, as shown in Figure 2, and as is readily apparent when SIFE and SOKE are both present in the same measurement, as in geometry 1.

The relaxation times have been extracted by fitting the measured SIFE and SOKE signals to equation (4). The bipolar shapes of the extracted SIFE and SOKE components for Ti and Pd suggest the presence of two contributions for each effect of opposite sign and different relaxation times. Hence, a superposition of two double convolutions in the equation (4) was used in the fitting. This has been further justified by the fitting process. For a particular metal, it was not possible to obtain a satisfactory fit with a model that contained a single double convolution, if the same value of the zero delay time was used for all four traces. The fitted SIFE and SOKE contributions are represented in Figure 6 by solid curves. The fitting parameters are summarised in Table 3.

The values presented in Table 3 fall into two categories. In the first, occupied by the noble metals and Ni, the difference of the two relaxation times for SIFE (or SOKE) is small. This suggests that the transient polarisation response results from excitation of transitions in one particular region of the band structure. The initial and final states occupy finite phase volumes that may be large for intraband transitions. The corresponding relaxation time will generally exhibit dispersion associated with the range of initial and final states that are involved. The use of two similar relaxation

Table 3. The parameters obtained by fitting the SIFE and SOKE contributions shown in Figure 6 to equation (4) are presented.

	SIFE		SOKE	
	τ_1, fs ($A_{\theta,1}$, $A_{\eta,1}$, au)	τ_2, fs ($A_{\theta,2}$, $A_{\eta,2}$, au)	τ_1, fs ($A_{\theta,1}$, $A_{\eta,1}$, au)	τ_2, fs ($A_{\theta,2}$, $A_{\eta,2}$, au)
Cu	**25** (20.6, -3.4)	**27.5** (-19.5, 5.0)	**22.5** (8.0, 44.9)	**23.5** (-8.5, -43.3)
Ag	**24.5** (11.10, 0.51)	**26** (-10.80, -0.09)	**22** (8.88, 15.68)	**22.5** (-8.93, -15.48)
Au	**27.5** (175.6, -54.6)	**28** (-173.9, 56.5)	**31.5** (56.6, 136.7)	**32** (-57.2, -136.0)
Ni	**25.5** (-24.2, -78.7)	**26.5** (22.7, 79.8)	**24.5** (-70.9, 11.4)	**25** (70.0, -12.3)
Pd	**30** (54.3, 72.5)	**31.5** (-56.6, -68.5)	**9.5** (-4.7, 2.6)	**44** (0.79, -2.30)
Ti	**9** (-11.8, 38.6)	**23** (11.4, -12.0)	**17** (203.4, 126.1)	**18** (-192.1, -125.0)
Zr	**10.5** (71.8, 158.8)	**14.5** (-39.8, -124.9)	**10** (150.9, -33.1)	**13** (-120.7, 18.8)
Hf	**11** (134.5, 82.6)	**13** (-103.1, -77.7)	**10.5** (162.4, 288.3)	**12** (-148.9, -265.4)

times in our model provides a crude representation of this dispersion, which may lead to a small shift in the position of the maximum in the signal. Alternatively, if the relaxation times differ significantly, as in Hf, Zr, Pd SOKE and Ti SIFE, then the signal may originate from states in two distinctly different places in the band structure. These states may even have the same energy, since time resolved two photon photoemission experiments have shown that hot electron lifetimes at a particular energy depend on whether excitation was the result of an interband or intraband transition [3].

After the initial coherence of the electron-hole pairs has been broken, the distribution of linear and angular momentum of the hot electrons and holes may still retain some memory of the pump polarisation. The action of a linearly polarised pump pulse reduces the symmetry of the momentum space electron distribution and induces an axial symmetry. The SOKE relaxation times characterise the recovery of the symmetry of the distribution, but only in the region of the band structure sensed by the probe beam, which may be a small fraction of the total phase space in the case of interband transitions.

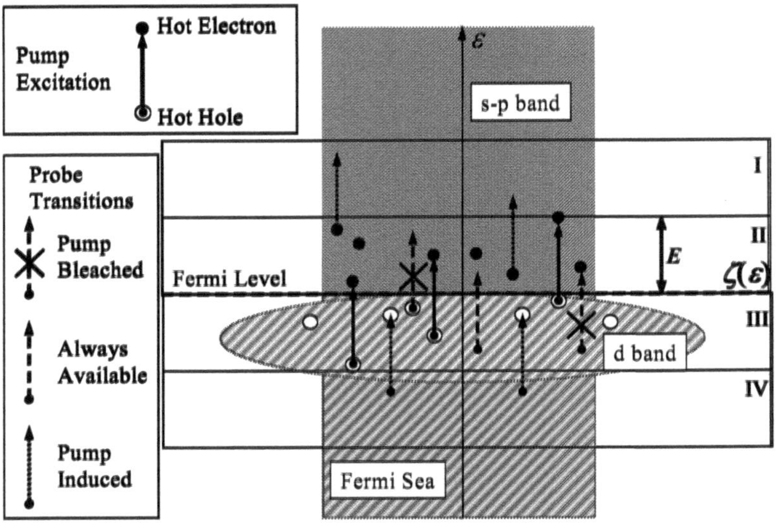

Figure 8. The effect of the pump excitation is schematically illustrated.

Because the linear momentum of the photon is small compared to that of the electrons and phonons, electronic linear momentum is conserved in interband transitions, and hence only vertical transitions are allowed, as shown in Figure 1. Then, due to energy conservation, the detection region is a narrow sub region of region II in Figure 8, for electrons, and a narrow sub region of region III, for holes. Therefore any inelastic scattering results in removal of an electron (hole) from the detection region and leads to a change in the measured linear momentum. Because a phonon, the linear momentum of which is finite, participates in an intraband transition, electronic linear momentum is not conserved, and hence indirect transitions are also allowed, as shown in Figure 1. Then the detection region may continuously span the entire region II in Figure 8, for electrons, and the entire region III, for holes. Inelastically scattered electrons and holes are then more likely to remain within the detection region and contribute to the transient polarisation response. This implies that experiments involving intraband transitions should provide a better estimate of the momentum relaxation time.

The action of a circularly polarised pulse leads to a repopulation of electrons and holes between states with a different angular momentum component in the direction of the pump pulse wave vector. The corresponding SIFE relaxation time describes the recovery of the symmetry of the initial angular momentum distribution. Again it is preferable that the probe response is dominated by intraband transitions or else inelastic scattering may obscure the relaxation of hot electron (hole) angular momentum.

Let us now discuss the precision and consistency of the measurements and analysis. The results of the fittings rely on an accurate determination of the zero time delay position in the measurement of each particular signal. The zero delay position was defined with respect to the onset of the reflectivity signal. Figure 9 presents transient reflectivity signals that were recorded simultaneously with the transient polarisation response of the Cu sample in two series of measurements. Panel (a) shows reflectivity signals recorded when the transient rotation was measured as function of the pump quarter wave plate orientation. The signals were recorded in random order, but are arranged in the panel in order corresponding to the orientation of the quarter wave plate. The bottom and top signals correspond to angles of 78.75° and -90° between the fast axis of the quarter wave plate and the plane of incidence, while the signals in between are equally spaced within this range. One can see that the signal onsets do not show any regular

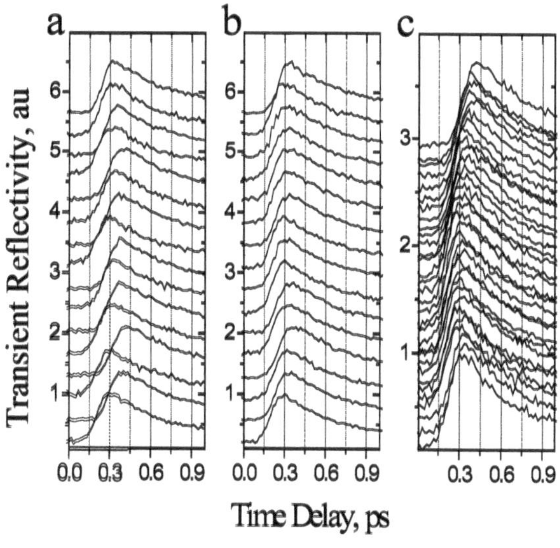

Figure 9. Transient reflectivity signals measured from the Cu film are presented. Panels (a) and (b) show reflectivity signals measured simultaneously with transient rotation and ellipticity, respectively, in order corresponding to the orientation of the quarter wave plate. The latter varies from 78.75° for the bottom trace to -90° for the top trace. Panel (c) shows the signals from panels (a) and (b) in the order that they were actually measured. The peak signal magnitude was normalised to unity in each case.

dependence upon the orientation of the quarter wave plate. Panel (b) shows the transient reflectivity for the same orientations of the pump quarter wave plate but in a different series of experiments in which the transient ellipticity was measured. The experimental configuration in this case was different to that used for the rotation measurement only in that a quarter wave plate was placed in front of the bridge detector, and so the optical paths of the pump and probe before the sample were not altered. However, the onsets of the reflectivity signals in the two series of measurements are not correlated. Panel (c) shows the signals from panels (a) and (b) in the order in which they were actually acquired during the measurements, so that about 8 hours time elapsed between the acquisition of the bottom and top signals. One can see that the onset of the signals shifts with time first towards negative and then towards positive time delays. This shows that the orientation of the quarter wave plate does not affect the time delay between the pump and probe pulses significantly. However, various drifts in other parts of the apparatus such as the laser or the sample holder may have a significant effect. Also, this shows that defining the temporal position of the transient polarisation signals relative to the reflectivity is a reasonable procedure.

Equation (4) was found to provide a good description of the observed SIFE and SOKE signal shapes. However, as one can see from Table 3, in cases where the percentage difference in the two fitted relaxation times for a particular effect was small (less than 10 percent), the values of the two associated amplitudes were found to almost cancel and hence the errors in the fitted parameters were comparable to or larger than the parameter values. This indicates a possible degeneracy in the fit and suggests that a more accurate model or a shorter optical pulse width may be needed. Any such model must be based on a solution of the Boltzmann kinetic equations with realistic assumptions about the band structure and interactions among quasi-particles. Nevertheless, our model is the simplest capable of reproducing the shape of the observed signals, including those with the bipolar shape, and it reveals the characteristic time scales of the underlying dynamics.

5. Summary

In summary, femtosecond ellipsometric measurements performed with an elliptically polarised pump have been performed to directly investigate the decay of optically induced transient linear and circular polarisation in metals. The response of different metals has been shown to fall into groups that correspond to the positions of the metals within the periodic table. A simple model has been introduced that successfully fits the shapes of the signals. In some cases two distinct relaxation times may be identified, that are interpreted in terms of scattering of the linear and angular momentum of the excited electrons. These relaxation times may be relevant to transport in hot electron devices such as the spin valve transistor or magnetic tunnel transistor. The capability of this new technique could be further enhanced through the use of shorter pulses and by separately tuning the wavelength of pump and probe, so that the mechanisms responsible for linear and angular momentum relaxation in metals may be better explored and understood.

Acknowledgments

The authors gratefully acknowledge the growth of the samples by Doctor M. Ali and Professor B.J. Hickey at the University of Leeds, the x-ray measurements performed by A.T.G. Pym and Professor B.K. Tanner at the University of Durham, the financial support of the UK Engineering and Physical Sciences Research Council (EPSRC), and the award of an Overseas Research Studentship (ORS) to V.V. Kruglyak.

References

[1] P.M.W. French, Rep. Progr. Phys. **58**, 169 (1995).
[2] R.J. Hicken, Philos. Trans. Roy. Soc. A **361**, 2827 (2003).
[3] H. Petek and S. Ogawa, Progr. Surf. Sci. **56**, 239 (1997).
[4] J. Hohlfeld et al., in *Nonlinear Optics in Metals*, Edited by K.H. Bennemann, Clarendon, Oxford, 1998.
[5] E. Beaurepaire et al., Phys. Rev. Lett. **76**, 4250 (1996).
[6] E. Beaurepaire et al., Phys. Rev. B **58**, 12134 (1998).
[7] J. Hohlfeld et al., Phys. Rev. B **65**, 12413 (2001).
[8] Y.I. Gorobets, Y.I. Dzhezherya, and A.F. Kravets, Metallofiz. Nov. Techn. **25**, 27 (2003).
[9] M. van Kampen et al., Phys. Rev. Lett. **88**, 227201 (2002).
[10] Q. Zhang et al., Phys. Rev. Lett. **89**, 177402 (2002).
[11] G. Ju et al., Phys. Rev. Lett. **82**, 3705 (1999).
[12] T. Rasing et al., Top. Appl. Phys. **87**, 213 (2003).
[13] R. Knorren, G. Bouzerar, and K.H. Bennemann, J. Phys.: Cond. Matt. **14**, R739 (2002).
[14] R. Jansen, J. Phys. D **36**, R289 (2003).
[15] R. Wilks and R.J. Hicken, J. Phys.: Condens. Matt. **16**, 4607 (2004).
[16] E. Knoesel, A. Hotzel, and M. Wolf, Phys. Rev. B **57**, 12812 (1998).
[17] S. Link, H.A. Durr, and W. Eberhardt, J. Phys.: Cond. Matt. **13**, 7873 (2001).
[18] S.V. Popov, Y.P. Svirko, and N.I. Zheludev, J. Opt. Soc. Am. B **13**, 2729 (1996); Y.P. Svirko and N.I. Zheludev, *Polarization of Light in Nonlinear Optics*, Wiley, New York, 1998.
[19] A.V. Kimel et al., Phys. Rev. B **63**, 235201 (2001).
[20] R. Wilks, PhD Thesis, Exeter, 2003.
[21] The SOKE components measured in geometry 2 were found to agree very well both in shape and magnitude with those extracted from measurements in geometry 1.
[22] G.A. Burdick, Phys. Rev. **129**, 138 (1967); N.E. Christensen and B.O. Seraphin, *ibid.* **4**, 3321 (1971); H. Eckardt, L. Fritsche, and J. Noffke, J. Phys. F: Met. Phys. **14**, 97 (1984).
[23] D.W. Lynch, C.G. Olson, and J.H. Weaver, Phys. Rev. B **11**, 3617 (1975).
[24] P.B. Johnson and R.W. Christy, Phys. Rev. B **9**, 5056 (1974).
[25] J.H. Weaver, Phys. Rev. B **11**, 1416 (1975); J.H. Weaver and R.L. Benbow, *ibid.* **12**, 3509 (1975).
[26] P.J Bennett et al., Opt. Comm. **147**, 148 (1998).

FERROMAGNETIC SHAPE MEMORY ALLOYS $Ni_{2+x}Mn_{1-x}Ga$

A. N. Vasiliev
Physics Faculty, Moscow State University, Moscow 119899, Russia
vasil@mig.phys.msu.ru

R. Z. Levitin
Physics Faculty, Moscow State University, Moscow 119899, Russia

V. V. Khovaylo
Institute of Radioengineering and Electronics of RAS, Moscow 125009, Russia
v-khovaylo@cplire.ru

Abstract: We present results of experimental studies of magnetic properties and phase transitions in a new class of smart materials, ferromagnetic shape memory alloys $Ni_{2+x}Mn_{1-x}Ga$. The phase diagram of these alloys was determined from magnetic, transport and calorimetric measurements in the concentration interval $0 \leq x \leq 0.36$. These measurements revealed a general tendency of the Curie temperature T_C to decrease and of the martensitic transformation temperature T_m to increase upon substitution of Mn for Ni. This tendency leads to the occurrence of three characteristic compositional intervals in which (i) a martensitic transformation occurs in the ferromagnetic matrix ($T_m < T_C$), (ii) a first-order magnetostructural phase transition from ferromagnetic martensite $\leftrightarrow$ to paramagnetic austenite takes place at $T_m \approx T_C$, and (iii) a martensitic transformation occurs in the paramagnetic state when ($T_m > T_C$).

Magnetic properties of $Ni_{2+x}Mn_{1-x}Ga$ alloys were studied systematically in a compositional interval $0 \leq x \leq 0.19$. As the results of these studies, compositional dependencies of the spontaneous magnetization $M_s(T)$, the jump of magnetization ΔM at the martensitic transformation, the paramagnetic Curie temperature Θ, the effective paramagnetic moment μ_{eff}, and the influence of a magnetic field on the martensitic transformation temperature T_m were clarified. Based on the experimental results obtained, the degree of delocalization of magnetic moments in the Ni and Mn magnetic subsystems was evaluated.

J. Franse et al. (eds.), Smart Materials for Ranging Systems, 223–243.

Keywords: ferromagnetism, martensitic transformations, Heusler alloys, phase diagram.

1. Introduction

Ni_2MnGa Heusler alloy is a prototype of a new class of smart materials, ferromagnetic shape memory alloys. The shape memory effect in these substances is related to a thermoelastic martensitic transformation, a reversible structural phase transition occurring with a small temperature hysteresis. Combination of ferromagnetic properties and martensitic transformation allows realization of a magnetically driven shape memory effect, expanding considerably the area of its technical applications. Besides Ni_2MnGa, several other Heusler alloys and the intermetallic compounds Fe-Pd, Fe-Pt, and Fe-Ni-Co-Ti have been studied in this connection (see for a review Ref. [1]). The most promising results, however, were obtained for the Ni_2MnGa-based alloys. This unique alloy system has made it possible to achieve giant (up to 10%) magnetic-field-induced strains upon application of a moderate magnetic field ~ 1 T.

In spite of a large number of recent investigations, many fundamental aspects of the Ni_2MnGa-based system are still not clearly understood. This concerns, first of all, the compositional stability of the martensitic phase in this system. Nowadays it is generally acknowledged that the martensitic transformation temperature T_m of Ni-Mn-Ga correlates with the electron concentration e/a [2]. Although recent experimental results [3] imply that the martensitic transformation can be observed even in the alloys with a large deviation from stoichiometry, the extent of non-stoichiometry has not been established yet. Since the ferromagnetic transition temperature T_C also depends on composition, this can lead to the coupling of T_C and T_m. Such a situation is realized in $Ni_{2+x}Mn_{1-x}Ga$ [4], and, presumably, in $Ni_{2+x}MnGa_{1-x}$ and $Ni_2Mn_{1+x}Ga_{1-x}$. Despite interesting magnetic and magnetocaloric properties of these alloys [5, 6, 7], compositional intervals of the coexistence of T_m and T_C have not been determined so far.

The magnetic properties of Ni-Mn-Ga alloys are also not sufficiently clarified. Even for the most intensively studied $Ni_{2+x}Mn_{1-x}Ga$ system, there are known only the compositional dependence of the Curie temperature and some information on the magnetocrystalline anisotropy and the saturation magnetization of these compounds. Little is known about the change of magnetic moment at the martensitic transformation, though this parameter determines the shift of the martensitic transition temperature in an external magnetic field, giving a basis for many technical applications. So far, the exchange interaction parameters of these alloys have not been estimated either.

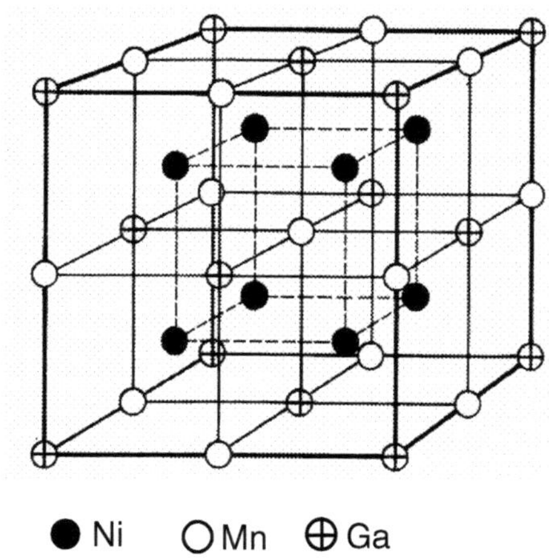

Figure 1. $L2_1$ crystal structure of Ni_2MnGa.

We conceive that a deeper understanding of physical mechanisms underlying the magnetically driven shape memory effect and other interesting phenomena observed in $Ni_{2+x}Mn_{1-x}Ga$ is difficult without knowledge of the phase diagram and the magnetic parameters of this system. This Chapter is organized as follows. Crystal structures and magnetic properties of $Ni_{2+x}Mn_{1-x}Ga$ are outlined in Section 2. Details regarding sample preparation and experimental techniques used in the study are presented in Section 3. The phase diagram of $Ni_{2+x}Mn_{1-x}Ga$ is discussed in Section 4. The results of a systematic study of the magnetic properties of $Ni_{2+x}Mn_{1-x}Ga$ are presented in Section 5. In Section 6 we summarize the main features of the phase diagram and the magnetic properties of the studied system.

2. Basic structural and magnetic characteristics of $Ni_{2+x}Mn_{1-x}Ga$ alloys

The high-temperature austenitic phase of $Ni_{2+x}Mn_{1-x}Ga$ alloys has a cubic structure of $Fm\bar{3}m$ space group (Fig. 1). A structural phase transition to a modulated tetragonal phase is observed in these alloys on cooling. It is worth noting that the crystal structure and the space group of the low-temperature phase is still a subject of controversy (see, for example, Refs. [8, 9]), which is aggravated by a compositional dependence of the crystal structure of martensite. Thus, for example, recent results of high-resolution neutron diffraction [10] give ground to conclude that for the stoichiometric composition the martensitic phase, being considered for a long time as a tetragonal, has actually the orthorhombic symmetry of Pnnm space group.

The structural (martensitic) transition in Ni_2MnGa-based Heusler alloys was described as driven by a band Jahn-Teller effect [11, 12]. In the framework of this model, the results of theoretical [12] and experimental [11] studies of the redistribution of magnetization upon martensitic transformation in Ni_2MnGa alloys have been discussed.

The martensitic transformation temperature, equal to $\approx$ 200 K for stoichiometric Ni_2MnGa, linearly increases with increasing x in substituted $Ni_{2+x}Mn_{1-x}Ga$ solid solutions, reaching approximately 320–330 K in the alloys with $x = 0.18 - 0.19$ [4] (alloys with a higher Ni excess have not been studied so far). Note that there exists a scatter in the data as for the martensitic transformation temperature, as well as for other relevant properties of the alloys, which is probably caused by a different degree of ordering [13, 14] and/or deviation from the nominal composition of the alloys.

The martensitic transformation does not exhaust the sequence of structural phase transitions in $Ni_{2+x}Mn_{1-x}Ga$, because phase transitions occurring above (premartensitic phase transition) and below (intermartensitic phase transition) the martensitic-transformation temperature were also found in these alloys. The premartensitic transition is observed only in near-stoichiometric Ni-Mn-Ga alloys. For stoichiometric Ni_2MnGa ($x = 0$) the temperature of this transition T_P is $\approx$ 260 K. In $Ni_{2+x}Mn_{1-x}Ga$ alloys the premartensitic transition is essentially independent of composition, $T_P \sim 260$ K, and disappears in a $Ni_{2.09}Mn_{0.91}Ga$ ($x = 0.09$) alloy [15]. An intermartensitic transition from a five-layered to a seven-layered martensitic structure is observed at $T_I = 283$ K during cooling of $Ni_{2.16}Mn_{0.84}Ga$ [16]. In this composition, the resistivity ρ exhibits a huge temperature hysteresis, which was attributed to the coexistence of both the martensitic phases over a wide temperature interval [16].

$Ni_{2+x}Mn_{1-x}Ga$ alloys are ferromagnetic at low temperatures. The Curie temperature, $T_C \approx 370$ K for the stoichiometric Ni_2MnGa, decreases approximately linearly with increasing Ni concentration, so that in the $x = 0.18 - 0.19$ alloys T_C coincides with the martensitic transformation temperature. Hence, these alloys undergo a structural (martensitic) transition from paramagnetic austenite to ferromagnetic martensite, whereas the magnetic state of alloys with a lower Ni concentration does not change during martensitic transformation and both the austenitic and the martensitic phases remains ferromagnetic. The martensitic transformation, however, affects the magnetic characteristics of these alloys and results in a sharp growth of the magnetic anisotropy energy and an increase in magnetization saturation [17].

Neutron diffraction measurements of stoichiometric composition [11, 18] revealed that the magnetic moment is localized mainly on Mn atoms. Reported values of the Mn magnetic moment range from 3.8 to 4.2 μ_B. The magnetic moment of Ni atoms is considerably smaller, about 0.2 – 0.4 μ_B. Compositional dependence of the magnetic moment in Ni_{2+x} $Mn_{1-x}Ga$ alloys has not been reported. It is known only that the magnetization saturation decreases with increasing x [19, 20].

3. Sample preparation and measurements

Polycrystalline samples of $Ni_{2+x}Mn_{1-x}Ga$ ($0 \leq x \leq 0.39$) alloys were prepared by a conventional arc-melting method in an atmosphere of spectroscopically pure argon gas. The samples were homogenized at 1050 K for nine days with subsequent quenching in ice water. The microstructure of the samples was examined by optical microscopy. It turned out that the sample with the highest Ni content, $Ni_{2.39}Mn_{0.61}Ga$, has a trace of a secondary phase and is, therefore, compositionally inhomogeneous. Because of this, phase transitions and magnetic properties were studied for $Ni_{2+x}Mn_{1-x}Ga$ ($0 \leq x \leq 0.36$) compositions. For the measurements we used only the samples whose weight loss during arc-melting did not exceed 0.2%.

Martensitic and magnetic phase transition temperatures were determined from temperature dependencies of magnetization and resistivity, and from differential scanning calorimetry (DSC) measurements. The temperature dependence of the resistivity, $\rho(T)$, was measured by a standard four-probe technique in a temperature interval 77 – 420 K with a heating/cooling rate 1 K/min. Differential scanning calorimetry measurements were performed by a Pyris-1 DSC equipment in a temperature interval 77 – 573 K with a heating/cooling rate 5 K/min. A high-temperature calorimeter NETZSCH-404 was used for the DSC measurements at temperatures above 573 K. The latent heat of the martensitic transition was determined from the results of differential scanning calorimetry measurements.

The magnetization M was measured in a temperature interval 5 – 700 K by a SQUID magnetometer in magnetic fields up to 5 T, by a vibrating sample magnetometer in magnetic fields up to 1.8 T, and in pulsed magnetic fields up to 10 T. At comparatively low temperatures the spontaneous magnetization M_s was determined by a linear extrapolation of the field dependence of the magnetization $M(H)$ from high fields to the saturation field. In the vicinity of the Curie temperature, where the $M(H)$ dependence is non-linear, M_s was determined by the Belov-Arrott method for second-order magnetic phase transitions. Using the Belov-Arrott method, we also determined the Curie temperature for all the alloys, except for the compositions for which the ferromagnetic-paramagnetic transition is a first-order phase transition. The paramagnetic susceptibility of the alloys was determined from $M(T)$ dependencies measured in a magnetic field of 0.2 T in a temperature interval from the Curie point T_C up to 700 K. The full cycle of the magnetic measurements was performed for $x = 0$, 0.04, 0.08, 0.12, 0.16, and 0.19

4. Phase diagram of $Ni_{2+x}Mn_{1-x}Ga$ ($0 \leq x \leq 0.36$) alloys

The martensitic transformation temperature T_m, determined from DSC measurements, revealed a monotonic increase with increasing x in the compositional interval $0 \leq x \leq 0.22$. Further increase in x from $x = 0.22$ to $x = 0.27$ does not affect T_m significantly, $T_m \approx 370$ K. A drastic jump of T_m is observed as the composition changes from $x = 0.27$ to $x = 0.30$. In the compositional interval $0.30 \leq x \leq 0.36$ the martensitic transformation temperature increases ~ 100 K, from $T_m = 528$ K ($x = 0.30$) to $T_m = 626$ K ($x = 0.36$).

Since previous studies of $Ni_{2+x}Mn_{1-x}Ga$ ($0 \leq x \leq 0.19$) alloys [4] have shown that the martensitic T_m and the ferromagnetic T_C transition temperatures merge in alloys with $x \geq 0.18$, the non-monotonic behaviour of T_m, evident from the DSC measurements, could be related to the coupling of T_m and T_C. In order to check this, we measured temperature dependencies of the magnetization M in a low magnetic field $H = 500$ Oe. Since the ferromagnetic-paramagnetic transition for the alloys with merged T_m and T_C was expected to be a first-order phase transition, the Curie temperature T_C was determined as a minimum on the temperature derivative of the magnetization curves measured upon heating, dM/dT. Results of these measurements revealed that $T_C \approx T_m$ in the interval $0.18 \leq x \leq 0.27$. In the alloys with $x \geq 0.30$, the Curie temperature is lower than the martensitic transformation temperature.

The hase diagram of $Ni_{2+x}Mn_{1-x}Ga$ in the studied compositional interval, constructed from the DSC and the low-field magnetization measurements, is shown in Fig. 2. Three distinct regions can be distinguished on this diagram. In the first region, characterized by the Ni excess $0 \leq x \leq 0.16$, $T_C > T_m$ and the martensitic transformation takes place in the ferromagnetic state. Alloys from the second region with Ni excess $0.18 \leq x \leq 0.27$ are characterized by a coupled magnetostructural transition, i.e. $T_m \approx T_C$. Finally, the third region is characterized

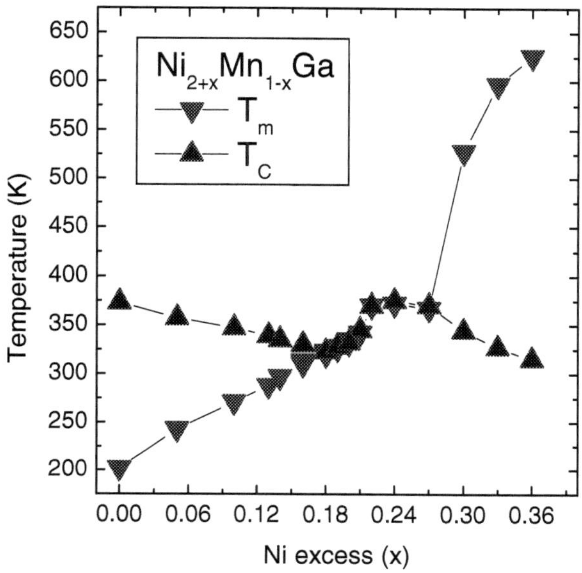

Figure 2. Temperatures of martensitic (T_m) and ferromagnetic (T_C) phase transitions determined from DSC and low-field magnetization measurements, as a function of Ni excess x in the studied $Ni_{2+x}Mn_{1-x}Ga$ alloys.

by a high martensitic-transformation temperature, $T_m > 550$ K, and a comparatively low Curie temperature, $T_C < 350$ K. In this region, the martensitic transformation takes place in the paramagnetic state. The occurrence of a martensitic transformation at high temperatures makes the alloys from this region attractive for application as high-temperature shape memory alloys.

As evident from Fig. 2, T_m and T_C of the alloys belonging to the second region exhibit a non-monotonic dependence on the Ni excess x. Approaching T_m and T_C results in enhancement of the magnetoelastic interaction [4], which reaches its maximum in the $x = 0.18$ alloy, when the magnetic and the structural subsystems couple. The Curie temperature, $T_C = 376$ K for Ni_2MnGa, decreases with increasing deviation from the stoichiometry and reaches a local minimum, $T_C = 323$ K, in the $Ni_{2.18}Mn_{0.82}Ga$ composition. The increase of T_C in the $0.18 \le x \le 0.22$ alloys is presumably caused by the coupling of the magnetic and structural subsystems and by the tendency of T_m to increase with Ni excess.

Since substitution of Mn for Ni results in dilution of the magnetic subsystem, the observed increase of T_C in the $0.18 \le x \le 0.22$ alloys discloses a strong interrelation between magnetic and structural subsystems of $Ni_{2+x}Mn_{1-x}Ga$. The almost constant temperature of the magnetostructural transition, $T_m \approx T_C \approx 370$ K, observed in the $0.22 \leq x \leq 0.27$ alloys, is probably caused by the strong magnetoelastic interaction and by a competition between increasing chemical pressure and further dilution of the magnetic subsystem.

At the critical $Ni_{2.30}Mn_{0.70}Ga$ composition T_m and T_C are no longer coupled, which results in a drastic increase of the martensitic transformation temperature up to ≈ 530 K, and in a decrease of T_C down to ≈ 350 K. In the alloys with the higher Ni excess, the martensitic transformation occurs at temperatures above 600 K, whereas the Curie temperature continues to decrease. Considering the empirical correlation between the electron concentration e/a and the martensitic transformation temperature T_m [2], it can be suggested that further increase in T_m of the $0.30 \leq x \leq 0.36$ alloys can be attained by the substitution of Ga for Ni or Mn.

The constructed phase diagram evidences that both T_m and T_C merge in a rather wide interval of compositions, from $x = 0.18$ to $x = 0.27$. The existence of a coupled magnetostructural phase transition for $\sim 10\%$ substitution of Mn for Ni in $Ni_{2+x}Mn_{1-x}Ga$ is the most distinct feature of the phase diagram of these alloys. Recent theoretical considerations [21] imply that the strong interrelation between the structural and the magnetic subsystems, observed in an extended range of compositions, can be accounted for by the influence of volume magnetostriction on the phase transitions of $Ni_{2+x}Mn_{1-x}Ga$ alloys.

5. Magnetic properties of $Ni_{2+x}Mn_{1-x}Ga$ ($0 \leq x \leq 0.19$) alloys

Temperature dependencies of the spontaneous magnetization M_s of the $Ni_{2+x}Mn_{1-x}Ga$ alloys are shown in Fig. 3. It is seen that M_s smoothly decreases with increasing temperature and exhibits a more pronounced change (a broad jump) when approaching a critical temperature T_m. This anomalous change in the magnetization is caused, as has been shown in numerous studies (see, for example, Refs. [18, 22, 23]), by a structural phase transition from martensite to austenite. As evident from our measurements, in the $x < 0.19$ alloys the austenitic phase remains ferromagnetic above T_m, whereas in the $x = 0.19$ composition the transformation from martensite to austenite is accompanied by a transition from the ferromagnetic to the paramagnetic state.

The compositional dependence of the martensitic transformation temperature T_m, determined from the magnetic measurements, is shown in Fig. 4. This figure also shows the dependence of the ferromagnetic or-

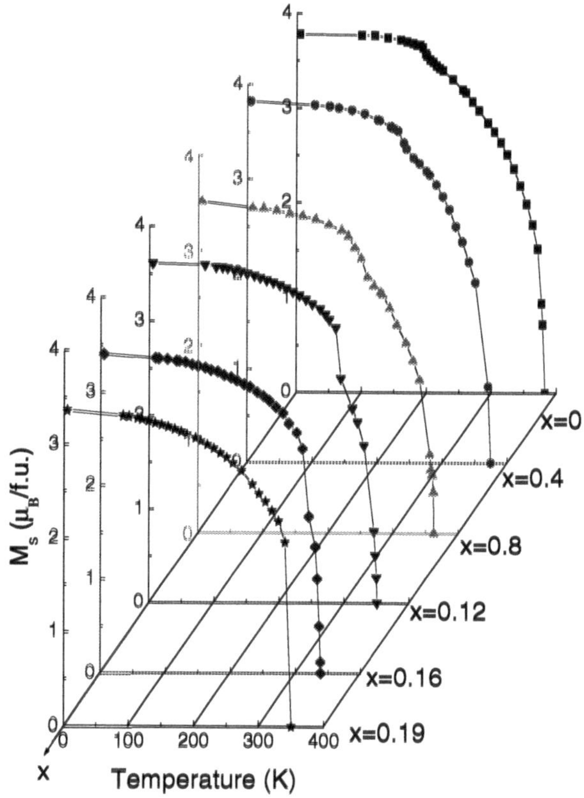

Figure 3. Temperature dependencies of spontaneous magnetization of $Ni_{2+x}Mn_{1-x}Ga$ ($0 \leqq x \leqq 0.19$) alloys.

dering temperature T_C on the Ni concentration x. It is seen that both dependencies are practically linear, T_m increasing and T_C decreasing with increasing Ni content x, so that the critical Ni concentration, at which these temperatures coincide, is in the interval $x = 0.18-0.19$. Our phase diagram of the $Ni_{2+x}Mn_{1-x}Ga$ ($0 \leq x \leq 0.19$) system, obtained from the magnetization measurements, is in good agreement with that obtained from DSC measurements (Fig. 2) and with previously reported ones [4, 24].

From the spontaneous magnetization data (Fig. 3) it follows that the magnetic moment of the alloys, μ, determined by extrapolation of the spontaneous magnetization to 0 K, decreases approximately in the same manner as T_C does with increasing degree of substitution of Mn by Ni (Fig. 5). The magnetic moment for stoichiometric Ni_2MnGa, determined

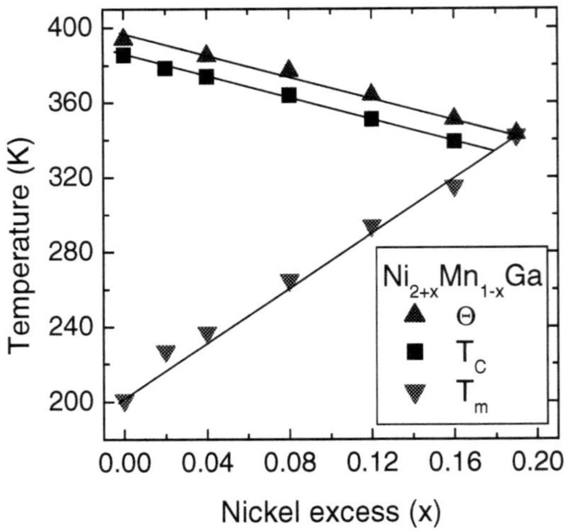

Figure 4. Compositional dependencies of the martensitic transformation temperature T_m, the the Curie temperature T_C and paramagnetic Curie temperature Θ determined from magnetic measurements.

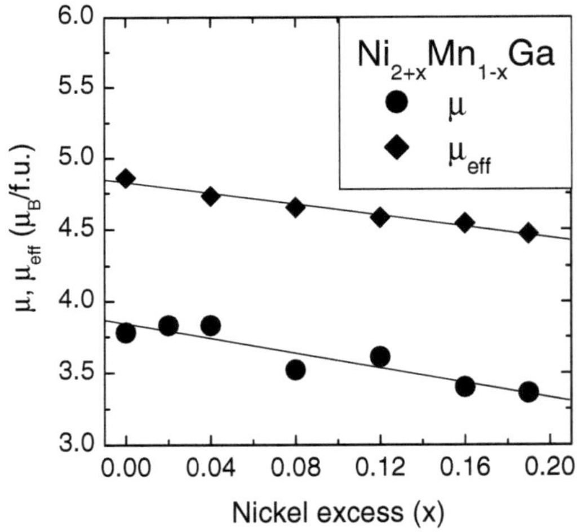

Figure 5. Compositional dependencies of the saturation magnetic moment $M_s(0)$ and the effective magnetic moment μ_{eff} of $Ni_{2+x}Mn_{1-x}Ga$ ($0 \leq x \leq 0.19$) alloys.

from our measurements, is close to that reported in others studies [11, 18, 19].

Our results of the spontaneous magnetization of the $Ni_{2+x}Mn_{1-x}Ga$ system, shown in Fig. 3, indicate that in the alloys studied the change of

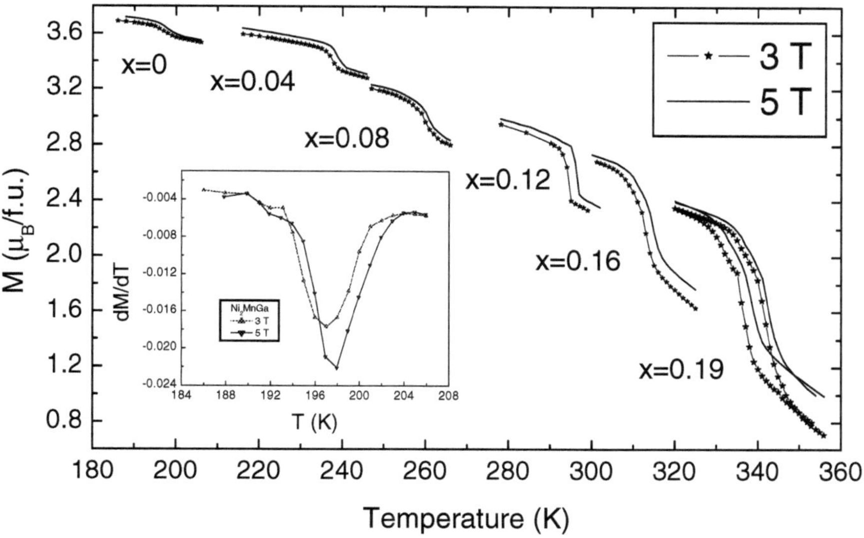

Figure 6. Magnetization jump at the martensitic transition in magnetic fields of 3 and 5 T. For $0 \leq x \leq 0.16$ compositions $M(T)$ dependencies were measured upon heating. For the $x = 0.19$ composition, a temperature hysteresis loop of the magnetization observed at martensitic transition is shown.

spontaneous magnetization at the structural transformation from martensite to austenite increases as the degree of substitution of Mn for Ni increases. A jump of magnetization at the structural transition is also observed in magnetic fields larger than the saturation field, as illustrated by Fig. 6, where $M(T)$ dependencies measured in fields of 3 T and 5 T are shown. The compositional dependence of the magnitude of the magnetization jump ΔM in various magnetic fields is illustrated by Fig. 7.

It is evident from Fig. 6 that in a higher magnetic field the magnetization jump is observed at a higher temperature, which is caused by the influence of the magnetic field on the martensitic transformation temperature. From our measurements it follows that the shift of T_m, caused by magnetic field, weakly depends on composition and is approximately $1.5 - 2$ K for the field increase from 3 to 5 T (see Table 1).

Literature data on the influence of a magnetic field on T_m for stoichiometric Ni_2MnGa [23] and for $x = 0.18$ and $x = 0.19$ compositions [5, 6, 24] show a considerable scattering. For the stoichiometric composition the shift of T_m by a magnetic field was estimated as $dT_m/dH \approx 0.2$ K/T [23]. For the $x = 0.18$ and $x = 0.19$ alloys, $dT_m/dH \approx 1$ K/T was reported in Refs. [6, 24], whereas in Ref. [5]

Table 1. Experimental values of latent heat of the martensitic transformation Q, the transformation temperature T_m, and the magnetization jump ΔM at the martensitic transformation for $Ni_{2+x}Mn_{1-x}Ga$ ($0 \leq x \leq 0.19$) alloys. ΔT is the shift of the martensitic transformation temperature T_m in a field $\Delta H = 2$ T calculated from the Clapeyron-Clausius relation (theory) and obtained from the magnetization measurements (experiment).

x	Q (J/mol)	T_m (K)	ΔM (μ_B)	ΔT (K) ($\Delta H = 2$ T) theory	ΔT (K) ($\Delta H = 2$ T) experiment
0	270	201	0.1	0.82 ± 0.2	0.8 ± 0.5
0.04	600	237	0.17	0.75 ± 0.2	0.95 ± 0.5
0.08	910	265	0.28	0.92 ± 0.2	0.95 ± 0.5
0.12	1250	294	0.41	1.07 ± 0.2	1.10 ± 0.5
0.16	1710	315	0.62	1.28 ± 0.2	1.30 ± 0.5
0.19	2260	342	0.96	1.62 ± 0.2	1.60 ± 0.5

this quantity was estimated as 3.5 K/T. It is worth noting that the shift of T_m is determined with a significant error. This is caused mainly by the fact that the jump of the magnetization at the martensitic transformation is broad, which makes a correct determination of T_m difficult. Besides, since the martensitic transformation is a first-order structural phase transition and has, therefore, a temperature hysteresis, the T_m temperature can be different from the temperature at which the jump of magnetization is observed. The most correct method to determine T_m from magnetic measurements is to determine this temperature as the average of temperatures, at which a magnetization jump is observed on cooling and heating, respectively. In our experiments we determined T_m as the temperature of a magnetization jump, observed during heating. Analysis of the experimental data for the $x = 0.19$ sample, for which a temperature hysteresis loop of the magnetization observed at the structural transition was measured (see Fig. 6), shows that this results in an error of $2 - 3$ K for the correct determination of T_m. At the same time, since the width of the temperature hysteresis loop is approximately the same in different magnetic fields, an additional error in the determination of the shift of T_m caused by a magnetic field does not exceed 0.3 K.

Temperature dependencies of the reciprocal paramagnetic susceptibility of the $Ni_{2-x}Mn_{1-x}Ga$ alloys are shown in Fig. 8. In the temperature

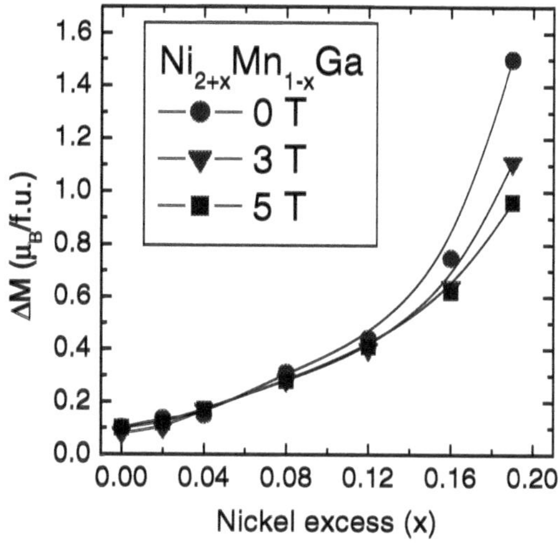

Figure 7. The magnetization jump at the martensitic transition in various magnetic fields as a function of Ni concentration in $Ni_{2+x}Mn_{1-x}Ga$ ($0 \leq x \leq 0.19$) alloys.

interval of the measurements, the susceptibility of all the samples studied follows a Curie-Weiss law.The resulting composition dependence of the paramagnetic Curie temperature Θ and that of the effective magnetic moment μ_{eff} are shown in Figs. 4 and 5, respectively. It is evident that both these parameters decrease monotonically with increasing Ni excess x. In the literature, paramagnetic susceptibility measurements were reported only for the stoichiometric Ni_2MnGa alloy [18, 25]. Our values of Θ and μ_{eff} are somewhat larger than the values reported previously. This difference could be accounted for by the fact that our measurements were performed in a larger temperature interval.

In Fig. 9 the compositional dependence is shown of the latent heat Q of the martensite-austenite phase transition. It is evident that Q strongly increases with increasing x. These results are in good agreement with a recently published one [26].

Using the compositional dependencies of the magnetic moment and the effective magnetic moment of the alloys (Fig. 5), we can determine the magnetic moments and the effective magnetic moments of Mn and Ni atoms (see Table 2) from the formulae

$$\mu = (1 - x)\mu_{\text{Mn}} + (2 + x)\mu_{\text{Ni}} \tag{1}$$

and

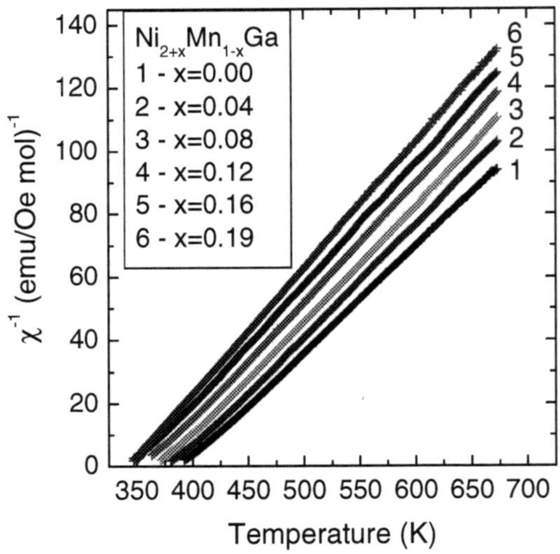

Figure 8. Temperature dependencies of the paramagnetic susceptibility of $Ni_{2+x}Mn_{1-x}Ga$ ($0 \leq x \leq 0.19$) alloys.

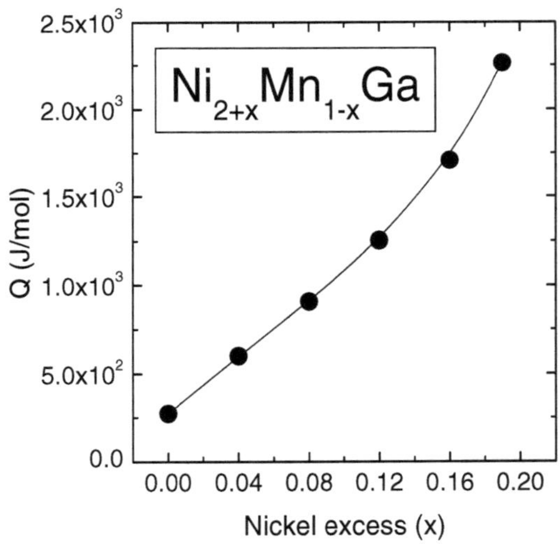

Figure 9. Compositional dependence of the latent heat of the martensitic transition for $Ni_{2+x}Mn_{1-x}Ga$ ($0 \leq x \leq 0.19$).

$$\mu_{\text{eff}} = \sqrt{(1-x)\mu^2_{\text{eff Mn}} + (2+x)\mu^2_{\text{eff Ni}}}\,, \tag{2}$$

Table 2. Magnetic moments μ and effective magnetic moments μ_{eff} of Mn and Ni atoms.

	μ (μ_B)	μ_{eff} (μ_B)	$\mu_{\text{eff}}^{\text{loc}}$ (μ_B)	$\mu_{\text{eff}}^{\text{loc}}/\mu_{\text{eff}}$
Mn	2.99 ± 0.32	4.43 ± 0.13	3.86 ± 0.14	0.87 ± 0.11
Ni	0.43 ± 0.14	1.35 ± 0.18	1.05 ± 0.21	0.77 ± 0.10

respectively. The obtained values of the magnetic moments of the constituting atoms of the alloys are in good accordance with the results of neutron diffraction and NMR studies of the stoichiometric composition [10, 11, 27], which showed that in Ni_2MnGa the Mn magnetic moment $\mu_{\text{Mn}} \approx 2.84 - 3.41\ \mu_B$, whereas the Ni magnetic moment $\mu_{\text{Ni}} \approx 0.3 - 0.41\ \mu_B$.

Note that in the calculation we proceeded from a simplified model and assumed that the magnetic moments of the constituting atoms do not change with deviation from stoichiometry and that the Ni atoms possess the same moment in different crystallographic positions. In general this is not the case, because magnetism of Heusler alloys is described by the band model, which means that the magnetic moments depend on the density of states near the Fermi level and on the exchange splitting parameter, i.e. on concentration- and structural-dependent parameters. As has been noted in Ref. [28], in Ni_2MnX Heusler alloys the distance between the atoms is sufficiently large so that the direct overlapping of electron shells is negligible and the delocalization effects are of secondary importance. Therefore, a localized moments model is applicable in the first approximation for the description of the magnetic properties of these alloys. From the results of magnetic and NMR measurements of Ni_2MnGa [27] it was concluded, however, that in this alloy only the Mn subsystem is localized, whereas the Ni magnetic moments are essentially delocalized.

The character of magnetism can be judged from the comparison of the magnetic moments of the constituting atoms and their effective magnetic moments (see Table 2).

In the model of localized magnetic moments for the spin-only state (orbital moment is quenched) the interrelation between the effective magnetic moment and the moment in the magnetically ordered state is given by

$$\mu_{\text{eff}}^{\text{loc}} = \sqrt{\mu(\mu + 2)} \tag{3}$$

whereas in the band model the calculated [from (3)] value of $\mu_{\rm eff}^{\rm loc}$ should be smaller than the experimental value of the effective moment due to the influence of delocalization effects. The values of $\mu_{\rm eff}^{\rm loc}$ for the Mn and Ni subsystems are given in Table 2. As evident from these data, for both the subsystems $\mu_{\rm eff}$ and $\mu_{\rm eff}^{\rm loc}$ are close to each other, although in both cases $\mu_{\rm eff}^{\rm loc}$ is slightly smaller than $\mu_{\rm eff}$. Within the experimental error of the measurements the $\mu_{\rm eff}^{\rm loc}/\mu_{\rm eff}$ ratio, which can be considered as a measure of the degree of localization, is the same for both the subsystems. Thus, it does not follow from our experimental data that the Ni subsystem is more delocalized that the Mn one.

It should be recognized, however, that the magnetic moments in the magnetically ordered state were determined in the martensitic phase, whereas the effective magnetic moments were calculated from the paramagnetic susceptibility measured in the austenitic state. It makes no difference if magnetism is described in the localized model, because in this case the magnitude of magnetic moment depends weakly on the crystallographic environment. In the band model, however, the magnetic moments depend on the degree of overlapping of electronic shells that changes upon structural transformation due to the changes in interatomic distances. Because of this, a possibility that magnitudes of magnetic moments will change upon structural transition must not be ruled out. The qualitative arguments given above are supported by the experimental results reported in Ref. [27] which indicate that the magnetic moment of Mn is the same in austenitic and martensitic phases, whereas magnetic moment of Ni in the austenitic phase is larger than that in the martensitic phase. The latter observation is conditioned by a higher density of states of Ni at the Fermi level in the austenitic state, as electronic structure calculations have revealed [12].

As evident from Fig. 4, the Curie temperature of the austenitic phase decreases upon substitution of Mn for Ni. This is due to the fact that upon such a substitution a number of Ni atoms with smaller, as compared to Mn, magnetic moment increases. A similar tendency presumably takes place for a hypothetical Curie temperature of the martensitic phase. This follows from the fact that, as is seen from Fig. 3, in the low-temperature martensitic phase the magnetization of the $Ni_{2+x}Mn_{1-x}Ga$ alloys with high Ni excess decreases more rapidly with increasing temperature.

The collected data allow us a rough estimation of the hypothetical Curie temperature of the martensitic phase. Figure 10 shows temperature dependencies of the reduced spontaneous magnetization $m = M_s(T)/M_s(0)$ of the alloys as a function of reduced temperature $t = T/T_C$. It is seen that the magnetization of the austenitic phase and

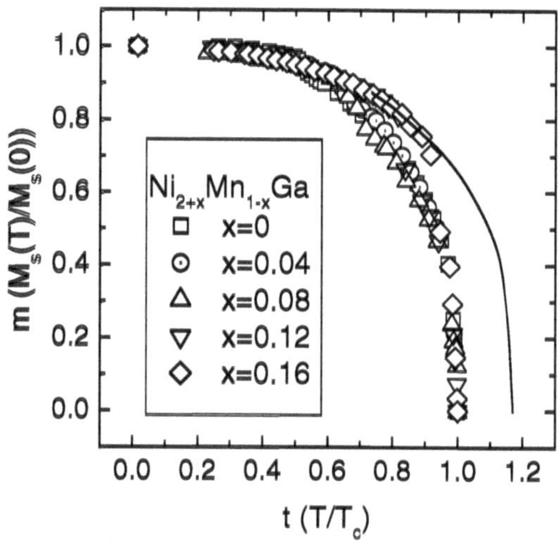

Figure 10. Reduced magnetization $m = M_s(T)/M_s(0)$ as a function of reduced temperature $t = T/T_C$ for $Ni_{2+x}Mn_{1-x}Ga$ ($0 \leq x \leq 0.19$) alloys. The solid line is the reduced magnetization of a virtual martensitic phase.

the magnetization of the martensitic phase change with temperature in a different way, whereas the reduced magnetization of these phases is approximately the same for different compositions. Comparing $m(t)$ dependencies of the martensitic and austenitic phases, and assuming that the difference between these dependencies arises because the Curie temperature of the martensitic phase is higher than the Curie temperature of the austenitic phase, it is possible to reconstruct a hypothetical reduced magnetization of the martensitic phase at high temperatures, above T_m, (shown by the solid line in Fig. 10) and to determine the relative Curie temperature of the martensitic phase. As evident from Fig. 10, this comparison indicates that the hypothetical Curie temperatures of the martensitic phase are $\approx 17\%$ higher than those of the corresponding alloys in the austenitic state. This estimation is twice as large as that obtained from phenomenological Landau theory [29].

The larger value of the Curie temperature of martensite as compared to the Curie temperature of the corresponding austenitic phase is determined, as it has been mentioned above, by the changes in interatomic distances and in the overlapping of electronic wave functions, taking place at austenite-martensite transformation, which results in a change of the exchange interactions. As evident from the analysis of the experimental results, this effect can not be attributed solely to a change in the unit cell volume at the martensitic transformation. Indeed, a study of the influence of hydrostatic pressure on the Curie temperature T_C and on the martensitic transformation temperature T_m of stoichiometric Ni_2MnGa [25] indicated that the exchange interaction of the austenitic phase increases with decreasing unit cell volume ,and yet the results of this study as well as the results of a X-ray diffraction study of the crystal structure at the stoichiometric composition [30] showed that the unit cell volume of the martensitic phase is larger than that of the austenitic phase. Therefore, it seems likely that the primary role in the martensitic transformations in Heusler alloys is played by crystal lattice distortions, leading to a change in magnitude and character of exchange interactions. Such a mechanism of the influence of a structural transition on exchange interactions in intermetallic compounds $Gd_5(Si_xGe_{1-x})_4$ was discussed recently in Ref. [31].

As evident from Fig. 7, the magnitude of the magnetization jump, ΔM at the structural transition drastically increases with increasing Ni content. This is caused by the fact that in the alloys with a high Ni content the structural transition occurs at higher temperatures, where the difference between the magnetization of martensitic and austenitic phases increases, which is dictated by the different Curie temperatures of these phases. It is also seen from Fig. 7 that for small x the magnitude of the magnetization jump weakly depends on the field, whereas in the $x = 0.16$ and $x = 0.19$ compositions the influence of the magnetic field becomes more pronounced and ΔM decreases with increasing field. The behavior of ΔM in the alloys with small x originates from the fact that the martensitic transformation in these alloys occurs at temperatures far below the Curie temperature T_C and therefore the influence of a magnetic field on the magnetization is weak. In the alloys with higher x, T_m is close to T_C of the austenitic phase and the external field strongly affects the magnetization of this phase, whereas magnetization of the martensitic phase having a higher hypothetical Curie temperature increases more weakly with increasing field, which explains the decrease of ΔM upon application of the magnetic field.

It has already been mentioned that the temperature of the structural transition shifts to higher temperatures upon application of a magnetic field. Such behavior is governed by the influence of the Zeeman energy, which stabilizes the martensitic phase with a larger magnetization. Experimental data on the shift of T_m are presented in Table 1. These results indicate that for the alloys studied the shift of T_m is rather small ($1-2$ K as the magnetic field changes with 2 T) and slightly increases with increasing Ni content. The table also contains a theoretical estimation of the shift of T_m in the magnetic field, derived from a thermodynamical relation for first-order phase transitions:

$$\Delta T = \Delta M H T_m / Q \ .$$

The agreement between experimental and theoretical values can be considered as satisfactory.

6. Conclusion

The results of experimental studies of $Ni_{2+x}Mn_{1-x}Ga$ ($0 \leq x \leq 0.36$) presented in this Chapter revealed the following main features in the phase diagram and the magnetic properties of these alloys. According to the experimental phase diagram of this system, ferromagnetic and martensitic phase transitions approach each other in $0 \leq x \leq 0.16$ compositions, couple in a compositional interval from $x = 0.18$ to $x = 0.27$ and move apart in the alloys with $x \geq 0.30$. A peculiar characteristic of the phase diagram is the observation of the coupled magnetostructural phase transition in rather a wide compositional range, for which the change in the Mn content is $\approx 10\%$.

Magnetic properties of $Ni_{2+x}Mn_{1-x}Ga$ alloys were studied systematically for $0 \leq x \leq 0.19$. From the measurements of the spontaneous magnetization $M_s(T)$, the jumps of the magnetization ΔM at the structural phase transitions were determined. The virtual Curie temperatures of martensite were estimated from the comparison of the magnetization in martensitic and austenitic phases. Both saturation magnetic moments in the ferromagnetic state and the effective magnetic moments in the paramagnetic state of Mn and Ni atoms were determined, and delocalization effects on magnetism in these alloys were discussed. The experimental results obtained show that the shift of the martensitic transition temperature is weakly dependent on composition. The values of this shift are in good correspondence with the Clapeyron-Clausius formalism taking into account the experimental data on latent heat of the martensite-austenite transformations.

Acknowledgments

The author thanks ABC for their interest, cooperation and support and many useful discussions. Financial support by ABCDE is appreciated.

References

[1] A. N. Vasil'ev, V. D. Buchel'nikov, T. Takagi, V. V. Khovailo, and E. I. Estrin, Physics - Uspekhi **46**, 559 (2003).

[2] V. A. Chernenko, Scripta Mater. **40**, 523 (1999).

[3] I. Takeuchi, O. O. Famodu, J. C. Read, M. A. Aronova, K.-S. Chang, C. Craciunescu, S. E. Lofland, M. Wuttig, F. C. Wellstood, L. Knauss, and A. Orozco, Nature Mater. **2**, 180 (2003).

[4] A. N. Vasil'ev, A. D. Bozhko, V. V. Khovailo, I. E. Dikshtein, V. G. Shavrov, V. D. Buchelnikov, M. Matsumoto, S. Suzuki, T. Takagi, and J. Tani, Phys. Rev. B **59**, 1113 (1999).

[5] V. V. Khovailo, T. Takagi, J. Tani, R. Z. Levitin, A. A. Cherechukin, M. Matsumoto, and R. Note, Phys. Rev. B **65**, 092410 (2002).

[6] D. A. Filippov, V. V. Khovailo, V. V. Koledov, E. P. Krasnoperov, R. Z. Levitin, V. G. Shavrov, and T. Takagi, J. Magn. Magn. Mater. **258–259**, 507 (2003).

[7] L. Pareti, M. Solzi, F. Albertini, and A. Paoluzi, Eur. Phys. J. B **32**, 303 (2003).

[8] B. Wedel, M. Suzuki, Y. Murakami, C. Wedel, T. Suzuki, D. Shindo, and K. Itagaki, J. Alloys Comp. **290**, 137 (1999).

[9] J. Pons, V. A. Chernenko, R. Santamarta, and E. Cesari, Acta Mater. **48**, 3027 (2000).

[10] P. J. Brown, J. Crangle, T. Kanomata, M. Matsumoto, K.-U. Neumann, B. Ouladdiaf, and K. R. A. Ziebeck, J. Phys.: Condens. Matter **14**, 10159 (2002).

[11] P. J. Brown, A. Y. Bargawi, J. Crangle, K.-U. Neumann, and K. R. A. Ziebeck, J. Phys.: Condens. Matter **11**, 4715 (1999).

[12] S. Fujii, S. Ishida, and S. Asano, J. Phys. Soc. Jpn. **58**, 3657 (1989).

[13] H. Hosoda, T. Sugimoto, K. Ohkubo, S. Miura, T. Mohri, and S. Miyazaki, Int. J. Appl. Electromagn. Mech. **12**, 9 (2000).

[14] M. Kreissl, K.-U. Neumann, T. Stephens, and K. R. A. Ziebeck, J. Phys.: Condens. Matter **15**, 3831 (2003).

[15] V. V. Khovailo, T. Takagi, A. D. Bozhko, M. Matsumoto, J. Tani, and V. G. Shavrov, J. Phys.: Condens. Matter **13**, 9655 (2001).

[16] V. V. Khovailo, K. Oikawa, C. Wedel, T. Takagi, T. Abe, and K. Sugiyama, J. Phys.: Condens. Matter **16**, 1951 (2004).

[17] R. Tickle and R. D. James, J. Magn. Magn. Mater. **195**, 627 (1999).

[18] P. J. Webster, K. R. A. Ziebeck, S. L. Town, and M. S. Peak, Philos. Mag. B **49**, 295 (1984).

[19] W. H. Wang, F. X. Hu, J. L. Chen, Y. X. Li, Z. Wang, Z. Y. Gao, Y. F. Zheng, L. C. Zhao, G. H. Wu, and W. S. Zan, IEEE Trans. Magn. **37**, 2715 (2001).

[20] F. Albertini, L. Pareti, A. Paoluzi, L. Morellon, P. A. Algarabel, M. R. Ibarra, and L. Righi, Appl. Phys. Lett. **81**, 4032 (2002).

[21] V. D. Buchelnikov (unpublished).

[22] V. V. Kokorin, V. A. Chernenko, V. I. Val'kov, S. M. Konoplyuk, and E. A. Khapalyuk, Fizika Tv. Tela **37**, 3718 (1995) [Phys. Solid State **37**, 2049 (1995)].

[23] A. Gonzàlez-Comas, E. Obradó, L. Mañosa, A. Planes, V. A. Chernenko, B. J. Hattink, and A. Labarta, Phys. Rev. B **60**, 7085 (1999).

[24] A. D. Bozhko, A. N. Vasil'ev, V. V. Khovailo, I. E. Dikshtein, V. V. Koledov, S. M. Seletskii, A. A. Tulaikova, A. A. Cherechukin, V. G. Shavrov, V. D. Buchel'nikov, Zh. Eksp. Teor. Fiz. **115**, 1740 (1999) [JETP **88**, 954 (1999)].

[25] T. Kanomata, K. Shirakawa, and T. Kaneko, J. Magn. Magn. Mater. **65**, 76 (1987).

[26] V. V. Khovailo, K. Oikawa, T. Abe, and T. Takagi, J. Appl. Phys. **93**, 8483 (2003).

[27] K. Ooiwa, K. Endo, and A. Shinogi, J. Magn. Magn. Mater. **104–107**, 2011 (1992).

[28] T. Moriya, *Spin fluctuations in itinerant electron magnetism* (Springer-Verlag, Berlin–Heidelberg–New-York–Tokyo, 1985).

[29] V. A. Chernenko, V. A. L'vov, S. P. Zagorodnyuk, and T. Takagi, Phys. Rev. B **67**, 064407 (2003).

[30] Y. Ma, S. Awaji, K. Watanabe, M. Matsumoto, and N. Kobayashi, Solid State Comm. **113**, 671 (2000).

[31] W. Choe, V. K. Pecharsky, A. O. Pecharsky, K. A. Gschneidner, Jr., V. G. Young, Jr., and G. J. Miller, Phys. Rev. Lett. **84**, 4617 (2000).

MAGNETOSTRICTION OF MANGANITES AND COBALTITES

H. Szymczak and R. Szymczak
Institute of Physics, Polish Academy of Sciences, Lotnikow 32/46, 02-668 Warsaw, Poland
szymh@ifpan.edu.pl

Abstract: A comparison of the magnetostrictive effects in manganite and cobaltite thin films and single crystals is presented. A detailed analysis of the magnetostriction has been performed for epitaxial thin films with the same level of doping, namely on $La_{0.67}Ca_{0.33}MnO_3$ and $La_{0.7}Sr_{0.3}CoO_3$ films grown on twins free substrates using laser deposition. The magnetoelastic stress was measured using a capacitance cantilever technique in a wide temperature range 4.2-300 K. It was shown that the magnetostriction of the investigated manganite films is typical in value for ferromagnetic 3d metals. From the temperature dependence of the magnetostriction constant it may be concluded that a single-ion mechanism dominates in this type of film. In the case of cobaltite films and single crystals, two mechanisms give a significant contribution to the magnetostrictive effects: magnetic-field induced increase of the volume of ferromagnetic clusters and magnetic-field induced spin state transitions. The first mechanism dominates at high temperatures.

Keywords: Magnetostriction, magnetocrystalline anisotropy, perovskites, manganites, cobaltites, phase separation.

1. Introduction

The discovery in 1986 of high-temperature superconductivity in ceramic cuprates of *perovskite* structure started a period of very intensive research of transition metal oxides. Soon afterwards, in 1993, the colossal magnetoresistance effect was discovered in manganite perovskites, again leading to an increasing research activity in the field of magnetic oxides. It is

J. Franse et al. (eds.), Smart Materials for Ranging Systems, 245–258.

now well established that this group of materials exhibits a wide range of exotic and still imperfectly understood properties, which are controlled by strong electron-lattice and electron-electron correlations [1-3].Recent experimental studies have shown that structural instabilities in the two families of *perovskite* systems play a key role in their characteristic superconducting and magnetoresistive properties.

Generally, materials with strong electron-lattice coupling are expected to exhibit giant magnetostriction. This paper will review the magnetostrictive effects in manganites and cobaltites characterized very often by a colossal value of the magnetoresistance.

The mixed-valence manganite and cobaltite oxides reveal unique magnetic and transport properties determined by competition of a variety of interactions, of which the most important are double exchange, superexchange and Jahn-Teller. The orbital degeneracy of Mn^{3+} and Co^{3+} ions leads to a variety of instabilities and is responsible for strong magnetoelastic coupling. [4,5]. In both families of compounds charge ordering, metal-insulator transitions and a high value of the magnetoresistance are observed. In the case of cobaltites, the various spin states of the cobalt ions makes this situation even more complicated than in manganites, leading to significant differences in their magnetic properties.

Because of surface contributions as well as due to interface strains, the magnetostriction of thin films is expected to be different from that observed in bulk materials. Motivated by this observation and by possible technical applications of thin films of manganites and cobaltites, we decided to present in this paper mostly results of investigations of magnetostrictive effects in thin films. It has to be mentioned that no data exist in the literature on magnetostriction of cobaltite thin films.

In order to obtain reliable results, magnetostriction should be measured in epitaxial films, free of twins, which are typical for materials of perovskite structure. In the case of films it is not easy, because usually the films are deposited on perovskite substrates, thus presenting difficulties in eliminating twin structures.

2 Experimental details

The magnetoelastic energy, simply related to the magnetostriction constants, can be written in the following form in the coordinate axes connected to the local anisotropy:

$$U_{me} = \sum B_{klmn}\, \alpha_k^{(i)} \alpha_l^{(i)}\, \varepsilon_{mn}^{(i)} \qquad (1)$$

Here, $\alpha_k^{(i)}$ are the direction cosines of the local magnetization $\boldsymbol{M}^{(i)}$ and $\varepsilon_{mn}^{(i)}$ are the components of the local strain tensor.

It has been shown, both experimentally and theoretically (see [5] for details), that the magnetoelastic tensor $[B]$ in nanoscale magnetic multilayers consists of two parts:

$$[B] = [B]_v + 2t^{-1}[B]_s \qquad (2)$$

where $[B]_v$ describes the bulk (volume) magnetoelastic coupling and $[B]_s$ describes the surface magnetoelastic coupling (t is the thickness of the magnetic layers).

Although we have performed magnetostriction measurements on thin films of various chemical compositions deposited on various substrates, we restrict ourselves here to a presentation of two typical examples. Both investigated cobaltite and manganite films were chosen to have the same level of doping (about 30%) and to have no structural instabilities in the temperature range from room temperature down to helium temperature. The last condition is very important, because in the vicinity of a phase transition the magnetostriction changes rapidly, making impossible any generalization.

The thin manganite $La_{0.67}Ca_{0.33}MnO_3$ (t = 880 Å) and cobaltite $La_{0.7}Sr_{0.3}CoO_3$ (t =650 Å) layers were grown by pulsed laser deposition without post annealing onto a low mismatched, **twins free,** monocrystalline $SAT_{0.7}CAT_{0.1}LA_{0.2}$ (001) (a_0=3.874 Å) substrate (where SAT means $SrAl_{0.5}Ta_{0.5}O_3$, CAT- $CaAl_{0.5}Ta_{0.5}O_3$, LA- $LaAlO_3$ single crystals). The substrate temperature was kept at 730°C during the deposition. The magnetic properties of the films were studied using a SQUID magnetometer. No magnetically detected structural phase transitions were observed in the temperature range 4.2-300 K. The magnetoelastic stress was investigated using a capacitance cantilever technique [6]. X-ray diffraction (Fig. 1) measurements showed that each film is single phase, epitaxial, with the pseudocubic lattice parameter a_0 = 3.848Å and a_0 = 3.839Å for manganite and cobaltite films, respectively. The samples were cut along appropriate crystallographic directions giving the samples dimensions of 14 mm by 4 mm. That allowed us to neglect any additional stresses appearing in the direction parallel to the clamped edge of the cantilever. The longer dimension was along the [100] crystalline axis. The $SAT_{0.7}CAT_{0.1}LA_{0.2}$ substrate was chosen because of the low lattice constant mismatch for both films of $La_{0.67}Ca_{0.33}MnO_3$ and $La_{0.7}Sr_{0.3}CoO_3$ composition and because of the absence of twin structures.

Single crystals of cobaltites $La_{0.8}Ca_{0.2}CoO_3$ were grown using the floating zone method with radiation heating. The X-ray diffraction analysis at 296 K confirmed that the samples were of single phase with a rhombohedral structure (space group R-3m). The point Laue patterns indicated some twinnings, therefore the grown crystals were cut into smaller pieces and a polarizing microscope was used to select samples with **the smallest** number of twins. The three terminal capacitance technique was applied in order to measure magnetostriction.

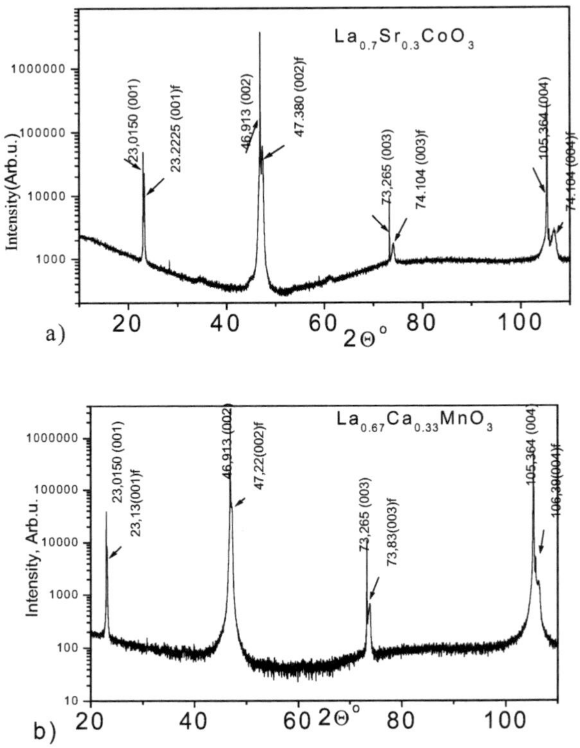

Figure 1. X-ray diffraction data of $La_{0.7}Sr_{0.3}CoO_3$ (a) and $La_{0.67}Ca_{0.33}MnO_3$ (b) films deposited onto $SAT_{0.7}CAT_{0.1}LA_{0.2}$ twins free substrate.

3. Results and analysis

First observations of the large magnetostrictive effects have been performed [7,8] very soon after the discovery of colossal magnetoresistance in this group of materials. It was suggested that two different mechanisms are responsible for magnetostriction in bulk manganites. Below the Curie temperature the magnetostriction is anisotropic and typical in value for a ferromagnetic *3d* metal. In the paramagnetic phase the magnetostriction should be related to the quenching of the charge localization mechanism under the application of an external magnetic field. Nevertheless, it was suggested that in $(La_{0.8}Ba_{0.2})_{0.93}MnO_3$ single crystals [9] magnetostriction is determined by a specific mechanism due to the presence of free-energy terms proportional to $M_iM_jM_kH_l$. According to [10], the magnetostriction in manganites is determined by the combined influence of the Jahn-Teller effect and exchange interactions. Interesting effects of irreversible giant changes of magnetostriction related to metamagnetic phase transitions have been observed in $Nd_{0.6}Ca_{0.4}MnO_3$ [11] and $La_{0.825}Sr_{0.175}MnO_3$ [12].

3.1 Manganite $La_{0.67}Ca_{0.33}MnO_3$ films

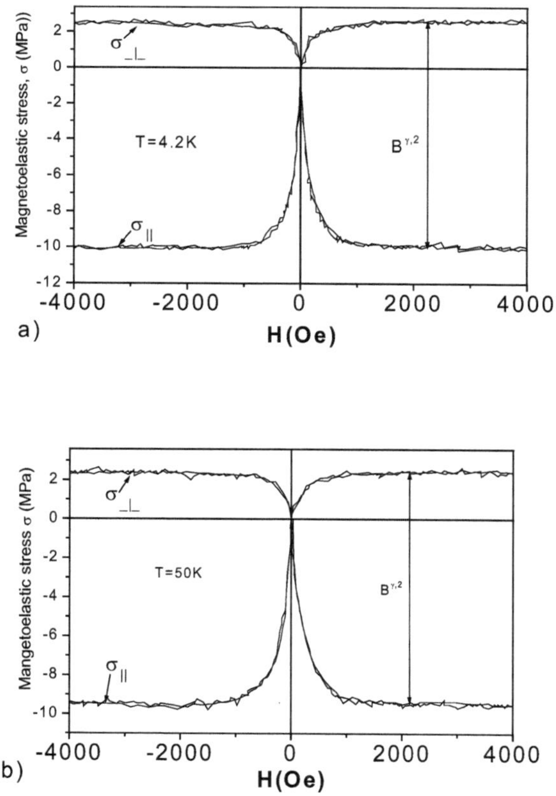

Figure 2. Magnetoelastic isotherms of a $La_{0.67}Ca_{0.33}MnO_3$ film on $SAT_{0.7}CAT_{0.1}LA_{0.2}$ substrate taken at 4.2 K (a) and 50 K (b).

Figure 2 shows both longitudinal and transversal magnetoelastic stresses in manganite $La_{0.67}Ca_{0.33}MnO_3$ films, with respect to the direction of the magnetic field.

The absolute value of the magnetoelastic-stress parameter $B^{\gamma,2}$ (see [13] for a detailed definition), determined from the difference between longitudinal and transversal magnetoelastic stresses, decreases with increasing temperature. The experimental points are fitted well with the Callen and Callen law [14], suggesting the domination of a single-ion mechanism in the magnetoelastic interaction (Fig. 3). The magnetostriction constant λ_{100}, using an expression [5] $B^{\gamma,2} = -(3/2)\lambda_{100}(c_{11} - c_{12})$, is calculated to be 83 ppm at 4.2 K. The elastic-constant difference c_{11} - c_{12} was evaluated to be 100 GPa [15].

It results from the performed measurements that the magnetostriction of ferromagnetic manganites, not having structural instabilities in the investigated temperature range, is typical in value for a ferromagnetic 3d metal. Similar values were obtained for bulk manganites with a similar level of doping. (see [5] for references). The origin of this unusual behavior of the system with e_g electrons interacting strongly with the lattice is still under discussion. The magnetostriction measurements performed for stoichiometric $La_{1-x}Sr_xMnO_3$ with $x = 0.11$, 0.13, and 0.165 indicated that large magnetostriction is observed only for the ferromagnetic phase [16]. It has to be mentioned that due to the magnetostriction the strain induced anisotropy arises in pseudomorphic grown epitaxial manganite films [17].

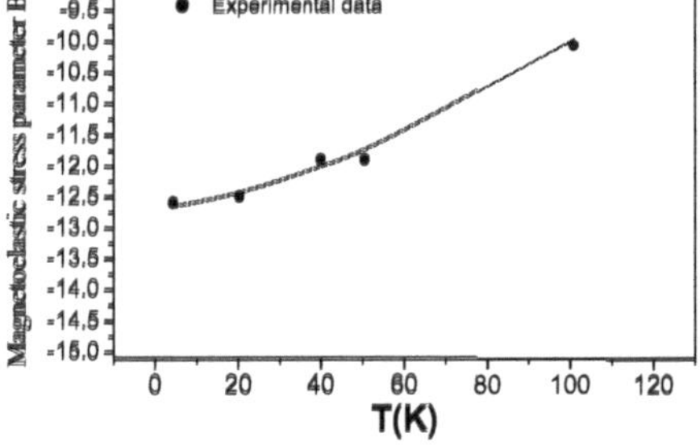

Figure 3. Temperature dependence of the magnetoelastic stress parameter $B^{\gamma,2}$ of $La_{0.67}Ca_{0.33}MnO_3$ film. The continuous line is a fit by the Callen and Callen law [5]: $B^{\gamma,2}(T) = B^{\gamma,2}(0)I_{5/2}L^{-1}[m(T)]$, where *m(T)*- reduced saturation magnetization and $B^{\gamma,2}(0)$=-12.7MPa .

The magnetostrictive effects are reviewed and analysed in detail by Koroleva [18]. It is interesting that the magnetostriction was shown to be very high in bilayered manganites [19]. Magnetoelastic phenomena have been shown [20] to be strongly correlated with magnetotransport. It is suggested that these effects are due to the field induced changes in the orbital-state occupancy in the 2D conduction band with temperature, magnetic field and doping level.

3.2 Cobaltite $La_{0.7}Sr_{0.3}CoO_3$ films

Magnetization measurements performed on cobaltite $La_{0.7}Sr_{0.3}CoO_3$ films indicate cluster spin-glass properties of the studied system [21]. It was also shown that an external magnetic field induces increase of the volume of ferromagnetically ordered clusters (embedded in the paramagnetic or antiferromagnetic matrix). Moreover, this field is responsible for the transition of Co^{3+} ions from a low-spin state (resulting spin S=0) to an intermediate spin state (S=1). It can not be excluded that the transition to the

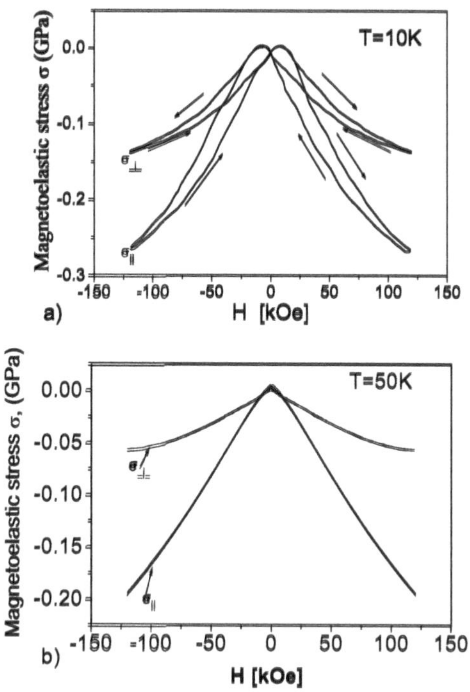

Figure 4. Magnetoelastic loops of the $La_{0.7}Sr_{0.3}CoO_3/SAT_{0.7}CAT_{0.1}LA_{0.2}$ cantilever taken at 10 K (a) and 50 K (b).

high-spin state (S=2) is also induced in the high-temperature region. This strong effect of the magnetic field on the magnetic properties manifests itself in the measured magnetostrictive properties of the cobaltite films. Figure 4 presents the longitudinal and transversal magnetoelastic stresses with respect to the direction of magnetic field measured at different temperatures. At low temperatures strong hysteretic behaviour is observed. This effect decreases in higher temperature ranges. A similar effect is observed in magnetic hysteresis loops (Figure 5). The unusual shape of the hysteresis loops at low temperature is typical for cluster glass systems. The narrowing effect observed at low fields is due to spin transitions induced by magnetic field. Both magnetoelastic and magnetic hysteresis loops indicate a decrease of the coercivity with increase of temperature.

The magnetostriction constant, λ_{100}, is calculated to be 1870 ppm at 50 K at the magnetic field of 12 T. To calculate this value the elastic constant difference c_{11} - c_{12} was evaluated to be 50 GPa using mechanical properties data [22]. The measured value of λ_{100} is very high, comparable to that observed in polycrystalline $La_{0.7}Sr_{0.3}CoO_3$ [4, 5]. This huge value of the magnetostriction has its origin in two phenomena:

- orbital instability of the Co^{3+} ions under the magnetic field inducing a transition from an orbitally non-degenerated low-spin state to an orbitally degenerate intermediate-spin state
- increase of the volume of ferromagnetic clusters with the increase of the magnetic field intensity

Due to the second effect the system may reach a percolation threshold and consequently a metal-insulator phase transition may be induced by the magnetic field. It has to be stressed that the contributions to the effective magnetostriction of cobaltites mentioned above have a different dependence on temperature. An increase of temperature induces low-spin to intermediate-spin transitions. At the same time the volume of ferromagnetic clusters decreases with increasing temperature. The competition of these mechanisms leads to the unusual dependence of the effective magnetostriction of cobaltites on temperature.

Figure 6 presents the thermal variation of the magnetoelastic stress parameter $B^{\gamma,2}$ of the $La_{0.7}Sr_{0.3}CoO_3$ film. This dependence is nonmonotonic. It is because the mentioned two mechanisms act in opposite directions. Since at high temperatures the magnetostriction tends to be smaller, it indicates that at least in this temperature range magnetostriction arising due to spin transitions is significantly suppressed and the dominating mechanism is related to the decrease of the volume of ferromagnetic clusters. In the low-temperature region both mechanism are of comparable importance.

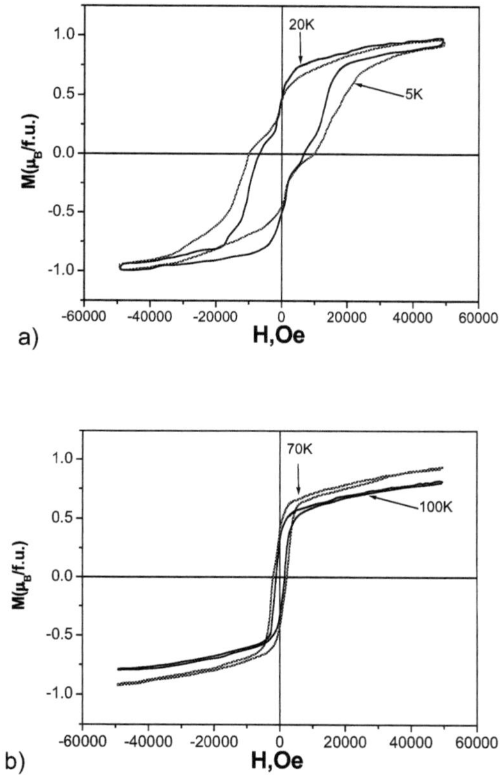

Figure 5. Magnetization loops of $La_{0.7}Sr_{0.3}CoO_3/SAT_{0.7}CAT_{0.1}LA_{0.2}$ films measured at different temperatures (20, 5 K (a)) and (70, 100 K (b)).

The giant magnetostrictive effects observed in cobaltite films suggest that these films should be anomalously sensitive to applied strain, such as would occur if a film is grown on a poorly lattice matched substrate. It offers a possibility of tuning the properties of cobaltite films with strain [23].

3.3 Cobaltite $La_{0.8}Ca_{0.2}CoO_3$ single crystals

Magnetization, magnetocrystalline anisotropy and magnetostriction measurements have been performed on La0.8Ca0.2CoO3 single crystals in the temperature range from 2K to 150 K in magnetic fields up to 12 T. Magnetization measurements indicate that the magnetic moments are frozen into a cluster glass state (Figs 6-7). A high magnetic field was shown to be able to transfer the system into a quasi-ferromagnetic one. This quasi-ferromagnetic state, typical for systems near the percolation threshold, is characterized by giant magnetostriction. Both magnetocrystalline anisotropy and magnetostriction was shown to be described well in a two-ion model. At

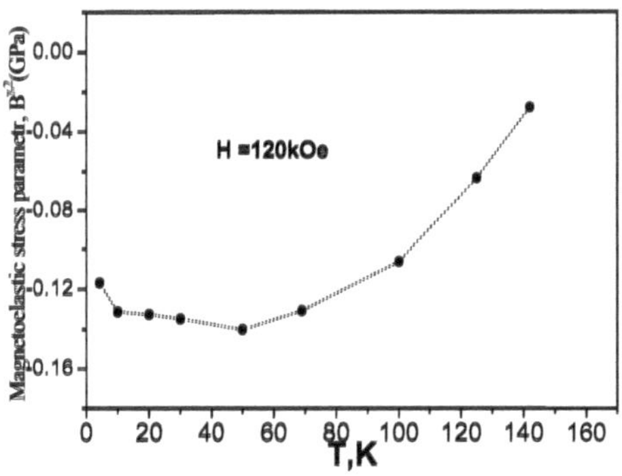

Figure 6. Thermal variation of the magnetoelastic stress parameter $B^{\gamma,2}$ of the $La_{0.7}Sr_{0.3}CoO_3$ film ,taken at 120 kOe.

the same time, the enhancement of the magnetostriction constants is due to the intrinsically inhomogeneous magnetic state of cobaltites in the low-doping regime.

At the same time the magnetoelastic properties of $La_{0.8}Ca_{0.2}CoO_3$ single crystals were shown to be typical for materials with a cluster glass magnetic ordering [24]. A giant magnetoelastic effect ($\lambda_{\bullet} \approx 1000$ ppm) has been found in the cobaltite being in the cluster glass state (Figs 8-9). This effect is due to a magnetic-field induced increase of the volume of the ferromagnetic phase. It leads to an increase of the magnetization and, as a consequence, of the magnetostriction. Additional contributions to the magnetostriction (as well as to the magnetocrystalline anisotropy) may arise due to the field induced spin state transition of the trivalent Co ion. As a result it may be shown that in cobaltite besides two basically magnetostriction mechanisms (single-ion, two-ion) two new mechanisms are expected to exist:

i) Mechanisms associated with electronic-phase separation,
ii) Mechanism related to the field-induced transitions between the low-spin state and the intermediate spin state.

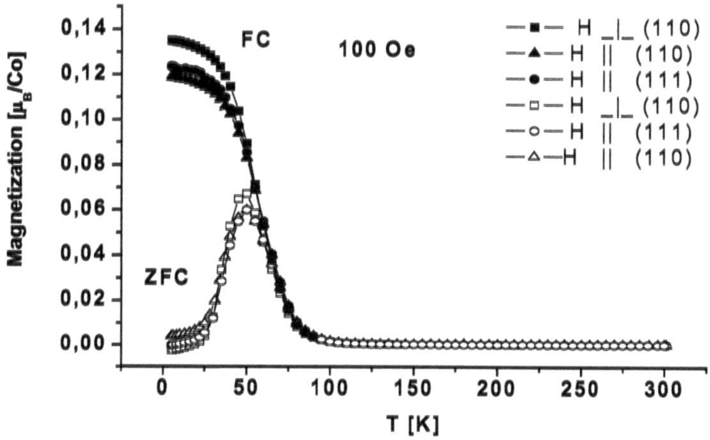

Figure 7. FC and ZFC magnetization as function of temperature for different magnetic-field orientations.

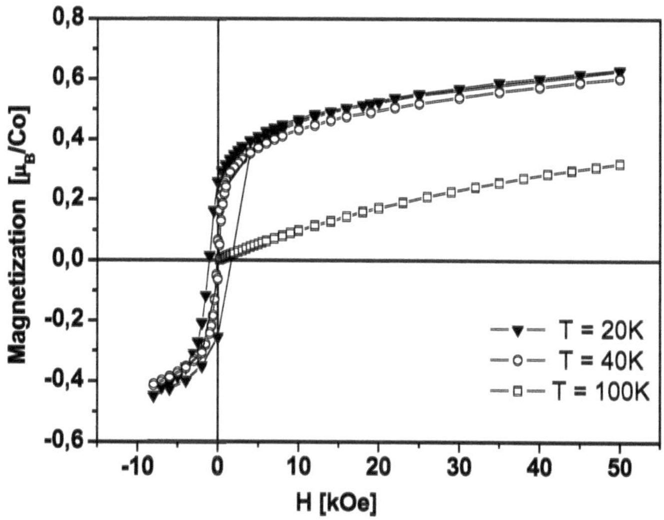

Figure 8. The magnetization of a monocrystalline sample as function of the magnetic field applied along the [111] crystallographic axis at different temperatures.

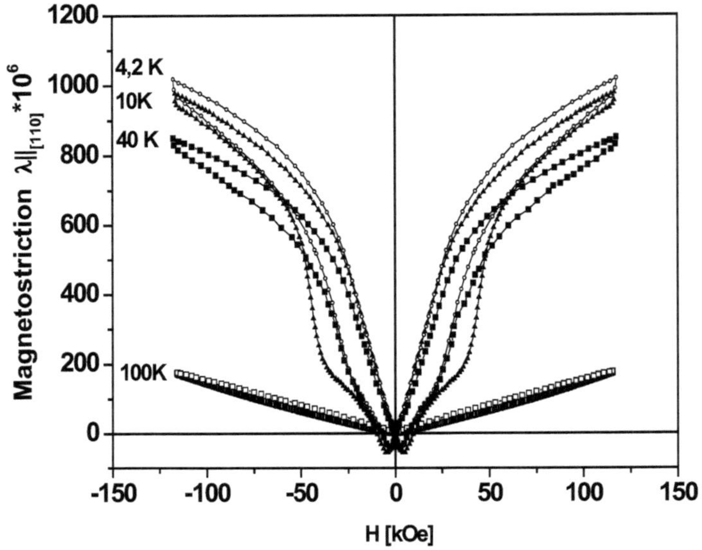

a)

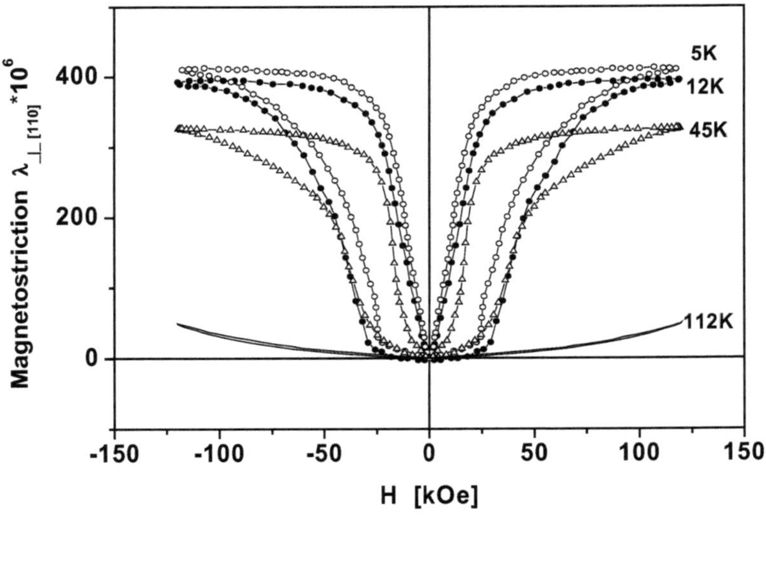

b)

Figure 9. Transversal (a) and longitudinal (b) magnetostriction of a $La_{0.8}Ca_{0.2}CoO_3$ single crystal along the [110] crystallographic axis at different temperatures.

4. Summary

The magnetostriction in manganite and cobaltite thin films was measured as a function of temperature and magnetic field. It was established that magnetostriction in manganite films is governed by a single-ion mechanism. In the case of cobaltite films, the situation is more complex because of the phase separation observed in the investigated films. It was shown that the giant magnetostriction in cobaltites arises due to two phenomena: orbital instability of the Co^{3+} ions under the magnetic field inducing a transition from an orbitally non-degenerated low-spin state to an orbitally degenerate intermediate-spin state, and increase of the volume of ferromagnetic clusters with increasing magnetic-field intensity. It seems that the last mechanism is common for all materials exhibiting a phase separation into ferromagnetic clusters embedded in a non-ferromagnetic matrix.

Acknowledgments

The work was supported in part by the Polish State Committee for Scientific Research (project 1 P03B 038 27)

References

[1] A.J. Millis, Nature **392** ,147 (1998).
[3] A.P. Ramirez, J.Phys.:Condens. Matter **9**,8171 (1997).
[3] J.M.D. Coey, M. Viret, and S. von Molnar, Adv. Phys. **48**,167(1999).
[4] M.R. Ibarra, R. Mahendiran, C. Marquina, B. Garcia-Landa, J. Blasco, Phys. Rev. **B 57**, R3217 (1998).
[5] H. Szymczak, J. Magn. Magn. Mater. **211**,186 (2000).
[6] B. Kundys, Yu. Bukhantsev, Vasiliev S, Kundys D, Berkowski M, Dyakonov VP, Rev. Scien. Instr. **75**, 2192 (2004).
[7] M.R. Ibarra, P.A. Algarabel, C. Marquina, J. Blasco and J. Garcia , Phys. Rev. Lett. **75**,3541 (1995).
[8] J.M. de Teresa, J. Blasco, M.R. Ibarra, J. Garcia, C. Marquina, P. Algarabel and A. del Moral, J. Appl. Phys. **79**,5175 (1996).
[9] Yu. Bukhatsev, Ya.M. Mukovskii and H. Szymczak, J. Magn. Magn. Mater. **272-273**,2053 (2004\0.
[10] I. Chaudhuri and S.K. Ghatak, J. Alloys and Compounds, **326**, 54 (2001).
[11] I.O. Troyanchuk, D.D. Khalyavin, N.V. Kasper, L.S. Lobanovsky, H. Szymczak and A. Nabialek, J. Solid State Chemistry **134**,215 (1997).
[12] Yu.F. Popov, A.M. Kadomtseva, G.P. Vorobev, V.Yu. Ivanov, A.A. Mukhin, A.K. Zvezdin and A.M. Balbashov, J.Appl.Phys. **83**,7160(1998).
[13] E. du Termolet de Lacheisserie, Magnetostriction: Theory and Application of Magnetoelastisity CRC Press, Boca Raton, FL, (1993).
[14] E. Callen and H.B. Callen, Phys. Rev. **A139**, 455 (1965).
[15] T.W. Darling, A. Migliori, E.G. Moshopoulou, S.A. Trugman, J.J. Neumeier, J.L. Sarrao, A.R. Bishop and J.D. Thompson, Phys.Rev. **B 57**, 5093 (1998).

[16] B. Dabrowski, L. Gladczuk, A. Wisniewski, Z. Bukowski, R. Dybzinski, A. Szewczyk, M. Gutowska, S. Kolesnik, C.W. Kimball and H. Szymczak, J. Appl. Physics **87**,3011 (2000).

[17] L. Ranno, A. Llobet, R. Tiron, and E. Favre-Nicolin, Applied Surface Science **188**, 170 (2002).

[18] L.I. Koroleva, Magnetic Semiconductors, Moscow State University,2003.

[19] H. Ogasawara, M. Matsukawa, M. Yoshizawa, M. Apostu, R. Suryanarayanan, G. Dhalenne, A. Revcolevschi, K. Itoh and N. Kobayashi , J. Magn. Magn. Mater. **226-230**.990 (2001).

[20] N. Orlovskaya, K. Kleveland, T. Grande, and M.-A. Einarsrud, J. Eur. Ceram. Soc. **20**, 51 (2000).

[21] H. Szymczak, R. Szymczak, M. Baran and J. Fink-Finowicki, J. Magn. Magn. Materials **272-276**, 1327 (2004).

[22] T. Kimura. Y. Tomioka, A. Asamitsu, and Y. Tokura , Phys. Rev. Lett. **81**, 5920 (1998).

[23] Zhi-Hong Wang, H. Kronmuller, O.I. Lebedev, G.M. Gross, F.S. Razavi. H.-U. Habermeier, B.G. Shen, Phys. Rev. **B65**, 054411 (2002).

[24] H. Szymczak, R. Szymczak, M. Baran and J. Fink-Finowicki, J. Magn. Magn. Mater. **272-276**, 1327 (2004).

STRUCTURAL STUDY OF MAGNETOSTRICTIVE SUPERCONDUCTING COMPOUNDS

V.V. Eremenko, S.B. Feodosyev, I.A. Gospodarev, V.A. Sirenko, M.Yu. Shvedun
Institute for Low Temperature Physics & Engineering National Academy of Sciences of Ukraine,
eremenko@ilt.kharkov.ua

M. Tovar
BENSC, Hahn-Meitner-Institute, Germany,

W. McCallum
Ames Lab., Iowa, USA

Abstract: It was shown recently, that high-temperature superconductors (HTSC) with perovskite-like structure are characterized by giant (up to 10^{-4}) magnetostriction. It means, that they are potential candidates for magnetostrictive applications. This observation attracted a number of experimental and theoretical studies of magnetostriction in the related materials. The reported activity is THE FIRST structural study of magnetostriction in superconductors. We studied layered superconductors, the high-temperature ones with rare earth substitutions, corresponding to different levels of doping (concentration of carriers) and layered dichalcogenides in order to explain the volume changes at low temperatures below the superconducting transition. Therefore dilatometric, neutron and X-ray diffraction studies above and below T_c, including measurements in magnetic field, were performed. Preliminary X-ray measurements have shown sufficient differences of the unit cell volumes in normal and superconducting state. The calculated lattice parameters revealed a pronounced change of the unit cell volume. Negative thermal expansion was observed and explained in detail.

Keywords: superconductors, perovskite-like structures, negative thermal expansion.

J. Franse et al. (eds.), Smart Materials for Ranging Systems, 259–270.

1. Introduction

Negative thermal expansion (NTE) has been known for some time in several compounds such as perovskite ferroelectrics [1] and layered superconductors [2]. Unusual lattice dynamics is responsible for this phenomenon. In such multilayer structures the lattice parameter in one of the crystallographic directions exceeds sufficiently the characteristic radius of the interatomic interaction. It weakens the influence of the regularity of the distribution of the atoms in the crystal on the phonon spectrum and vibration properties of the crystal. The structures of this type exhibit a state intermediate between crystals and disordered systems. We consider this effect for the HTSC perovskite like system $Eu_{1+x}(Ba_{1-y}R_y)_{2-x}Cu_3O_{7-d}$ with Ba substitution by the light rare-earth La. It was shown recently, that the structural distortions in $Eu(Ba_{1-x}R_x)Cu_3O_{7+d}$ caused by rare-earth atoms (R = La, Nd, Pr) on Ba sites include a crystallographic transition from orthorhombic to tetragonal symmetry, changes of "buckling angle" at the CuO_2 plane layer, and changes in ionic band lengths [4]. It results in suppression of high-temperature superconductivity in these compounds and in a number of specific effects in the vicinity of the superconducting transition, which ask for a comprehensive study. The presented work is part of such a study.

2. Experimental part

2.1 Experiment details

We studied the HTSC with rare-earth substitutions for Ba, corresponding to different levels of doping (concentration of carriers) [3] in order to check the connection of volume changes with phase transitions at low temperatures, below the superconducting transition. The crystal structure contains layers of Eu, CuO_2, BaO, and CuO_{1-d}. Sample characterization has shown that the preparation procedure provided perfect location of the substituting La atoms in Ba positions. A neutron diffraction patterns (NDP) study was performed above and below T_c, including measurements in magnetic fields up to 4 T, together with low-temperature X-ray measurements. The X-ray spectra $I(\Theta)$ were registered with a DRON-2.0 diffractometer, using the Bragg-Brentano scheme of focusing (scheme Θ-2Θ). The complete spectra of the diffracted radiation were obtained, with a subsequent identification of the unit cell system and parameters. The accuracy of the intensity measurements was 0.5%. The lattice parameters were determined to accuracy within ±0.01%. Error bars are shown on the measured curves. The DRON-2.0 diffractometer was also used in the cryogenic experiments: in this case a small-size cryostat for X-ray powder measurements in the temperature range 4.2÷300K was installed. In the experiment an X-ray tube with a copper anode was used ($\lambda_{Cuk\alpha} = 1.54178$ Å). The spectra were registered in the angle range $2\Theta = 20° \div 60°$.

2.2. Experimental results

Typical measured diffraction patterns are shown in Fig. 1. The calculated lattice parameters are summarized in Table 1 and illustrated by Fig. 2 for $Eu(Ba_{0.85}La_{0.15})_2Cu_3O_{7-d}$ as example . The space groups and discrepancy factors for these compounds are: $EuBa_2Cu_3O_{7-d}$: Pmmm, R_p=4.2%, R_{wp}=5.4%, R_B=6.2%; $Eu(Ba_{0.85}La_{0.15})_2Cu_3O_{7-d}$: P4/mmm, R_p=4.6%, R_{wp}=5.6%, R_B=6.4%; $Eu(Ba_{1-x}Nd_x)_2Cu_3O_{7-d}$: Pmmm, R_p=4.4%, R_{wp}=5.6%, R_B=6.2%. The data in Table 1 show a weak orthorhombicity in Pmmm structures.

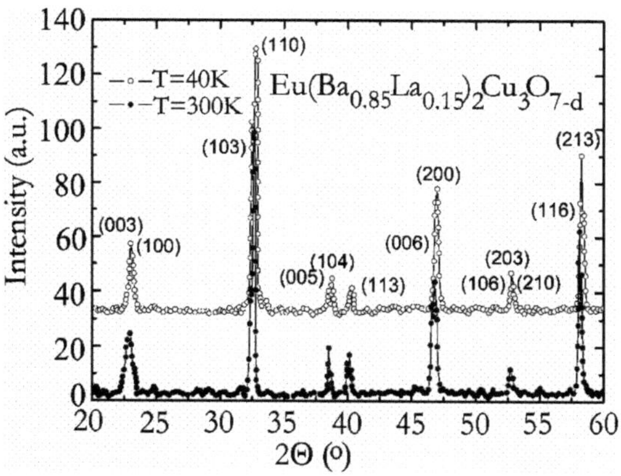

Figure 1. X-ray diffraction patterns for $Eu(Ba_{0.85}La_{0.15})_2Cu_3O_{7-d}$.

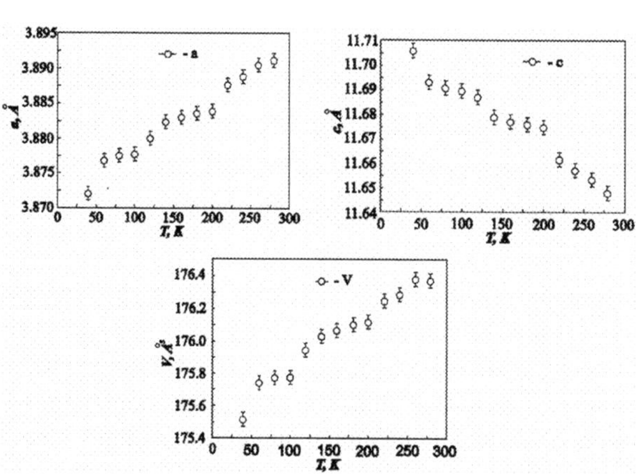

Figure 2. Temperature dependence of the unit cell parameters for $Eu(Ba_{0.85}La_{0.15})_2Cu_3O_{7-d}$.

Here we restrict ourselves to the problem of anisotropic negative expansion, illustrated by Fig. 2 for $Eu(Ba_{0.85}La_{0.15})_2Cu_3O_{7-d}$ as example. The same tendency is observed for the non-substituted compound $Eu_1Ba_2Cu_3O_{7-d}$, as our X-ray measurements revealed. The corresponding numerical data are presented in Table 1.

Table 1.

T,K	***a*, Å**			***b*, Å**		***c*, Å**		
	$EuBa_2Cu_3O_{7-d}$	$Eu(Ba_{0.85}La_{0.15})_2Cu_3O_{7-d}$	$Eu(Ba_{1-x}Nd_x)_2Cu_3O_{7-d}$	$EuBa_2Cu_3O_{7-d}$	$Eu(Ba_{1-x}Nd_x)_2Cu_3O_{7-d}$	$EuBa_2Cu_3O_{7-d}$	$Eu(Ba_{0.85}La_{0.15})_2Cu_3O_{7-d}$	$Eu(Ba_{1-x}Nd_x)_2Cu_3O_{7-d}$
300	3,8313		3.8186	3,8900	3.8868	11,6699		11.5687
280		3,8912					11,6481	
260		3,8905					11,6534	
240	3,8305	3,8888		3,8893		11,6764	11,6571	
220		3,8877					11,6617	
200		3,8839					11,6753	
180	3,8295	3,8836		3,8873		11,6791	11,6761	
160		3,8830					11,6774	
140		3,8824	3.8187		3.8904		11,6789	11.5851
120	3,8243	3,8800		3,8855		11,6838	11,6871	
100		3,8777					11,6894	
80	3,8243	3,8775		3,8841		11,6862	11,6908	
60		3,8768					11,6930	
40		3,8721	3.8184		3.8892		11,7064	11.5871

3. Discussion: "Membrane effect" as possible mechanism, responsible for negative expansion in strongly anisotropic layered and multilayered Structures. (Microscopic theory)

Strong anisotropy of thermal expansion when, at positive volumetric expansion, the lattice is compressed along any crystal axis (or plane), indicates, first of all, a strong anisotropy of the interatomic interactions.

The possibility of negative thermal-expansion coefficient (TEC) values along a direction of strong coupling in layered or chain structures (the so-called "membrane effect") was suggested for the first time by Lifshitz [4] for strongly anisotropic compounds. In the phonon spectra of such compounds the so-called bending modes with a quadratic dispersion law $\omega(k)$ in the long-wave limit (at $k \to 0$) are distinctly seen.

Negative TEC were found experimentally both (see, e.g., [5]) for strongly anisotropic layered crystals, characterized by the modes in the vibration spectrum with a specific "quasi-bending" curvature in a range of small k (e.g. graphite), and for compounds where the manifestation of the corresponding bending in the phonon spectra was extremely weak, or even was not detected at all (such as GaSe, GaS, InSe). In [6,7] the reasons for negative TEC along strong-coupling directions in crystals with a pronounced anisotropy of interatomic interaction and of elastic properties was proposed on the basis of a microscopic approach. In such strongly anisotropic crystals, the amplitude of the atomic vibrations along directions of weak coupling (that is perpendicular to layers or chains) is much higher than in directions of strong coupling. Displacement of atoms along a direction of weak coupling leads to an increase in distance between atoms inside a layer (or chain) of magnitude Δl, proportional to the square of this displacement. Consequently, in a layer a compressing force proportional to Δl arises, which is the reason for the anisotropic negative expansion alike the «membrane effect». We suggest the following microscopic description for these compressing forces.

If the potential energy of a lattice Φ is expanded in a number of components of small displacements of atoms from their equilibrium position u_i up to a cubic term, then the temperature dependence of the principal values of the LTEC tensor $\alpha_{ii}(T)$ are, in view of the translational invariancy of the lattice, described by the following expression:

$$D_i \alpha_{ii}(T) = -\frac{1}{2} \sum_{\substack{r,r' \\ k,l}} \Phi_{ikl}(r,r') \left\langle u_k(r) \;\; u_k(r') \right\rangle_T , \qquad (1)$$

Here, $D_i = \sum_{r} \Phi_{ik}(r)|x_k|$, $\Phi_{ik}(r) = \Phi_{ik}(r - r') = \dfrac{\partial^2 \Phi}{\partial u_i(r) \partial u_k(r')}$ are the elements of a force matrix.

$\Phi_{ikl}(r,r') = \Phi_{ik}(r - r'', r' - r'') = \dfrac{\partial^3 \Phi}{\partial u_i(r) \partial u_k(r') \partial u_l(r'')}$ x_i are the components of the radius - vectors of the equilibrium positions of atoms r; the symbol $\left\langle u_i(r) \;\; u_k(r') \right\rangle_T$ is the correlator of atom displacements, calculated in harmonic approximation.

Considering nonlinear effects in the vibrations of a crystal lattice (see, for example, [8]) it is necessary to take into account anharmonicity only in the terms connected to the largest interatomic forces, while the potential energy of weak forces of interlayer (or interchain) interactions, as well as non-central forces should be considered in the harmonic approach. Therefore; in (1) it is possible to neglect the summands, containing correlators of the atom displacements from various layers or chains, i.e. the correlators of $\left\langle u_x(\boldsymbol{r})\ u_y(0)\right\rangle_T$ type. Besides, as in the majority of crystals the central interaction between atoms decreases with distance quickly enough, as that in (1) it is possible to neglect the summands containing correlators between the second and more distant neighbors of the same layer or a chain. Correlators of $\left\langle u_\mu(\boldsymbol{r})\ u_\mu(0)\right\rangle_T$ type for nearest neighbours (a direction μ is chosen along a strong-coupling one) grow with temperature not faster than the root-mean-square displacement $\left\langle u_\mu^2(\boldsymbol{r})\right\rangle_T$, and both these temperature dependences are similar (that follows, for example, from the expressions given in [9]). Therefore the behaviour of the temperature dependence of LTEC along a direction of strong coupling of a layered or chained crystal can be described by a simple enough expression:

$$\alpha_{\|}(T)=\alpha_{\mu\mu}(T)=A\frac{\partial}{\partial T}\left\langle u_\mu^2\right\rangle_T\left\{\delta-\frac{\frac{\partial}{\partial T}\left\langle u_\perp^2\right\rangle_T}{\frac{\partial}{\partial T}\left\langle u_\mu^2\right\rangle_T}\right\}. \tag{2}$$

Here $\left\langle u_\perp^2\right\rangle_T$ is the root-mean-square displacement of atoms along a direction of weak coupling, that is $\left\langle u_\perp^2\right\rangle_T=\left\langle u_c^2\right\rangle_T$ for layered crystals (if ***ab*** is a layer plane) and $\left\langle u_\perp^2\right\rangle_T=\left\langle u_a^2\right\rangle_T+\left\langle u_b^2\right\rangle_T$ for chained ones, built with chains along the ***c*** axis). The parameters A and δ, included in (2), are obviously expressed with the anharmonic constants Φ_{ikl} ***(r, r')***, but the explicit form of the corresponding expressions is not presented here because of their bulkiness.

Let's analyse the temperature dependence of the quantity

$$\Delta(T) = \frac{\partial}{\partial T}\left\langle u_{\perp}^{2}\right\rangle_{T} \Big/ \frac{\partial}{\partial T}\left\langle u_{\mu}^{2}\right\rangle_{T} \,. \tag{3}$$

In strongly anisotropic structures, the mean square displacements $\left\langle u_{\perp}^{2}\right\rangle_{T}$ not only exceed $\left\langle u_{\mu}^{2}\right\rangle_{T}$, but also approach the classical limit at much lower temperatures. Thus, near to a temperature T⊥ (the temperature at which $\left\langle u_{\perp}^{2}\right\rangle_{T}$ approaches the classical limit) the function $\Delta(T)$ has a maximum, which size is proportional to exp ($\gamma^{-1/2}$), where γ is the ratio of weak interlayer or interchain interaction to strong interaction along layers or chains, respectively. Therefore near this temperature the LTEC $\alpha_{\|}$ is anomalously small and may adopt negative values. Values of temperatures of maxima obtained in [6, 7] for the quantities $\Delta(T)$ have shown good agreement with temperatures of minima on the measured dependences LTEC $\alpha_{\|}(T)$ of strongly anisotropic layered crystals [5]. The microscopic model, developed in [6,7], may be used almost without any modification for the description of thermal expansion of such compounds as MgB_2 or $NbSe_2$, which are also layered crystals, characterized by high anisotropy of both the force interatomic interaction and the elastic properties. Fig. 3 presents $\Delta(T)$,

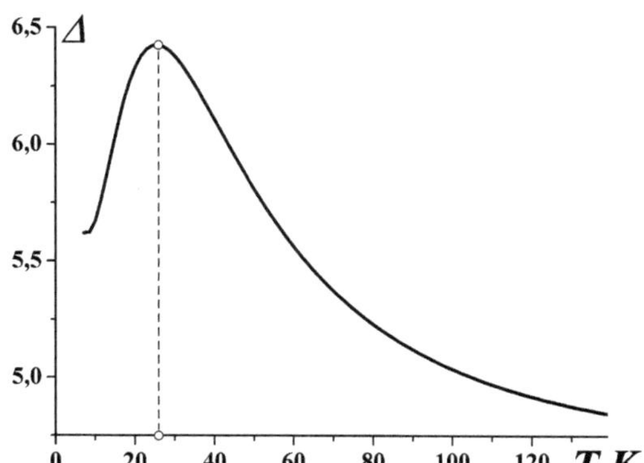

Figure 3. Ratio of the mean square root displacements derivatives along directions of weak and strong coupling, calculated in the model of a highly anisotropic layered crystal. Its anisotropy of interatomic interaction and elastic properties correspond to those of $NbSe_2$. The pronounced maximum on this curve corresponds to a minimum on the temperature dependence of the thermal expansion along the layers.

calculated in the framework of the model of strongly anisotropic layered crystals with anisotropy of elastic constants corresponding to the elastic anisotropy of $NbSe_2$ [10].

The temperature of the maximum on this curve should correspond to the temperature of a minimum on the temperature dependence of the TEC along an in-plane crystal direction (normal to the hexagonal axis) of this compound. Near this temperature, negative values of TEC are possible along the specified directions (α_a , α_b<0). It should be noted that in [5-7] crystal structures were considered, in which the strong anisotropy of the interatomic interaction at long distances remains and manifests itself, in particular, in an anisotropy of the elastic moduli. In crystals such as HTSC 1-2-3 the anisotropy of the elastic modules is insignificant and, moreover, has «the other sign», as the sound velocity along the layers is higher than in a direction normal to the layers [11]. At the same time, in such multilayered compounds the lattice period along the ***c*** axis is not only larger than that along the ***a*** and ***b*** axes, but also essentially exceeds the characteristic radius of the interatomic interaction. It leads to a weakening of the effect of the crystal structure regularity on the character of quasi-particle excitations. In the spectral characteristics of such crystals, features common to disordered systems are revealed. In particular, for phonon spectra of multilayered structures the presence of quasi-local (weakly dispersed) modes [12-14] is typical. Consequently, the excitation propagation along the ***c*** axis in such systems will be complicated, in comparison with their propagation in layers.

In multilayered lattices, even in such ones, for which the macroscopic characteristics are not distinguished by an appreciable anisotropy (as, for example, HTSC type 1-2-3), the interaction between separate atoms or atomic groups can be strongly anisotropic. The "damping" interaction propagation between layers inherent in substances of the specified class may result in appreciable manifestation of such local anisotropy both in the phonon spectrum [15] and in the behaviour of some vibrationary characteristics, in particular the root-mean-square displacement of atoms from separate layers along various crystal directions.

Some multilayered HTSC, for example $Bi_2Sr_2CaCu_2O_x$, show an anisotropy of the elastic moduli inherent for layered crystals, and negative thermal expansion in a direction within the layer [16], which can be described by formula (2). At the same time for multilayered structures such as HTSC 1-2-3, where the interlayer interaction between all layers is of the same order, the intralayer interaction essentially varies from one layer to another layer. Local anisotropy of «chain type» is characteristic for layers with weak intralayer interaction (a layer of the rare earth and a layer of chains Cu-O). In these layers the root-mean-square displacement of atoms in a direction within the plane is beyond the classical limit at lower temperatures, and is appreciably higher than the root-mean-square displacement of the same atoms in a direction normal to the layers. Therefore, the existence of temperature intervals is possible, where the interlayer distance Cu_2O- Eu - Cu_2O and BaO - CuO - BaO decreases with increasing temperature. This might be the reason for the negative values of LTEC along the ***c*** axis (Fig. 4).

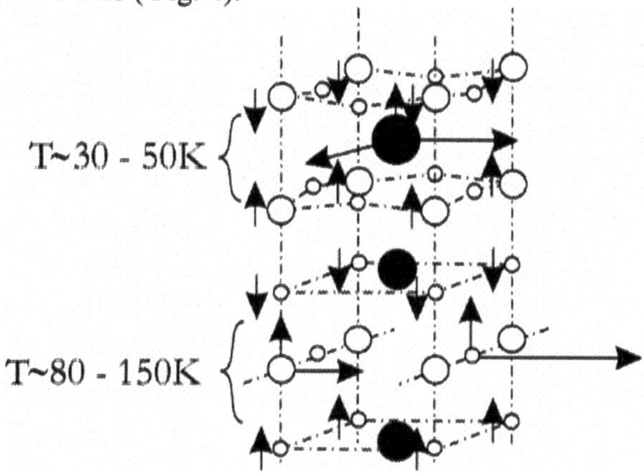

Figure 4. Schematic presentation of the causes of negative temperature expansion along the ***c*** axis in the $EuBa_2Cu_3O_{7-d}$ structure. The thick arrows indicate relative amplitudes of atomic vibrations in different crystallographic directions. The thin arrows demonstrate compression of the related corresponding interlayer spacings .

The temperature dependence of each spacing change can be described by some local thermal expansivity $\alpha_c^{(l)}(T)$ for which a relation analogous to eq. 2 holds, namely:

$$\alpha_c^{(l)}(T) = A\frac{\partial}{\partial T}\left\langle \left(u_c^l\right)^2\right\rangle_T \left\{\delta^{(l)} - \Delta^{(l)}(T)\right\}, \tag{4}$$

where the index l numbers atomic layers (R or CuO respectively). The functions $\Delta^{(l)}(T)$ are defined similarly to eq. (3) by addition of index l for the corresponding root-mean-square displacements. Then $\left\langle \left(u_i^l\right)^2\right\rangle_T$ (*i*=***a***, ***b***, ***c***) means either the root-mean-square displacement $\left\langle u_i^2\right\rangle_T^R$ itself for the layers of rare-earth elements, or a half-sum $\frac{1}{2}\left\{\left\langle u_i^2\right\rangle_T^{Cu} + \left\langle u_i^2\right\rangle_T^{O}\right\}$ - for the CuO layers. The temperature dependencies $\Delta^{(l)}(T)$ were calculated by a method of ***J***-matrices (see, for example, [17-19]). The data on the force constants are obtained on the basis of the results given in [20-23]. The calculation results are presented in Fig. 5.

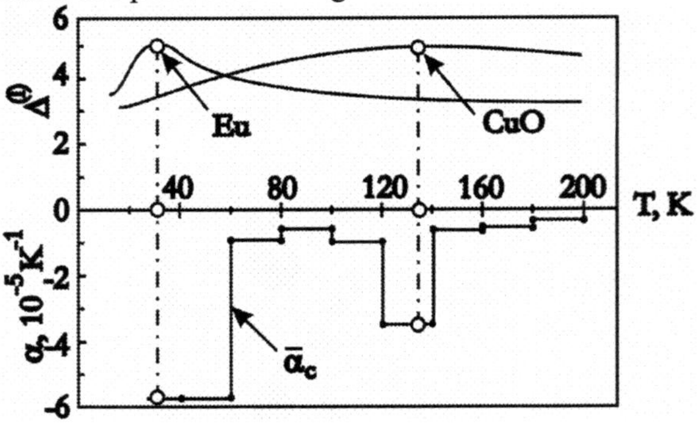

Figure 5. Temperature dependencies $\Delta^{(l)}(T)$ for Eu and CuO layers in $EuBa_2Cu_3O_7$, and the LTEC temperature dependence $\overline{\alpha}_c(T)$ averaged over temperature intervals between experimental points.

In the same figure the temperature dependence of the average values of the LTEC along the ***c*** axis, on each interval between temperatures of experimental points (see Tab. 1 and Fig. 2), is presented.

$$\overline{\alpha}_c\left(T \in [T_i, T_{i+1}]\right) = \frac{2}{T_{i+1} - T_i} \cdot \frac{c(T_{i+1}) - c(T_i)}{c(T_{i+1}) + c(T_i)}$$

Temperatures of maxima on the $\Delta^{(l)}(T)$ curves are in intervals of the least values $\overline{\alpha}_c(T)$, where the speed of decrease of the lattice parameter ***c*** is the greatest (see Tab. 1). Partial replacement of Ba by La does not lead to significant changes in the phonon spectrum and in oscillatory characteristics considered here. The perfect agreement between the calculated results and experiment unambiguously supports the mechanism proposed here for negative thermal expansion in the ***c*** direction, the mechanism being based on crystal lattice dynamics in a quasi-anharmonic approach. At temperatures $T \leq$ 50 K, negative thermal expansion is caused by compression of interlayer spacings $Cu_2O \leftrightarrow Eu \leftrightarrow Cu_2O$, and at 80K $\leq T \leq$ 150K by the $BaO \leftrightarrow CuO \leftrightarrow BaO$ ones (see Fig. 4). At temperatures above 200 K an even faster linear decrease of the parameter ***c*** with temperature is observed, which is not explained by the proposed model (in the temperature range 260 ÷ 280 K the crystal size decreases), as in this temperature range the quasi-harmonic lattice dynamics model becomes non-applicable. It should be noted, that it follows from the experimental curves, that at $T \leq$ 180 K the LTEC practically vanishes. This feature suggests that independent mechanisms are responsible for negative thermal expansion at high (more than 200 K) and at low temperatures $T \leq$ 140 K, or at least that there is only a weak effect of the "high temperature" mechanism on the thermal expansion at $T \leq 180$ K. The results do not show a significant contribution of BaO layers to the negative expansion.

Acknowledgments

This work was supported by grants of the BENSC MAT-01-1271 and MES of Ukraine under contract #M/257-2004.

References

[1] G. Shirane and S. Hoshina, *J. Phys. Soc. Jpn.*, **6**, 265 (1951).

[2] N.V. Anshukova, B.M. Bulychev, A.I. Golovashkin, L.I. Ivanova, I.B. Krynetskii, A.A. Minakov, A.P. Rusakov, *Zh. Éxp. Teor. Fiz.* **124**, 80 (2003).

[3] Youwen Xu., Kramer MJ., Dennis KW., Wu H., O'Connor A., McCallum RW., Malik SK. Yelon WB., Substitution for Ba by light rare-earth in Eu123 solid solutions. *Physica C*, **341-348**, pt.1, Nov. 2000, pp.613-14.

[4] I.M. Lifshits, *Zh. Éxp. Teor. Fiz.* **22**, 475 (1952).

[5] G.L. Belenkii, R.A.Suleimanov, N.A.Abdullaev, V.Ya.Shteinshraiber, *Fiz. Tverd. Tela*, **26**, 3560 (1984).

[6] S.B. Feodosyev, I.A. Gospodarev, E.S. Syrkin, *Phys. Stat/ Sol (b)*, **150**, K19 (1988).

[7] S.B. Feodosyev, E.S. Syrkin, I.A. Gospodarev, V.P. Popov, A.A. Gurskas, N.M. Nesterenko, *Fiz. Tverd. Tela*, **31**, 186 (1989).

[8] A.M. Kossevich, *The Crystal Lattice (Phonons, Solitons, Dislocations)*, WILLEY-VCH Verlag Berlin GmbH, Berlin, 1999.

[9] S.B. Feodosyev, I.A. Gospodarev, M.A. Strzhemechny, R.J. Hemley, *Physica B*, **300**, 186 (2001)

[10] H.G. Smith, N. Wakabayashi, in *Dynamics of Solids and Liquids by Neutron Scattering,*, S.W. Lovesey and T. Springer (eds.), Springer Verlag, New York (1977).

[11] Ming Lei, J.L. Sarrao, W.M. Visssher, T.M. Bell, J.D. Thompson, A. Migliory, U.W. Welp, B.W. Veal, *Phys. Rev. B*, **47**, 6154 (1993).

[12] E.S. Syrkin, S.B. Feodosyev, *Fiz. Nizk. Temp.* **17**, 1055 (1991). [*Sov. J. Low Temp. Phys.*, **17**, 539 (1991)].

[13] A.M. Kosevich, E.S. Syrkin, S.B. Feodosyev, *Phys.Lett.*, **A167**, 94 (1992).

[14] I.A. Gospodarev, A.M. Kosevich, E.S. Syrkin, S.B. Feodosyev, *Fiz. Nizk. Temp.*, **22**, 593 (1996) [*Low Temp. Phys.*, **22**, 457 (1996)].

[15] S.B. Feodosyev, I.A. Gospodarev, A.M. Kosevich, E.S. Syrkin, *Phys. Low Dim. Str.*, **10/11**, 209 (1995).

[16] I.A. Gospodarev, A.P. Isakina, A.I. Prokhvatilov, E.S. Syrkin, S.B. Feodosyev, *Fiz. Nizk. Temp.*, **16**, 673 (1990) [*Sov. J. Low Temp. Phys.*, **16**, 396 (1990)].

[17] V.I. Peresada, in *Physics of Condensed State* [in Russian], B.Verkin Inst. of Low Temp. and Eng., Kharkov (1968) **2**, 172.

[18] V.I. Peresada, V.N. Afanas'ev, V.S. Borovikov, *Fiz. Nizk. Temp.*, **1**, 461 (1975) [*Sov. J. Low Temp. Phys.*, **1**, 227 (1975)].

[19] R. Haydock, in *Solid State Physics*, **35**, 129, H. Ehrenreich, F. Seitz, D. Turnbull (eds.), Academic Press, New York (1980).

[20] S. Mass, T. Jasuda, Y. Horie, M. Kusada, T. Fukami, *Journ. Phys. Soc. Jpn.*,**57**, 1024 (1988).

[21] V.K. Fedotov, A.I. Kolesnikov, V.I. Kulakov, E.G. Ponyatovskiî, I. Natkanets, Yu. Maîer, and Ya. Kravszyk, *Fiz. Tverd. Tela (St. Petersburg)*, **35**, 310 (1993) [*Phys. Solid State*, **38**, 156 (1993)].

[22] L. Pintschovius, N. Pyka, W. Reichard, A.Yu. Rumiantsev, N.L. Mitrifanov, A.S. Ivanov, G. Golin, A. Bourges, *Physica B*, **174**, 323 (1991).

[23] J.J. Rhyne, D.A. Neumann, J.A. Gitaas, F. Beech, *Phys. Rev. B*, **36**, 2294 (1991).

Subject index: keywords

Keyword	paper nr	page nr
Anderson model	8	153
Antiferromagnet	4	67
Cobaltites	14	245
Colossal magnetoresistance	9, 14	165, 245
Computer simulation	2	27
Copper metaborate CuB_2O_4	3, 5	50, 101
De Haas-van Alphen effect	4	67
Dichalcogenides (layered)	15	259
Domain walls	10	179
Exchange		
bias	10	179
coupling (interlayer)	10	179
interaction	9	165
Ferromagnetism	13	224
Glass		
Oxide glass	2	27
Heavy fermions	7	129
Heusler alloys	13	224
High T_c superconductor	6	115
Intermediate state	4	67
Iron	2	27
Kondo lattice	8	153
Magnetic -		
anisotropy	1	1
microscope	11	195
multilayers (see multilayers)		

recording	11	195
resonance	2	27
phase transitions (see also Phase)	9	165
roughness	10	179
susceptibility	8	153
Magnetization		
reversal	10	179
Magnetocrystalline anisotropy	14	245
Magnetostriction	14	245
in superconductors	15	259
Magnetodynamics	1	1
Manganites	9, 14	163, 245
Martensitic transformation	13	224
Mixed state	4	67
Multilayers (magnetic)	10	179
Nanoparticles	1, 2	1, 27
Negative thermal expansion	15	259
NMR	5	101
Noble metals	12	203
Nonlinear optics	12	203
Optical pumping	12	203
Orbital moment	1	1
Orbital ordering	9	165
Oxide glass: see Glass (oxide glass)		
Perovskites (see also Cobaltites or Manganites)		
perovskite like structure	15	259
Phase		
boundary		
commensurate phase	3, 5	50, 101
diagram	13	224
incommensurate phase	3, 5	50, 101
magnetic phase diagram	3	50
separation	9, 14	165, 245
transitions (successive)	4, 5	67, 101
polarized neutron reflectivity	10	179

Quantum critical phenomena	7	129
Shape memory alloys	13	224
SIFE	12	203
Size effect	1	1
SOKE	12	203
Spin dynamics	12	203
Spin-flop transition	4	67
SQUID	11	195
Strong correlation	8	153
Superconductivity		
superconducting carrier	11	195
superconductors	15	259
Superparamagnetism	1, 2	1, 27
t-J model	6	115
Thermal expansion		
Negative thermal expansion	15	259
Time resolved optical response	12	203
$U(Pt,Pd)_3$	7	129
X-ray reflectivity	10	179

9 781402 046445

MIX
Papier aus verantwortungsvollen Quellen
Paper from responsible sources
FSC® C105338

If you have any concerns about our products,
you can contact us on
ProductSafety@springernature.com

In case Publisher is established outside the EU,
the EU authorized representative is:
Springer Nature Customer Service Center GmbH
Europaplatz 3, 69115 Heidelberg, Germany

Printed by Libri Plureos GmbH
in Hamburg, Germany